Weigang Zhang
Signals and Systems
De Gruyter Graduate

Also of Interest

Signals and systems
G. Li, L. Chang, S. Li, 2015
ISBN 978-3-11-037811-5, e-ISBN (PDF) 978-3-11-037954-9, e-ISBN
(EPUB) 978-3-11-041684-8

Energy harvesting
O. Kanoun (Ed.), 2017
ISBN 978-3-11-044368-4, e-ISBN (PDF) 978-3-11-044505-3, e-ISBN
(EPUB) 978-3-11-043611-2, Set-ISBN 978-3-11-044506-0

Pulse width modulation
S. Peddapelli, 2016
ISBN 978-3-11-046817-5, e-ISBN (PDF) 978-3-11-047042-0, e-ISBN
(EPUB) 978-3-11-046857-1, Set-ISBN 978-3-11-047043-7

Wind energy & system engineering
D. He, G. Yu, 2017
ISBN 978-3-11-040143-1, e-ISBN 978-3-11-040153-0, e-ISBN (EPUB)
978-3-11-040164-6, Set-ISBN 978-3-11-040154-7

Weigang Zhang

Signals and Systems

Volume 1: In continuous time

清華大学出版社
TSINGHUA UNIVERSITY PRESS

DE GRUYTER

Author
Prof. Weigang Zhang
Chang' an University
Mid South 2nd Ring Road
Shaanxi Province
710064 XI 'AN China
wgzhang@chd.edu.cn; 648383177@qq.com

ISBN 978-3-11-041754-8
e-ISBN (PDF) 978-3-11-041953-5
e-ISBN (EPUB) 978-3-11-042650-2

Library of Congress Cataloging-in-Publication Data
A CIP catalog record for this book has been applied for at the Library of Congress.

Bibliographic information published by the Deutsche Nationalbibliothek
The Deutsche Nationalbibliothek lists this publication in the Deutsche Nationalbibliografie;
detailed bibliographic data are available on the Internet at http://dnb.dnb.de.

© 2018 Walter de Gruyter GmbH, Berlin/Boston
Cover image: Creatas/Creatas/thinkstock
Typesetting: le-tex publishing services GmbH, Leipzig
Printing and binding: CPI books GmbH, Leck
♾ Printed on acid-free paper
Printed in Germany

www.degruyter.com

Preface

The course of *Signals and systems* is an important professional, fundamental course for undergraduates majoring in electronics, information and communication, and control, etc. The course has a profound influence on the cultivation of students' overall capabilities, such as independent learning, scientific thinking in problem solving, practical skills, etc. The course is not only compulsory for undergraduates, but it is also necessary for postgraduate entrance examinations in related majors. The course plays a critical role in undergraduate education and it is the theoretical foundation to information theory and information technology. It is even regarded as the key to opening the door to information science in the twenty-first century.

Main contents in the course

From the content aspect, the course of Signals and Systems is more of a mathematics course integrated into professional characteristics than a specialized course. The so called signal is actually the function in mathematics, given "voltage", "current" or other physical background. System is considered as a module that can transfer (process) a signal.

In essence, the contents of the course can be summarized as the study of the relation between before and after a signal is transformed by a given system or a function is processed by a given operation module. Here, the signal before being transformed is called the input or the excitation, and the signal after being transformed is called the output or the response. The excitation is the cause (or the independent variable in mathematical description), and the response is the consequence (or the dependent variable in mathematical description). Mathematically, it can be described as after an independent variable (excitation) is calculated by a module (system) a dependent variable (response) is the result (plotted in ► Figure 1). A real physical system (a transformer and the relation between voltages on its two ports) is shown in ► Figure 1a and the mathematical description or the equivalent model of this physical system is given

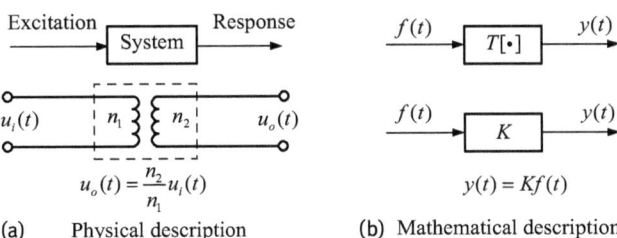

(a) Physical description (b) Mathematical description

Fig. 1: Excitation, response and system.

https://doi.org/10.1515/9783110419535-201

by ▶ Figure 1b. From the figures it can be seen that, in fact, Signals and Systems is just a course that abstracts a physical system as a mathematical model, and then studies the performance of the system through analyzing the model, namely solving the relationship between the excitation and the response. Please note that the symbol $T[\cdot]$ in ▶ Figure 1 shows a processing method or a transformation to the element in brackets [ ]. It is obvious that the signal is an object processed by the system, and the system is the main carrier to process the signal; both complement each other.

Herein, the relationship between signal and system is described by a mathematical model (mathematical analysis formula), such as $y(t) = T[f(t)]$, therefore, solving the mathematical model or solving the equation in layman's terms will run through the course. The various solutions for the model are main knowledge points of the book.

Since the signal and the system are the two key points of this course, all research is concentrated around them.

The analysis of signals covers the following points:
(1) Modeling. Various physical signals in the real world can be mathematically abstracted into mathematical models, and the process is known as "mathematical modeling." The aim of the modeling is to change a physical signal into a function that can be analyzed theoretically on paper.
(2) Decomposition and composition of signals. One signal can be decomposed into a linear combination of other signals; or, a set of signals can be used to represent a signal with their linear combination.

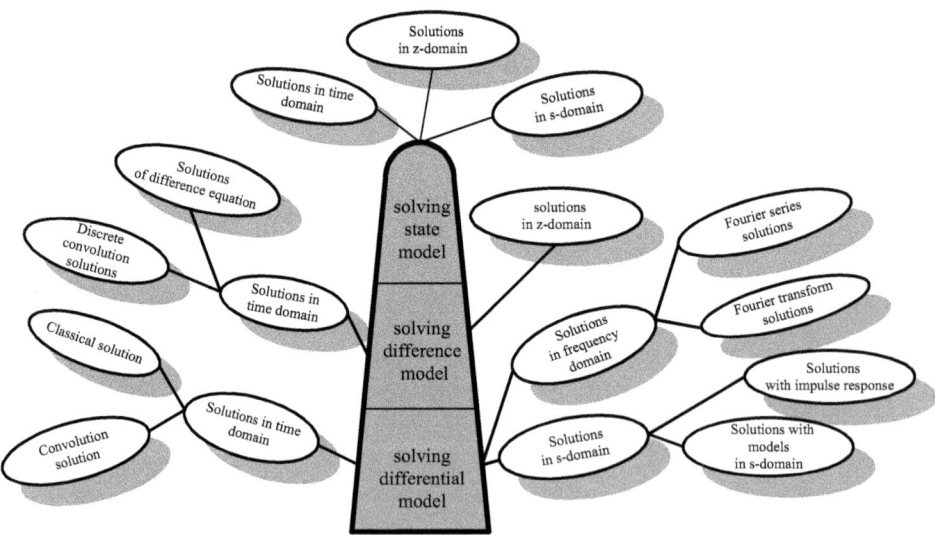

Fig. 2: Tree of content in Signals and Systems.

The analysis of systems focuses on the study of the response of a given system to an arbitrary excitation (▶ Figure 2), or, analyzes the transform characteristics of a system to signals known under the system constitution.

Briefly, this book consists of two parts, signal analysis and system analysis, and discusses how to solve differential or difference equations in the time and transform domains (real frequency domain, complex frequency domain and z domain).

Features of the course

This course has three main features:
(1) A sound theoretical basis. Various mathematical methods to solve the differential or difference equations in the time and transform domains are introduced.
(2) A strong specialty. Building up mathematical models of various systems from the real world must rely on the fundamental laws and theorems of related fields.
(3) Wide applications. The research results can be generalized to real applications in nature and society, even to nonlinear system analysis.

The aims of learning

After thinking carefully, we find that the real world is constituted by various systems. For example, the human body includes the nervous, blood and digestion systems, etc.; and then there is transport, lighting, water supply, finance, communication, control systems and so on in daily life. The functions of systems can be summarized as processing or transforming an input. So, the relationships between the inputs and the outputs of these systems are exactly the main topic studied in this book.

To facilitate research, real physical systems can be abstracted into mathematical models. Further, these models can be classified into two types of linear and nonlinear systems according to their characteristics. Thus, the aims of learning here are to master methods to analyze the relationship between the excitation and the response of a linear system and to apply these analytical results to the analysis of nonlinear systems and then to solve various practical problems in real systems.

Learning the course can help us to establish a correct, scientific and reasonable approach to analyzing and solving problems, and to improve the treatment ability of various problems and difficulties encountered in study, work and life, while at the same time, to master how to solve practical problems using basic knowledge, especially mathematical knowledge.

Research route of this book

The contents of this book can be divided into two layers: the lower layer which is signal analysis and the upper layer which is system analysis. The lower layer is the basis of the upper layer, while the upper one is the achievement of the lower one. Based on ► Figure 2, the research route of this book is shown in ► Figure 3.

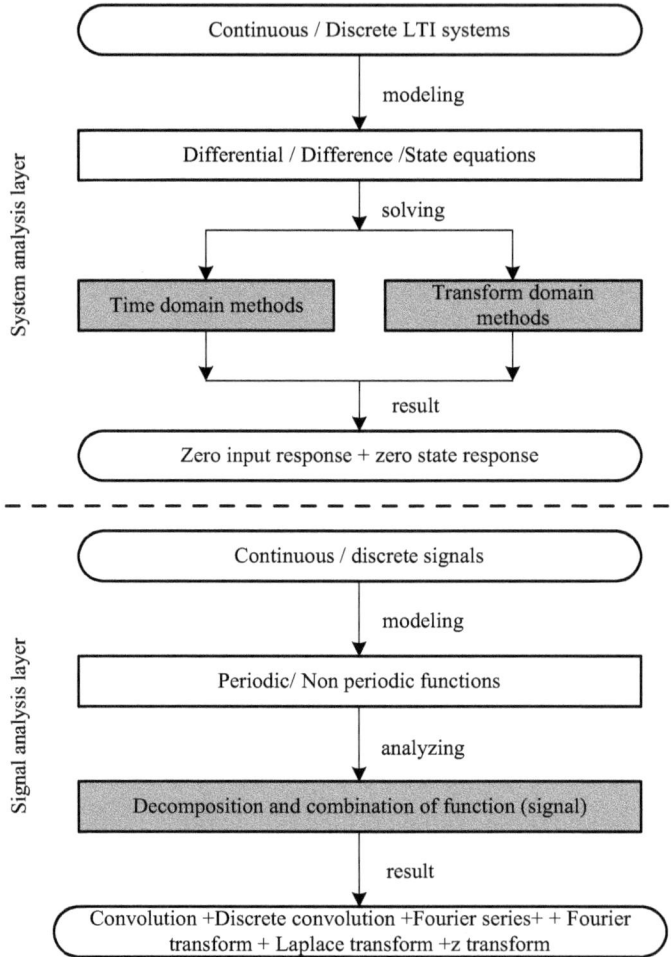

Fig. 3: The roadmap of the book.

Relationship between the course and other basic courses

There is no doubt that mathematics is an important basic course for signals and systems. The mathematics knowledge herein includes expansion and summation of series, solutions of differential/difference and algebraic equations, partial fraction expansion, basic calculus operations and linear algebra. In addition, this book, which takes electric systems as the objects to be studied, also involves professional knowledge of circuit analysis, analog circuits and digital circuits. Among of these fields, circuit analysis relates closely to the work herein; it is the precursor of the course, and the course is an expansion of content and the improvement in methodologies of the former. The similarities and differences between them are listed below:

(1) Both of the objects to be studied are circuits or networks consisting of electronic components.
(2) Both of the main aims are to find circuit variables such as voltage and current.
(3) The main analysis method in circuit analysis is to obtain the responses (node voltages and branch currents) of a circuit to excitations by constructing algebraic equations where excitations are direct or alternating current signals.
(4) The main analysis method for signals and systems is to obtain the responses (output voltages or output currents) of a system to excitations by building differential/difference equations where excitations are periodic or nonperiodic signals.

For instance, in ▶ Figure 4, if u_S and i are, respectively, the excitation and the response of a system or a circuit, how can we obtain the response i of the system under different excitations? From circuit analysis, the currents in ▶ Figure 4a and b are $i = \frac{u_S}{R}$ and $\dot{I} = \frac{\dot{U}_S}{R+j\omega L}$, respectively. Because excitations in ▶ Figure 4c and d are non-sinusoidal periodic and nonperiodic signals, respectively, the currents cannot be obtained. Fortunately, these problems can be solved by Fourier series and Fourier transform in signals and systems.

(5) Circuits Analysis includes solution methods for algebraic equations obtained by laws and theorems of circuits and the phasor analysis method for alternating current circuits.

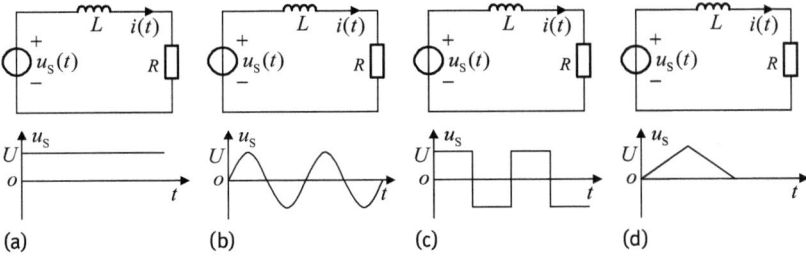

Fig. 4: Circuits Analysis examples.

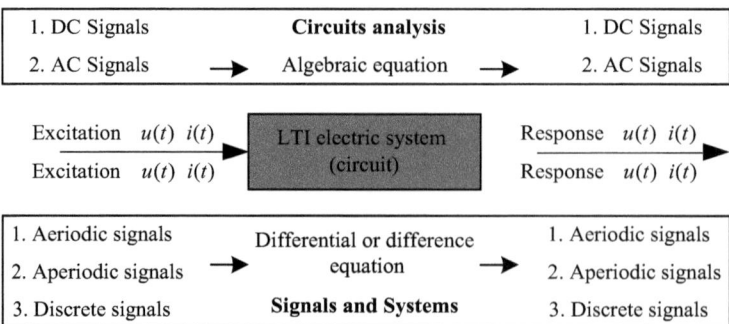

Fig. 5: Main differences between Signals and Systems and Circuits Analysis.

(6) Signals and Systems include solution methods for differential equations in the time, frequency and complex frequency domains and difference equations in the time and z domains.

The main differences and similarities between the two courses are shown in ▶ Figure 5. Note that the points in the figure only focus on electricity technology, in fact, the contents of Signals and Systems can be also used in mechanical systems and other analogous systems.

Status of the course

In conclusion, Signals and Systems is a professional basic course, which takes the signal and the system as the core, has system performance analysis as the purpose and employs mathematics as the tool to establish the mathematical model as the premise and solve the model as the means.

Features of this textbook

The Signals and Systems course is not only a specialized course with professional concepts, but also a mathematical one with massive computations. To help readers master the contents in a better way, we deliberately increased the number of examples (about 140 sets) to expand insights and to improve understanding by analogy. At the same time, we also arranged more than 160 problems with answers to help readers grab and consolidate knowledge learned. In addition, to deepen readers' understanding of the contents and to improve problem solving skills, a section called Solved Questions was added at the end of each chapter, which includes about 50 exam questions and solutions selected from other universities.

Due to space limitations, this book has been split into two volumes. Volume 1 mainly discusses problems concerning continuous-time signals and systems analysis. Volume 2 focuses on issues about discrete time signals and systems analysis.

I would like to express my sincere thanks to Associate Professor Wei-Feng Zhang, the coauthor of the books, and Associate Professors Tao Zhang and Shuai Ren, and lecturers Xiao-Xian Qian and Jian-Fang Xiong for preparing the Solved Questions and translating the Chinese manuscript into English. I also wish to thank Dan-Yun Zheng, Jie-Xu Zhao, Xiang-Yun Li, Pei-Cheng Wang, Juan-Juan Wu, Jing Wu and Run-Qing Li for proofreading all examples, exercises and answers and helping to translate the manuscript. Finally, thanks are due to authors and translators of reference books in the books.

The books are the summary of the authors' teaching experience of many years; any suggestions for improvements to the book would be greatly appreciated. Please feel free to contact us about any problems encountered during reading; we would be grateful for your comments.

June 2016   Weigang Zhang

Contents

Contents of Volume 2

1 Signals

Questions: How do we analyze the various signals that we often come across in the process of productive practice? Can we find some general methods?

How to proceed: Seek the similarities of different signals and classify them → select and analyze the basic signals → find out the analytical methods adaptive to most signals.

Results: Definitions and properties of 11 basic signals such as sinusoidal, complex exponential, step, impulse signals and so on. Study of the plotting method and 8 basic operations such as arithmetic, operations among even and odd signals, time shifting, reversal, time scaling, differential/integral, decomposition/synthetic and convolution integral.

1.1 The concept of signals

1. Definition of signals

We know that the aim of communication is to transfer information, which is based on the transmission of signals. In other words, a communication task can be completed only after a signal has been transferred. The signals used in communication are usually generated or controlled by people and are called artificial signals, such as the beacon fire in ancient China, the semaphores on ships, the voice of people, etc. In addition, there is another category of signals, which is usually seen in measurement and control fields. Because these signals are created by nature, they are called natural signals, such as the vibration of the earth's crust, temperature and humidity, the human pulse, etc. All signals in real life are either artificial or natural. Signals are the physical carriers of information and play an important role in communication, measurement and control systems.

According to live examples, signals can be defined as

all physical phenomena or quantities that can carry information and be perceived by humans or instruments are called signals.

Signals and systems is a foundational course of communication principles, automatic control principles, etc., so the information characteristic of signals is not the focus of discussion here. Whereas the signals in this book are only considered as mathematical functions which can be realized physically and are useful in productive practice, and furthermore the functions can be taken as general, abstract forms which characterize physical signals with both representativeness and universality. As a result, we define it as

a kind of physical quantity varying with time, such as the voltage and current in circuits.

https://doi.org/10.1515/9783110419535-001

To facilitate the research of physical signals in theory, we must abstract a mathematical model from them, and this process is called the modeling to signal; this mathematical model is just the function that is familiar to us. Therefore, in theory signal and function are equivalent herein, so the terms signal and function can be used interchangeably if we do not indicate otherwise.

2. Ways of describing signals
Usually, there are three methods to describe a signal in the familiar *time* domain:
1. A mathematical expression or model. This is a function of time.
2. A graph. This is a changing waveform related to the function (dependent variable) and time (independent variable), which can be plotted from the expression or the measurement.
3. A table. This is frequently used to describe those signals that cannot be expressed by a mathematical expression.

We also need to know and obtain the other three forms of signals, i.e. the expression, the graph and the table in transform domains, respectively. In this case, the independent variable of a signal is not time but real frequency, complex frequency or z. The forms to express signals in transform domains are the supplements for the *time* domain and the auxiliary tools for the study of signals, and are usually only used in the equation solving process or in specific circumstances. Of course, the different forms of signals have also different models.

3. Basic properties of signals
Generally, a signal has the four basic properties of time, frequency, energy and information.

The time feature includes changing relations of size (amplitude) and speed (frequency) and delay (phrase) with time variables. The frequency feature consists of changing relations of amplitude and phase with frequency variables. The energy feature involves changing the rules of the energy or the power of a signal with time and frequency variables. The information feature entails that a signal can carry the information that exists in a type of changing waveform of a signal in all cases.

4. Contents of signal research
Herein, the research on signals consists of two aspects: signal analysis and signal processing.

Signal analysis refers to the concepts, theories and methods to decompose a signal into several components, such as the convolution integral and convolution sum, Fourier series, Fourier transform, Laplace transform and z transform, etc.

Signal analysis can be applied in time and transform domains. Moreover, in the transform domain it can be regarded as the supplement and extension to concepts and methods for time domain analysis. For example, the spectrum concept obtained from *frequency* domain analysis is important in the application and the processing of signals.

Signal processing refers to the process, operation, modification and transformation to a signal based on special requirements or aims. Filter, amplification, modulation and demodulation, coding and decoding, encryption and decryption, equilibrium, smoothness and sharpening are all examples of signal processing.

The signal is the object being processed, whereas the main body executing the process to the signal is just the system.

5. Means of signal analysis

Analysis methods mainly include arithmetic, time shifting, time reversal, time scaling, differentiation/integration, decomposition/synthetic, convolution integration (sum) and plotting.

6. The purpose of signal analysis

Because a signal is employed by excitation and response of a system, the purpose of signal analysis is to facilitate analysis of transferring or processing features of the system to various signals, namely characteristics of the system. In a nutshell, signal analysis is the foundation of system analysis.

1.2 Classification of signals

Signals can be classified into the following categories according to their characteristics.

1.2.1 Continuous time and discrete time signals

The continuous-time signal and the discrete-time signal play leading roles in this book; they correspond to the analog signal and the digital signal.

1. Continuous-time signals

Usually, a signal takes time as the independent variable, so we can give a definition as:

A signal whose independent variable time has all values in the domain of definition is called a continuous-time signal or, simply, a continuous signal.

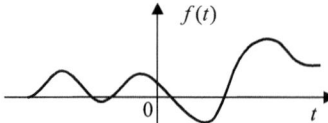

Fig. 1.1: A continuous-time signal (analog signal).

Two examples this type of signal are shown in ▸ Figure 1.1 and ▸ Figure 1.6. Note that values of the dependent variable of a continuous signal can be continuous or discontinuous.

The analog signal is always discussed together with the continuous signal, which can be defined as

A signal whose dependent variable changes continuously with continuous variations of independent variable time is called an analog signal.

The graph of an analog signal is a continuous curve (▸ Figure 1.1). Note that an analog signal must be a continuous signal, but the reverse is not true; for example, the continuous digital signal plotted in ▸ Figure 1.5b is not an analog signal.

Usually, a continuous signal can be denoted as $f(t)$, $x(t)$, $y(t)$, $s(t)$ and so on.

2. Discrete-time signals

A signal whose independent variable time only takes discrete values in the domain of definition is called a discrete-time signal or, simply, a discrete signal.

A typical signal where the independent variable is not time is the frequency spectrum of a periodic signal, which takes frequency as the independent variable; this will be introduced in Chapter 4. In this book, discrete signals usually refer to discrete-time signals if no other explanation is given.

The $f[t_n]$ in ▸ Figure 1.2 is a discrete signal that only has values at instants t_n. Assuming an interval between t_n and t_{n+1} is T_n, which is a constant T or a function of n, if $T_n = T$, the discrete signal $f[t_n]$ only has values at points at equal intervals

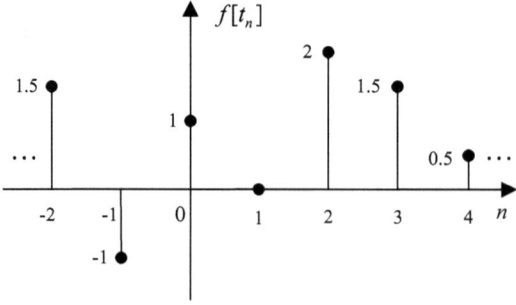

Fig. 1.2: A discrete-time signal.

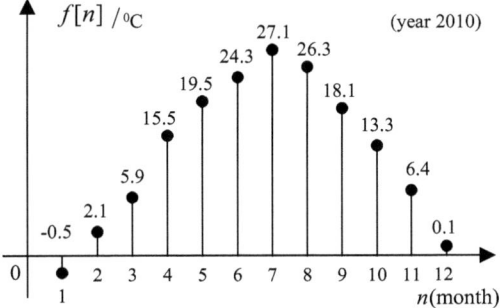

Fig. 1.3: Average monthly temperatures in a city in 2010.

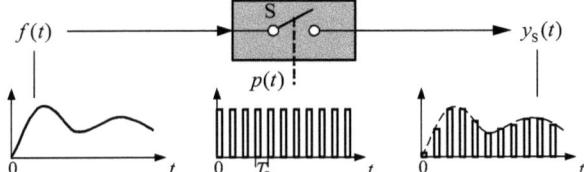

Fig. 1.4: The sampling process.

$t_n = \cdots - 2T, -T, 0, T, 2T \ldots$, and it is expressed as $f[nT]$ or as $f[n]$ if $T = 1$. Usually, this discrete signal whose value distributes uniformly on the time axis is also called a sequence.

To distinguish between continuous and discrete signals, we use square brackets to express a discrete signal, such as $f[n]$, $x[n]$, $y[n]$ and $s[n]$. At the same time, we stipulate that the independent variable of sequence n is only an integer, and sequence values at noninteger points of the independent variable are either undefined or zero.

Discrete signals can be divided into two types from their sources; one is a natural discrete signal of which the independent variable itself is discrete, such as the average temperatures of months in 2010 of a city $f[n]$ (► Figure 1.3). The other is an artificial discrete signal, which can be obtained by sampling to a continuous signal; for example, if a continuous signal $f(t)$ applies to a switch S controlled by a sampling pulse sequence $p(t)$, the output $y_S(t)$ of S is a discrete signal as shown in ► Figure 1.4.

Sampling is a kind of operation or process to acquire orderly instantaneous values of a continuous signal with a certain time interval.

Looking at the waveform, a discrete signal obtained by sampling has lost much content (information) carried by the original continuous signal; so readers are likely to ask: what is the purpose to make such a discrete signal?

In fact, it is an important part in a communication principles course to change a continuous or an analog signal into a discrete signal by sampling; the aim of such A/D conversion is to achieve digital communication. The sampling theorem is intended to ensure that all information carried by the original continuous signal can be recovered from the discrete.

Sampling theorem: A bandlimited or lowpass continuous signal $f(t)$ with frequency band $[f_L, f_H]$, if it is sampled with sampling frequency $f_S \geq 2f_H$ (the sampling interval is $T_S \leq \frac{1}{f_S}$), will be entirely determined by a sampled signal, which is represented as $y_S(t) = \{f(nT_S)\}$. Alternatively, the original continuous signal $f(t)$ can be restored from the sampled signal $y_S(t)$ without any distortion.

The sampling interval and the sampling frequency are called the Nyquist interval and the Nyquist frequency, respectively.

Note that lowpass signals are a kind of signal whose bandwidths must meet relation $f_H - f_L > f_L$, whereas signals that met relation $f_H - f_L < f_L$ are called bandpass signals.

The sampling theorem is a bridge between the continuous (analog) and discrete (digital) signals and is the theoretical basis of time division multiplexing technology. Therefore, it is widely used in the computer and the communication fields.

The main characteristics of discrete signals are as follows:
(1) Although the independent variable is discrete, the dependent variable (amplitude of the discrete signal) can be continuous (has infinite possible values) or discrete.
(2) The graph of a discrete signal is a series of vertical line segments located at discrete independent variable points.

Other details regarding discrete signals can be found in Chapter 8. A digital signal is often discussed accompanied with a discrete one and is defined as:

A discrete or continuous signal whose dependent variable is discrete and takes limited values is called a digital signal.

In general, a digital signal of which the independent variable is discrete is also called a digital sequence; this is shown as $f[n]$ in ▶ Figure 1.5a. The dependent variable values of the sequence are only 0 and 1, and the independent variable n is an integer, so $f[n]$ can be expressed as

$$f[n] = \begin{cases} 0 & (n < 0) \\ 1 & (n \geq 0) \end{cases}. \tag{1.2-1}$$

To change a discrete signal into a digital one, the discrete signal needs to be quantified as usual.

A process or a method changing infinite possible values of the dependent variable into finite ones is called quantization.

As a result, a digital signal is also said to be a signal with quantified amplitude.

In computer and communication fields, the common digital signal is a kind of continuous time signal as shown in ▶ Figure 1.5b, and is called a baseband digital

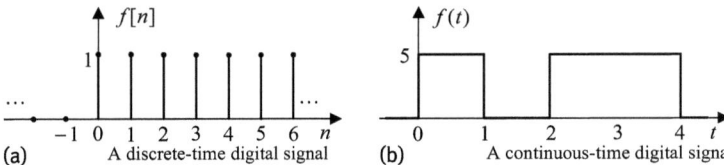

(a) A discrete-time digital signal   (b) A continuous-time digital signal

Fig. 1.5: Digital signals.

signal. From the communication standpoint, the digital signal is also considered as: *a signal using limited values or states of dependent variable to carry messages.*

In conclusion, the terms continuous and discrete describe the change of the independent variable of a signal. The terms analog and digital, however, depict the change of the dependent variable of a signal.

1.2.2 Periodic and aperiodic signals

The periodic signal is a type of signal which waveform arises periodically with a certain time interval and without beginning and end, as shown in ▶ Figure 1.6a, and satisfies the equation

$$f(t) = f(t + nT) \quad n = 0, \pm1, \pm2, \ldots \quad \text{(arbitrary integer)} . \tag{1.2-2}$$

The minimum value of T satisfying the equation is known as the period of a periodic signal. If the expression or a complete period waveform of a periodic signal is given, all values of the signal for all time can be determined. The trigonometric function $f(t) = A\sin(\omega t + \theta)$ is the best known example of a periodic signal to our knowledge.

A signal that is not periodic is called aperiodic such as the exponential signal $f(t) = Ae^{at}$ shown in ▶ Figure 1.6b.

A periodic signal will become aperiodic if its period T tends to infinity so that its waveform no longer arises repeatedly.

This concept is very important because it reveals the inner relationship between a periodic signal and an aperiodic one; this will be used to introduce the Fourier transform in Chapter 5.

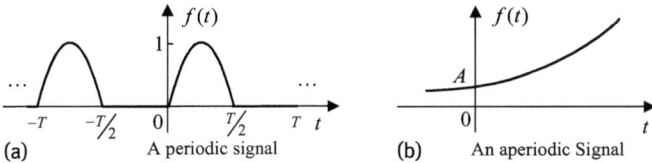

(a) A periodic signal   (b) An aperiodic Signal

Fig. 1.6: A periodic signal and an aperiodic signal.

1.2.3 Energy and power signals

To understand the energy feature of a current or a voltage signal, we need to study the energy or power consumption of a signal through a unit resistor. If the instantaneous power of signal $f(t)$ consumed by a unit resistor is $|f(t)|^2$, then following equations are true:

(1) The consumed energy of $f(t)$ over interval $[-T, T]$ is defined as

$$\int_{-T}^{T} |f(t)|^2 dt .$$ (1.2-3)

The average consumed power of $f(t)$ is defined as

$$\frac{1}{2T} \int_{-T}^{T} |f(t)|^2 dt .$$ (1.2-4)

(2) The consumed energy E of $f(t)$ over range $(-\infty, \infty)$ is defined as

$$E \stackrel{\text{def}}{=} \lim_{T \to \infty} \int_{-T}^{T} |f(t)|^2 dt \quad \text{(in J)} .$$ (1.2-5)

(3) The consumed power P of $f(t)$ over range $(-\infty, \infty)$ is defined as the average power

$$P \stackrel{\text{def}}{=} \lim_{T \to \infty} \frac{1}{2T} \int_{-T}^{T} |f(t)|^2 dt \quad \text{(in W)} .$$ (1.2-6)

Some conclusions based on the above definitions are as follows:
(1) If $0 < E < \infty$ (at this time, $P = 0$), the signal is called a limited energy signal or, simply, an energy signal. Its features are limited amplitude, limited duration time and aperiodicity, for example, a single rectangular pulse.
(2) If $0 < P < \infty$ (at this time, $E = \infty$), the signal is called a limited power signal or, simply, a power signal. Its features are limited amplitude and limitless duration time, for example, a direct current signal, a periodic signal and a random signal. A periodic signal must be a power signal, but a power signal is not necessarily a periodic signal.
(3) A signal cannot be both an energy signal and a power signal, but it can be neither an energy signal nor a power signal. For $f(t) = e^{-2|t|}$, its energy $E = \lim_{T \to \infty} \int_{-T}^{T} \left| e^{-2|t|} \right|^2 dt = 2 \int_{0}^{\infty} e^{-4t} dt = \frac{1}{2}$ and power $P = 0$, therefore it is an energy signal. For the example $f(t) = e^{-2t}$, its energy

$$E = \lim_{T \to \infty} \int_{-T}^{T} (e^{-2t})^2 dt = \lim_{T \to \infty} \left[\frac{1}{4} \left(e^{4T} - e^{-4T} \right) \right] = \infty ,$$

and power

$$P = \lim_{T \to \infty} \frac{E}{2T} = \lim_{T \to \infty} \frac{e^{4T} - e^{-4T}}{8T} = \lim_{T \to \infty} \frac{e^{4T}}{8T} = \lim_{T \to \infty} \frac{4e^{4T}}{8} = \infty .$$

Obviously, $f(t)$ is neither a power nor an energy signal. The classification for signals based on the energy and the power of signal cannot include all signals theoretically.

It is very meaningful for research on energy or power spectrums of a signal in communication principles work to classify signals into energy signals and power signals.

Note the following:

(1) Above T is not the period of signal but a certain moment; $f(t)$ can represent a voltage or a current signal.

(2) If $f(t)$ is a complex signal, the symbol $|f(t)|^2$ represents that $f(t)$ is first taken by magnitude computation and then squared.

1.2.4 Deterministic and random signals

Common signals in engineering are also divided into two types of deterministic and random signals from their change rules. Signals whose future values can be described accurately by a certain mathematical function are deterministic signals, for example, whole values of a sine signal can be determined by a sine function. If the value of a signal is random at any moment, namely, future values of the signal cannot be determined by an accurate function of time, the signal is called an uncertain or a random signal, such as speech signals from the human body. Because future values of this type of signal change randomly with time, it is necessary to describe them by a probability distribution or statistical average values, these signals are also called statistical time signals. Moreover, not only the amplitude of a random signal can change randomly, but also frequency and phase can also change randomly.

Strictly speaking, signals in the objective world are mainly random, for example, audio signals, image signals, biological electric signals, earthquake signals, etc. Only the basic signals for analyses and tests, for example, sines, square waves, triangle waves, exponential signals, etc., are exactly deterministic. The diagrams of deterministic and random signals are shown in ▶ Figure 1.7.

From the perspective of communication, it is meaningful to transmit random signals, because random signals always carry contents or information that receivers do not know but want to know. However, mainly deterministic signals will be considered herein. The reasoning here is that although deterministic signals cannot be used in communication tasks, they can be used to analyze the characteristics of a system as basic signals, and whose analysis methods and results can be directly generalized or applied to the analysis of random signals. This is the motivation for the analysis of

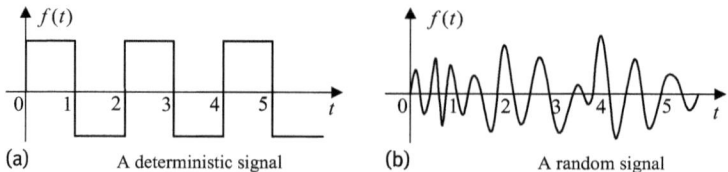

(a) A deterministic signal   (b) A random signal

Fig. 1.7: A deterministic signal and a random signal.

deterministic signals. On the other hand, the analysis of random signals often needs the additional help of stochastic processes to be completed, falling under the topic of communication principles and so will not be discussed herein.

1.2.5 Causal and anticausal signals

According to change features of a signal before and after an observed time point, signals can be divided into causal and anticausal types.

If time $t = 0$ or $n = 0$ is set as the time we start observing a signal, a causal signal satisfies

$$f(t) = 0 , \ t < 0 \quad \text{or} \quad f[n] = 0 , \ n < 0 , \qquad (1.2\text{-}7)$$

and an anticausal signal satisfies

$$f(t) = 0 , \ t > 0 \quad \text{or} \quad f[n] = 0 , \ n > 0 . \qquad (1.2\text{-}8)$$

1.3 Basic continuous-time signals

It is impossible to study various signals one by one in real life, so some basic signals have been selected to be researched for specific reasons as follows:

(1) Basic signals can be used to represent other signals accurately or approximately by mathematical means. For example, the basic form of Fourier series only includes sine and cosine functions, but it can represent most different periodic signals, which will be seen in Chapter 4.

(2) Responses of a system to basic signals play leading roles in system analysis and are of general significance. As examples, impulse or step responses are produced by systems in response to impulse or step signals.

Research on the characteristics and performance of basic signals will be the basis for the analysis of other signals and systems.

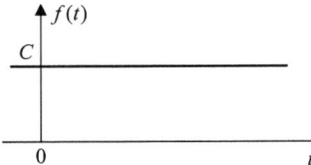

Fig. 1.8: A direct current signal.

1.3.1 Direct current signals

In general, a current or voltage signal with only one changing direction is called a DC signal (direct current signal). However, a DC signal which is often mentioned refers to one whose magnitude and direction are both changeless with time, namely, constant voltage or a constant current signal, which is expressed as

$$f(t) = C \quad \text{(a constant)}, \tag{1.3-1}$$

and is plotted in ▶ Figure 1.8. Obviously, the constant term which is familiar to us in a mathematical expression has a physical meaning in the electric engineering field, it represents a DC signal.

1.3.2 Sinusoidal signals

Sinusoidal signals generally refer to sine and cosine signals that are familiar to us. A cosine signal is actually a sine signal with a shifted phase $\frac{\pi}{2}$, namely, the mathematical relation between of them is $\cos(\omega t + \varphi) = \sin(\omega t + \varphi + \frac{\pi}{2})$. Therefore, sine and cosine signals are collectively called the sinusoidal signal to facilitate research, which is written as

$$f(t) = A \sin(\omega t + \varphi) \tag{1.3-2a}$$

or

$$f(t) = A \cos(\omega t + \varphi). \tag{1.3-2b}$$

Where A is the amplitude, ω the angular frequency and φ the initial phase. The waveform of equation (1.3-2a) is shown in ▶ Figure 1.9.

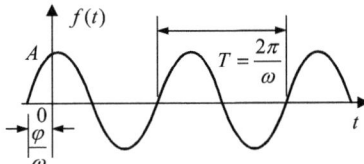

Fig. 1.9: A sinusoidal signal.

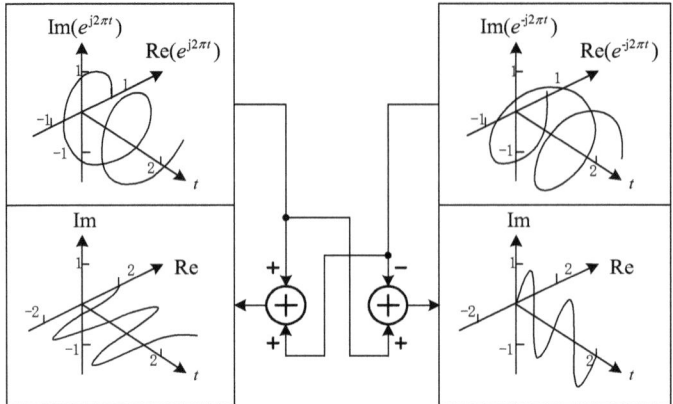

Fig. 1.10: Graphical interpretation of Euler's relation.

In the real world, the natural response (resonant wave) from a LC circuit, simple harmonic vibration in a mechanical system and the sound pressure wave of a single tone from an instrument are all examples of sinusoidal signals.

In circuits analysis, signals and systems, and communication principles courses, sinusoidal signals are often expressed by imaginary exponential signals with Euler's relation, e.g.

$$e^{j\omega t} = \cos \omega t + j \sin \omega t , \tag{1.3-3}$$

$$e^{-j\omega t} = \cos \omega t - j \sin \omega t , \tag{1.3-4}$$

or

$$\sin \omega t = \frac{e^{j\omega t} - e^{-j\omega t}}{2j} , \tag{1.3-5}$$

$$\cos \omega t = \frac{e^{j\omega t} + e^{-j\omega t}}{2} . \tag{1.3-6}$$

The significance of Euler's relation is that it is not only a bridge between trigonometric functions and exponential functions but also a link between real functions and imaginary functions. It has provided an effective way for sinusoidal AC circuit or sinusoidal signal analysis. The graphical explanation for Euler's relation when $\omega = 2\pi$ is shown in ▸ Figure 1.10.

1.3.3 Exponential signals

The exponential signal is also known as the real exponential signal and is of the form

$$f(t) = Ke^{at} , \tag{1.3-7}$$

where K is a constant and a is a real number. It is sketched in ▸ Figure 1.11a.

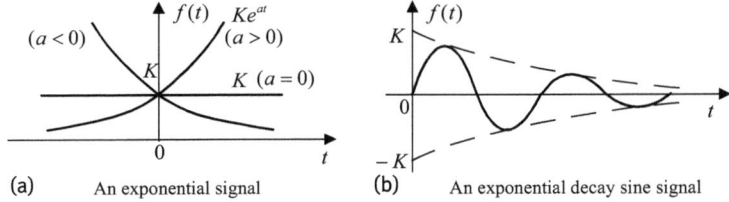

(a) An exponential signal   (b) An exponential decay sine signal

Fig. 1.11: An exponential signal and an exponential decay sine signal.

If $a > 0$, the signal size increases with time t; but if $a < 0$, the size decreases with time t; and if $a = 0$, the signal becomes a DC signal. Sometimes we also discuss those sinusoidal signals multiplied by a growing exponential or a decaying one. The sinusoidal signal multiplied by a decaying exponential is usually called a damped sinusoid and is expressed as

$$f(t) = \begin{cases} 0 & (t < 0) \\ Ke^{-at} \sin \omega t & (t \geq 0) \end{cases} . \tag{1.3-8}$$

It is plotted in ▶ Figure 1.11b.

1.3.4 Complex exponential signals

A complex exponential signal is expressed as

$$f(t) = Ke^{st} , \tag{1.3-9}$$

where $s = \sigma + j\omega$ is called a complex frequency, K is a real constant in general and both σ and ω are real numbers.

Based on Euler's relation, a complex exponential signal can also be expressed as

$$f(t) = Ke^{st} = Ke^{(\sigma+j\omega)t} = Ke^{\sigma t} \cos \omega t + jKe^{\sigma t} \sin \omega t .$$

This expression shows that a complex exponential signal can be decomposed into a real signal and an imaginary signal. Moreover, the real component contains a cosine signal; the imaginary part includes a sine signal. The ω in complex frequency s is the angular frequency of the sine and cosine signals, and σ represents the amplitude change of the sine and cosine functions with time. Therefore, five points are now given from the above:
(1) When $\sigma > 0$, the sine and cosine components are grown sinusoids.
(2) When $\sigma < 0$, the sine and cosine components are damped sinusoids.
(3) When $\sigma = 0$ and $K = 1$, the complex exponential signal has only the imaginary part and becomes an imaginary exponential signal $f(t) = e^{j\omega t}$.

(4) When $\omega = 0$, the complex exponential signal becomes a real exponential signal.
(5) When $s = 0$, the complex exponential signal becomes a DC signal.

Although the complex exponential signal cannot be generated in real practice, it can include changing situations of various signals from the above discussions. Therefore, it is frequently used to describe some basic signals, such as DC signals, exponential signals, sine and cosine signals, grown and damped sinusoids, so it is a very important basic signal. Another important feature is that the response of a linear time invariant system to it is still a complex exponential signal, which only has a different amplitude [as shown in equation (6.8-1)].

1.3.5 Signum signal

The signum signal is defined as

$$\text{sgn}(t) \overset{\text{def}}{=} \begin{cases} -1, & (t < 0) \\ 1, & (t > 0) \end{cases} \qquad (1.3\text{-}10)$$

and is plotted in ▶ Figure 1.12. Its meaning is obvious from the waveform, that is, when $t > 0$, the signal values are positive; otherwise, they are negative. Note that the signum signal can be also written as the sign signal.

1.3.6 Unit step signal

The unit step signal is defined by

$$\varepsilon(t) \overset{\text{def}}{=} \begin{cases} 0, & (t < 0) \\ 1, & (t > 0) \end{cases}, \qquad (1.3\text{-}11)$$

and is illustrated in ▶ Figure 1.13. It is to be noted that it has no definition at time $t = 0$.

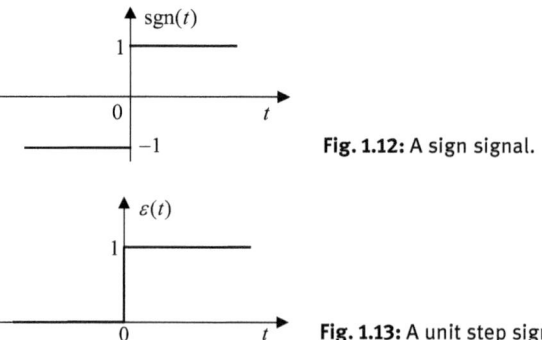

Fig. 1.12: A sign signal.

Fig. 1.13: A unit step signal.

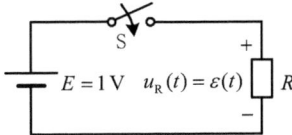

Fig. 1.14: An example of the unit step signal.

The unit step signal is abstracted from practical applications. For example, in ▶ Figure 1.14 a switch S is closed at $t = 0$, and voltage on a resistor R is $u_R(t) = E\varepsilon(t) = \varepsilon(t)$ in ideal condition. This shows that the unit step signal can be used to control other signals as a switch.

Note that the unit step signal is represented by $U(t)$ or $u(t)$ in some books, but $U(t)$ and $u(t)$ are more often used to express a voltage signal in electronics, and therefore, to distinguish it from the voltage symbol, $\varepsilon(t)$ is employed to represent the unit step signal in this book.

The unit step signal can be also combined with a sign signal and a DC signal

$$\varepsilon(t) = \frac{1}{2} + \frac{1}{2}\,\mathrm{sgn}(t)\,. \qquad (1.3\text{-}12)$$

Similarly, the sign signal is also described with a step signal and a DC signal

$$\mathrm{sgn}(t) = 2\varepsilon(t) - 1\,. \qquad (1.3\text{-}13)$$

The unit step signal has four main purposes:
(1) To indicate the starting time of an input or an output signal of a system or to control another signal as a switch. For example, a signal $f(t)$ acting in a system at time $t = 0$ can be denoted by $f(t)\varepsilon(t)$, and the zero-state response of a system can be expressed as $y_f(t) = Eu_C(t)\varepsilon(t)$.
(2) To describe the causality of a system or a signal, namely, the righted side property.
(3) To describe other signals by a linear combination of it and its shifted signals, which can simplify the expressions and analysis courses of these signals as well as the analysis course of system.
(4) As the integral signal of the unit impulse signal, it is used to generate the step response, which is closely related to the impulse response.

Example 1.3-1. Please write math expression of the signal in ▶ Figure 1.15.

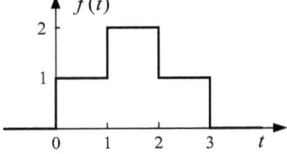

Fig. 1.15: E1.3-1.

Solution. This signal must be divided into five segments to show in the general method, but now from the unit step signal it can be concisely expressed as

$$f(t) = \varepsilon(t) + \varepsilon(t-1) - \varepsilon(t-2) - \varepsilon(t-3).$$

1.3.7 Unit ramp signal

The unit ramp signal is defined as

$$r(t) \overset{\text{def}}{=} \begin{cases} t & t > 0 \\ 0 & t \le 0 \end{cases} \tag{1.3-14}$$

and is sketched in ▶ Figure 1.16. It is also written as $r(t) = t\varepsilon(t)$.

A unit ramp signal and a unit step signal can be related by

$$r(t) = \int_{-\infty}^{t} \varepsilon(\tau)d\tau \tag{1.3-15}$$

or

$$\frac{dr(t)}{dt} = \varepsilon(t) \tag{1.3-16}$$

Obviously, the derivative of a unit ramp signal is a unit step signal, and this is the main purpose for introducing the ramp signal.

1.3.8 Unit impulse signal

The unit impulse signal (or Delta signal) can be defined in various ways.

1. Evolution from a rectangular pulse signal
As shown in ▶ Figure 1.17a, a rectangular pulse $p_\tau(t)$ with width τ height $\frac{1}{\tau}$ is given. Keeping its area $\tau \cdot \frac{1}{\tau} = 1$ the same, when the width τ approaches zero, the height $\frac{1}{\tau}$ will tend to infinity; in this case, $p_\tau(t)$ becomes a unit impulse signal and can be expressed as

$$\delta(t) \overset{\text{def}}{=} \lim_{\tau \to 0} p_\tau(t). \tag{1.3-17}$$

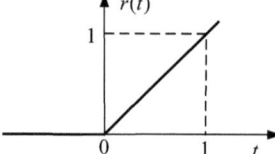

Fig. 1.16: A unit ramp signal.

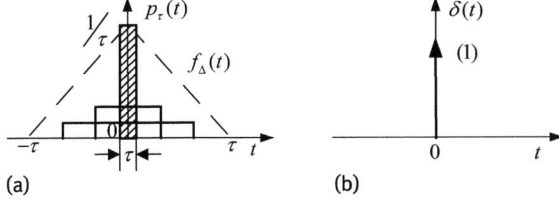

Fig. 1.17: A unit impulse signal.

It can be represented by a vertical arrow (as shown in ▶ Figure 1.17), and the pulse weight 1 (the area under the pulse) is labeled in parentheses next to the arrow. If the area of $p_\tau(t)$ is a constant A, then the impulse signal with weight A can be expressed as $A\delta(t)$, and in the graph A is placed in parenthesis next to the arrow.

2. Dirac's definition

Another definition was given by the British theoretical physicist Dirac (1902–1984), which is now also a common definition, as

$$\begin{cases} \delta(t) = 0, & t \neq 0 \\ \int_{-\infty}^{\infty} \delta(t)\mathrm{d}t = 1, & -\infty < t < +\infty \end{cases} . \tag{1.3-18}$$

It can be shown that values of a unit impulse signal are zero at all time points except the origin, but the area surrounded by the signal waveform is 1.

Of course, besides above two definitions, the unit impulse signal can be also defined by other methods, such as the evolution process of a triangle pulse in Example 1.3-4 and the definition from the screening characteristics.

The unit impulse signal $\delta(t)$ can be thought of as an idealized model of physical phenomena which can generate a lot of energy in a very short time. For example, lightning and earthquakes in the natural world, high voltage sparks in industrial production and the sound of a nail being hit by a hammer in real life, etc.

3. The main properties of unit impulse signal

(1) Sifting

Suppose signal $f(t)$ is continuous at $t = 0$ and bounded everywhere, then we have

$$f(t)\delta(t) = f(0)\delta(t) . \tag{1.3-19}$$

Equation (1.3-19) indicates that the product of $f(t)$ and $\delta(t)$ is a unit impulse signal with weight $f(0)$. Moreover, it can be also generalized as $f(t)\delta(t - t_0) = f(t_0)\delta(t - t_0)$.

Integrating on both sides of equation (1.3-19), we have

$$\int_{-\infty}^{\infty} f(t)\delta(t)\mathrm{d}t = \int_{-\infty}^{\infty} f(0)\delta(t)\mathrm{d}t = f(0) \int_{-\infty}^{\infty} \delta(t)\mathrm{d}t = f(0) . \tag{1.3-20}$$

Equation (1.3-20) can be extended as

$$\int_{-\infty}^{\infty} f(t)\delta(t - t_0)dt = \int_{-\infty}^{\infty} f(t_0)\delta(t - t_0)dt = f(t_0) . \qquad (1.3\text{-}21)$$

The sifting property shows that after we take the integral to the product of a continuous bounded function $f(t)$ and a unit impulse function $\delta(t)$ over range $(-\infty, \infty)$, the value of $f(t)$ at $t = 0$ can be obtained, that is, the value $f(0)$ is sifted. If $\delta(t)$ is delayed to t_0, $f(t_0)$ will be sifted. The term "sifting" can be also replaced by "sampling".

Note that equation (1.3-20) can also be treated as a definition of the unit impulse function, which means that if $x(t)$ holds the expression $\int_{-\infty}^{\infty} f(t)x(t)dt = f(0)$, $x(t)$ is a unit impulse function, or $x(t) = \delta(t)$. This definition is stricter than equation (1.3-17) in the mathematical sense.

(2) Even property
$\delta(t)$ is an even function because it meets

$$\delta(-t) = \delta(t) . \qquad (1.3\text{-}22)$$

Proof. Considering $\int_{-\infty}^{\infty} f(t)\delta(-t)dt$ and letting $t = -\tau$, we have

$$\int_{-\infty}^{\infty} f(t)\delta(-t)dt = -\int_{\infty}^{-\infty} f(-\tau)\delta(\tau)d\tau = \int_{-\infty}^{\infty} f(-\tau)\delta(\tau)d\tau = f(0) .$$

Combining expression $\int_{-\infty}^{\infty} f(t)\delta(t)dt = f(0)$, we obtain $\delta(-t) = \delta(t)$. $\square$

Example 1.3-2. Find the value of $\int_{1}^{3} \cos[\omega(t - 3)]\delta(2 - t)dt$.

Solution. From the even property of $\delta(t)$ we have

$$\int_{1}^{3} \cos[\omega(t - 3)]\delta(2 - t)dt = \int_{1}^{3} \cos[\omega(t - 3)]\delta(t - 2)dt .$$

From the sifting property, we have

$$\int_{1}^{3} \cos[\omega(t - 3)]\delta(t - 2)dt = \int_{1}^{3} \cos[\omega(2 - 3)]\delta(t - 2)dt$$

$$= \int_{1}^{3} \cos(-\omega)\delta(t - 2)dt$$

$$= \cos \omega \int_{1}^{3} \delta(t - 2)dt$$

Because values of $\delta(t-2)$ equal zero everywhere, except time $t = 2$, and the integral interval is $[1, 3]$, we have

$$\int_1^3 \cos[\omega(t-3)]\delta(2-t)\mathrm{d}t = \cos\omega .$$

(3) Time scaling

If a is a real number, we have

$$\delta(at) = \frac{1}{|a|}\delta(t) . \tag{1.3-23}$$

(4) Integral property

From Dirac's definition, we know that

$$\int_{-\infty}^t \delta(\tau)\mathrm{d}\tau = \begin{cases} 0 & (t < 0) \\ 1 & (t > 0) \end{cases} .$$

Referring to the definition [equation (1.3-10)] of the unit step signal, it can be obtained by

$$\int_{-\infty}^t \delta(\tau)\mathrm{d}\tau = \varepsilon(t) . \tag{1.3-24}$$

This shows that the unit step signal is a running integral of the unit impulse signal. Accordingly, the derivative of a unit step signal is a unit impulse signal

$$\delta(t) = \frac{\mathrm{d}\varepsilon(t)}{\mathrm{d}t} . \tag{1.3-25}$$

The expression explains that a unit impulse signal can describe the changing rate of which a unit step signal occurs a step at time $t = 0$.

There are four purposes of the impulse signal $\delta(t)$:

(1) To make the unit step signal derivable for all time.

(2) As a typical signal, it is used to produce the impulse response in the system analysis.

(3) To express a common signal with a linear combination of the signal and its time-shifted signals. This can be seen from the expression $f(t) = \int_{-\infty}^{+\infty} f(\tau)\delta(t-\tau)\mathrm{d}\tau$ in Section 1.4.8.

(4) For the modeling of a sampling system, i.e. it can be considered as an ideal sampling pulse train, like $p(t)$ in Section 1.2.1.

Because the impulse and the step signals are different from some familiar common signals, they are also known as singularity signals or generalized signals.

Example 1.3-3. The waveform of signal $f(t)$ is shown in ▶ Figure 1.18a, please write expressions of $f(t)$ and $y(t) = \frac{\mathrm{d}}{\mathrm{d}t}f(t)$, and draw waveform of $y(t)$.

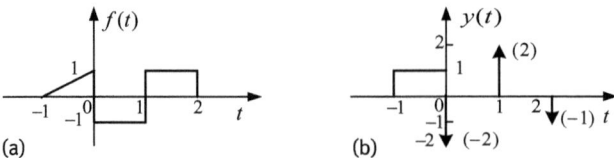

(a)   (b)

Fig. 1.18: E1.3-2.

Solution. From the unit step signal, $f(t)$ can be expressed as

$$f(t) = (t + 1)[\varepsilon(t + 1) - \varepsilon(t)] - \varepsilon(t) + 2\varepsilon(t - 1) - \varepsilon(t - 2).$$

Taking the derivative to $f(t)$, we have

$$y(t) = \frac{d}{dt}f(t) = \varepsilon(t + 1) - \varepsilon(t) + (t + 1)[\delta(t + 1) - \delta(t)] - \delta(t) + 2\delta(t - 1) - \delta(t - 2)$$

$$= \varepsilon(t + 1) - \varepsilon(t) - 2\delta(t) + 2\delta(t - 1) - \delta(t - 2).$$

▶ Figure 1.18b shows the plot of $y(t)$.

1.3.9 Unit doublet signal

The unit doublet signal is defined as the first order derivative of the unit impulse signal; it is written as

$$\delta'(t) = \frac{d\delta(t)}{dt} \quad \text{or} \quad \delta(t) = \int_{-\infty}^{t} \delta'(\tau)d\tau. \tag{1.3-26}$$

Its waveform is composed of two impulse signals whose weights are $-\infty$ and $+\infty$, respectively, as shown in ▶ Figure 1.19. Needless to say, it is also a singularity signal.

Example 1.3-4. Proof weights of a unit doublet signal are $\pm\infty$, as shown in ▶ Figure 1.18.

Proof. Because an impulse signal can be expressed as

$$\delta(t) = \lim_{\tau \to 0} \frac{1}{\tau} \left[\varepsilon\left(t + \frac{\tau}{2}\right) - \varepsilon\left(t - \frac{\tau}{2}\right) \right],$$

from $\delta(t) = \frac{d\varepsilon(t)}{dt}$ the unit doublet signal can be deduced by

$$\delta'(t) = \left\{ \lim_{\tau \to 0} \frac{1}{\tau} \left[\varepsilon\left(t + \frac{\tau}{2}\right) - \varepsilon\left(t - \frac{\tau}{2}\right) \right] \right\}' = \lim_{\tau \to 0} \frac{1}{\tau} \left[\delta\left(t + \frac{\tau}{2}\right) - \delta\left(t - \frac{\tau}{2}\right) \right].$$

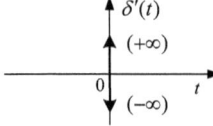

Fig. 1.19: A unit doublet signal.

Obviously, $\pm\frac{1}{\tau}$ are weights of the two impulse signals; when $\tau \to 0$, they tend to $\pm\infty$.

Another way, $\delta(t)$ can be considered as a limit of a triangle pulse $f_\Lambda(t)$ with width 2τ and height $\frac{1}{\tau}$. When $\tau \to 0$, we have $\delta(t) = \lim_{\tau\to 0} f_\Lambda(t)$, which is plotted with dotted lines in ▶ Figure 1.16a). So, $\delta'(t) = [\lim_{\tau\to 0} f_\Lambda(t)]' = \lim_{\tau\to 0} f_\Lambda'(t)$ is a limit of two rectangular pulses with width τ and different heights $\frac{1}{\tau^2}$ and $-\frac{1}{\tau^2}$, which are located at left and right sides of the point of origin when $\tau \to 0$. Their weights (areas) are limits of $\pm\frac{1}{\tau}$ when $\tau \to 0$, that is, $\pm\infty$.

Note: In the above proofs, because two impulse signals $\frac{1}{\tau}\delta(t+\frac{\tau}{2})$ and $-\frac{1}{\tau}\delta(t-\frac{\tau}{2})$ or two rectangular pulses $\frac{1}{\tau^2}$ and $-\frac{1}{\tau^2}$ are located separately on both sides of the origin, when $\tau \to 0$ they will be close to each other infinitely but cannot meet at the same time point after all, and, therefore, the effects of the two pulses cannot cancel each other out. □

The unit doublet signal has the following main features:
(1) Odd property

$$\delta'(-t) = -\delta'(t) . \tag{1.3-27}$$

(2) Integral property

$$\int_{-\infty}^{\infty} \delta'(t)\mathrm{d}t = 0 . \tag{1.3-28}$$

This feature can be seen from ▶ Figure 1.19; that is, since the areas under the positive and the negative impulse signals can be offset by each other, the integral should by zero. In addition, we have

$$\int_{-\infty}^{\infty} \delta'(t)f(t)\mathrm{d}t = -f'(0) , \tag{1.3-29}$$

where $f(t)$ is continuous at $t = 0$, and $f'(0)$ is first-order derivative value of $f(t)$ at $t = 0$.

Proof.

$$\int_{-\infty}^{\infty} \delta'(t)f(t)\mathrm{d}t = \int_{-\infty}^{\infty} f(t)\mathrm{d}\delta(t) = f(t)\delta(t)|_{-\infty}^{\infty} - \int_{-\infty}^{\infty} \delta(t)f'(t)\mathrm{d}t$$

$$= -f'(0) \int_{-\infty}^{\infty} \delta(t)\mathrm{d}t = -f'(0) .$$

Similarly,

$$\int_{-\infty}^{\infty} f(t)\delta^{(n)}(t)\mathrm{d}t = (-1)^n f^{(n)}(0) , \tag{1.3-30}$$

$$\int_{-\infty}^{\infty} f(t)\delta^{(n)}(t - t_0)\mathrm{d}t = (-1)^n f^{(n)}(t_0) . \tag{1.3-31}$$

Note: Equation (1.3-30) can also be considered as a definition of the unit doublet signal just like one of the unit impulse signal. □

(3) Multiplication property

$$f(t)\delta'(t - t_0) = f(t_0)\delta'(t - t_0) - f'(t_0)\delta(t - t_0) . \qquad (1.3\text{-}32)$$

(4) Timing scaling

$$\delta'(at) = \frac{1}{|a|}\frac{1}{a}\delta'(t) . \qquad (1.3\text{-}33)$$

It can be generalized as

$$\delta^{(n)}(at) = \frac{1}{|a|}\frac{1}{a^n}\delta^{(n)}(t) . \qquad (1.3\text{-}34)$$

(5) Convolution property

$$f(t) * \delta'(t) = \frac{d}{dt}f(t) . \qquad (1.3\text{-}35)$$

Example 1.3-5. Obtain the value of integral $\int_{-\infty}^{\infty} A \sin t\delta'(t)dt$.

Solution. According to equation (1.3-29), we obtain

$$\int_{-\infty}^{\infty} A \sin t\delta'(t)dt = -\left.\frac{dA \sin t}{dt}\right|_{t=0} = -A \cos t|_{t=0} = -A .$$

1.3.10 Unit gate signal

The unit gate signal $g_\tau(t)$ is a rectangular pulse signal located at the origin with width τ and height 1, and is defined as

$$g_\tau(t) \overset{\text{def}}{=} \begin{cases} 1 & |t| < \frac{\tau}{2} \\ 0 & |t| > \frac{\tau}{2} \end{cases} . \qquad (1.3\text{-}36)$$

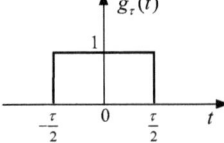

Fig. 1.20: A unit gate signal.

It is illustrated in ▶ Figure 1.20. Note that it is undefined at $t = -\frac{\tau}{2}$ and $\frac{\tau}{2}$. From ▶ Figure 1.20, it can be also described by two shifted step signals

$$g_\tau(t) = \varepsilon(t + \frac{\tau}{2}) - \varepsilon(t - \frac{\tau}{2}) . \qquad (1.3\text{-}37)$$

The gate signal is important in communication principles and has the following main functions:
(1) Representation of the frequency characteristic of an ideal lowpass filter.
(2) From the time shifting property, a series of shifted gate signals can be employed to represent a binary digital signal.
(3) It and the sampling signal, which is also an important signal, form the Fourier transform pair.

1.3.11 Bell shaped pulse signal

The bell shaped pulse signal is defined as

$$f(t) \overset{\text{def}}{=} Ee^{-(\frac{t}{\tau})^2} . \qquad (1.3\text{-}38)$$

It is illustrated in ▶ Figure 1.21 and is a monotone attenuation even function. The parameter τ represents the time width occupied by $f(t)$ when it declines from the maximum value E to $0.78E$. The bell shaped pulse signal is very useful in random signal analysis.

After careful observation, it is found that most of the above signals relate to complex exponential signals or impulse signals, so *complex exponential and impulse signals are considered as two core signals herein*.

To sum up, although there are various signals, those that are analyzed here are analyzed are continuous or discrete, periodic or aperiodic deterministic signals.

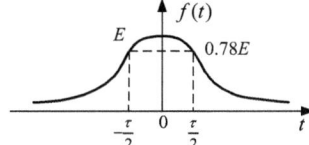

Fig. 1.21: The bell shaped pulse signal.

1.4 Operations of continuous signals

1.4.1 Arithmetic operations

A new continuous signal can be obtained by the addition (subtraction, multiplication or division) of two continuous signals, its value at any time being consistent with the sum (difference, product or quotient) of these two continuous signals at that time.

Example 1.4-1. The waveforms of $f_1(t)$ and $f_2(t)$ are shown in ▶ Figure 1.22a nd b, plot $f_1(t) + f_2(t)$ and $f_1(t) - f_2(t)$ and write their expressions.

Solution. According to the waveforms, the expressions of $f_1(t)$ and $f_2(t)$ are written as

$$f_1(t) = \begin{cases} 0 & (t < 0) \\ t & (0 \le t \le 1) \\ 1 & (t > 1) \end{cases} \quad \text{and} \quad f_2(t) = \begin{cases} 0 & (t < 0) \\ 1 & (0 \le t \le 1) \\ 0 & (t > 1) \end{cases}.$$

Moreover, the expressions of $f_1(t) + f_2(t)$ and $f_1(t) - f_2(t)$ are separately

$$f_1(t) + f_2(t) = \begin{cases} 0 & (t < 0) \\ t + 1 & (0 \le t \le 1) \\ 1 & (t > 1) \end{cases} \quad \text{and} \quad f_1(t) - f_2(t) = \begin{cases} 0 & (t < 0) \\ t - 1 & (0 \le t \le 1) \\ 1 & (t > 1) \end{cases}.$$

So, they are plotted in ▶ Figure 1.22c and d.

It is especially important to bear in mind that when two signals with different varying frequency are multiplied, the lower frequency signal will constitute the envelope of the product signal, and the envelope can reflect the change trend of the product signal. It is also the mathematical foundation of the amplitude modulation technology in a "communication principles" course.

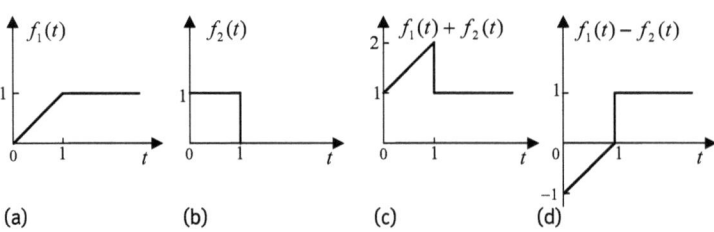

(a) (b) (c) (d)

Fig. 1.22: E1.4-1.

1.4.2 Operations of even and odd signals

Even and odd signals
If a signal $f(t)$ holds the equation $f(t) = -f(-t)$, $f(t)$ is an odd signal.
  If a signal $f(t)$ holds the equation $f(t) = f(-t)$, $f(t)$ is an even signal.
  The definitions are also suitable in discrete time.

Operations of even signals
(1) The sum or difference of two even signals is even. It is also suitable for discrete signals.
(2) The product of two even signals is even. It is also suitable for discrete signals.
(3) The derivative of an even signal is odd.

Proof. If $f(t)$ is even, then the derivative of $f(-t)$ is $f'(-t) = -f'(t)$, and obviously, $f'(t) = -f'(-t)$, and so the derivative of an even signal is odd.   □

Operations of odd signals
(1) The sum or difference of two odd signals is odd. It is also suitable for discrete signals.
(2) The product of two odd signals is even. It is also suitable for discrete signals.
(3) The derivative of an odd signal is even.

Operations of an odd signal and an even signal
(1) The product of an odd signal and an even signal is odd.
(2) The sum of an odd signal and an even signal is neither odd nor even.

The two points are also suitable for discrete signals.

1.4.3 Time shifting

The waveform of shifted signal $f(t + t_0)$ or $f(t - t_0)$ leads or lags behind one of $f(t)$ with a time t_0, that is, the waveform of $f(t + t_0)$ or $f(t - t_0)$ results by moving one of $f(t)$ to the left or to the right with t_0. Both are shown in ▶ Figure 1.23.

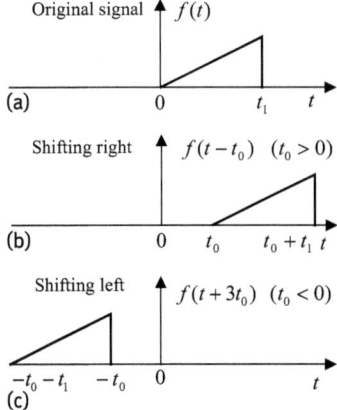

Fig. 1.23: Time Shifting Operation.

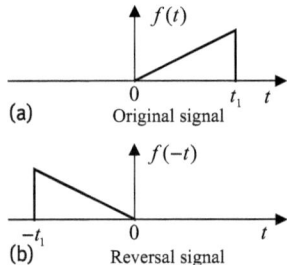

Fig. 1.24: The reversal operation.

1.4.4 Time reversal

The waveform of $f(-t)$ is the mirror image of $f(t)$ about the vertical axis, as is shown in ▶ Figure 1.24.

The waveform of $f[-(t + t_0)]$ can be obtained by moving $f(-t)$ (▶ Figure 1.25a) to the left with t_0 as shown in ▶ Figure 1.25b, and the waveform of $f[-(t - t_0)]$ results by moving $f(-t)$ to the right with t_0, as shown in ▶ Figure 1.25c.

1.4.5 Time scaling

The time scaling operation needs to replace the independent variable t with at in $f(t)$, and $f(at)$ can be obtained, where a is a constant and is called the scale coefficient.

(1) If $a > 1$, the waveform of $f(at)$ can be obtained by that the waveform of $f(t)$ (▶ Figure 1.26a) is compressed to $\frac{1}{a}$ times the original scale in time (▶ Figure 1.26b).

(2) If $0 < a < 1$, the waveform of $f(at)$ can be obtained by the waveform of $f(t)$ being linearly expanded to $\frac{1}{a}$ times of the original scale in time (▶ Figure 1.26c).

(3) If $a < 0$, the waveform of $f(at)$ can be obtained by the waveform of $f(t)$ being inversed first and then compressed or expanded to $\frac{1}{|a|}$ times the original scale in time.

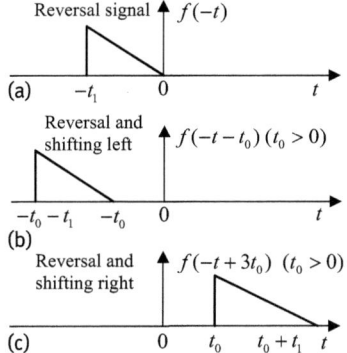

Fig. 1.25: The reversal operation with time shifting.

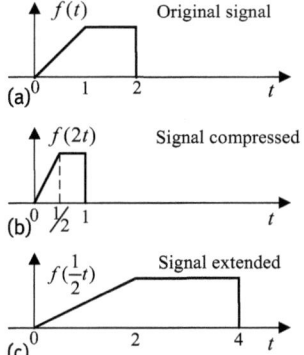

Fig. 1.26: Time scaling.

Example 1.4-2. Please give the waveform of $y(t) = f(-2t + 3)$ the gate signal $f(t)$ is given in ▶ Figure 1.27a.

Solution. The waveform of $y(t)$ can be obtained after $f(t)$ by scaling and shifting and reversal. The order of the first two operations can be changed, that is, there are two methods to solve the problem. Here, we use the way of shifting first and scaling to solve it.

First, $f(t + 3)$ is obtained by moving $f(t)$ to the left by 3 units; it is plotted in ▶ Figure 1.27b. Then, $f(2t + 3)$ is obtained by compressing the waveform of $f(t + 3)$ to half of the original one; it is shown as ▶ Figure 1.27c. Last, overturning the wave-

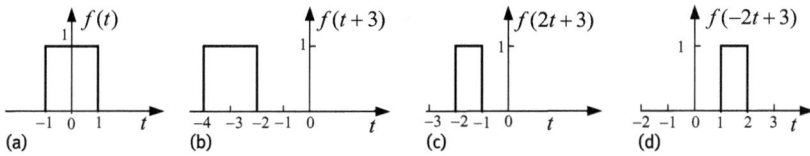

Fig. 1.27: E1.4-2.

form of $f(2t + 3)$ to obtain $y(t)$ (the mirror image with the vertical axis), it is sketched in ▶ Figure 1.27d.

Note: If there are time scaling and time shifting and reversal operations in an exercise at the same time, a suggested approach would be first time shifting, then scaling and, finally, reversal.

1.4.6 Differentiation and integration

If $f(t)$ is known, its differentiation and integration operations are shown, respectively, by

$$f'(t) = \frac{\mathrm{d}f(t)}{\mathrm{d}t} ,$$ (1.4-1)

$$f^{(-1)}(t) = \int_{-\infty}^{t} f(t)\mathrm{d}t .$$ (1.4-2)

1.4.7 Decomposition and synthesis

Generally speaking, the process or the method where one signal is broken into finite or infinite other signals is called decomposition. The process or the method where one signal is formed by several or infinite other signals is called synthesis.

Just as in Example 1.3-1, $f(t)$ in ▶ Figure 1.15 can be expressed as the algebraic sum of several step signals, and this expression is the decomposition to $f(t)$. For instance, again, any signal $f(t)$ can be broken into an algebraic sum of even and odd signals.

Proof. If $f_e(t)$ and $f_o(t)$ are, respectively, even and odd signals. we have

$$f_e(t) = \frac{1}{2}f(t) + \frac{1}{2}f(-t) ,$$ (1.4-3)

$$f_o(t) = \frac{1}{2}f(t) - \frac{1}{2}f(-t) .$$ (1.4-4)

It can be seen that

$$f(t) = f_e(t) + f_o(t) .$$

Equations (1.4-3) and (1.4-4) are usually used to find the odd and the even components in any signal. □

There are several signal decomposition methods. Fourier series, Fourier transform, Laplace transform and z transform, which will be introduced later, can reflect the concept of signal decomposition. *It can be said that decomposition is the essence of signal analysis.* Since synthesis is the inverse operation of decomposition, we have nothing more to say about it.

1.4.8 Convolution integral

The convolution integral, or simply convolution, is a kind of special operation, it is defined as follows:

The convolution of two given functions $f_1(t)$ and $f_2(t)$ is written as

$$f_1(t) * f_2(t) \overset{\text{def}}{=} \int_{-\infty}^{+\infty} f_1(\tau)f_2(t - \tau)d\tau , \qquad (1.4\text{-}5)$$

where, the operator symbol "*" represents the convolution operation.

Example 1.4-3. the signals $f_1(t) = (3e^{-2t} - 1)\varepsilon(t)$ and $f_2(t) = e^t\varepsilon(t)$ are known. Find their convolution.

Solution.

$$f_1(t) * f_2(t) = \int_{-\infty}^{+\infty} f(\tau)f(t - \tau)d\tau = \int_{-\infty}^{+\infty} \left(3e^{-2\tau} - 1\right)\varepsilon(\tau)e^{t-\tau}\varepsilon(t - \tau)d\tau .$$

Since $\varepsilon(\tau) = 0$ for $\tau < 0$, $\varepsilon(\tau) = 1$ for $\tau > 0$, $\varepsilon(t - \tau) = 0$ for $\tau > t$, and $\varepsilon(t - \tau) = 1$ for $\tau < t$, we have

$$f_1(t)*f_2(t) = \int_0^t (3e^{-2\tau} - 1)e^{t-\tau}d\tau = e^t \int_0^t 3e^{-3\tau}d\tau - e^t \int_0^t e^{-\tau}d\tau$$

$$= -e^t \cdot e^{-3\tau}\Big|_0^t + e^t \cdot e^{-\tau}\Big|_0^t$$

$$= 1 - e^{-2t} \quad t \geq 0 .$$

Equation (1.4-5) is a high level abstract mathematical equation, so we will interpret the concept of convolution by means of plotting to understand it better. According to the definition, the convolution operation can be divided into five steps, namely change of variables, reversal, shifting, multiplication, and integration, which will be explained by graphic transformation as follows.

Example 1.4-4. The waveforms of $f_1(t)$ and $f_2(t)$ are shown in ▶ Figure 1.28a and b, respectively. Find $y(t) = f_1(t) * f_2(t)$.

Solution. (1) Change of variables. Changing variable t of $f_1(t)$ and $f_2(t)$ into τ, we obtain $f_1(\tau)$ and $f_2(\tau)$ as shown in ▶ Figure 1.28a and b respectively.
(2) Reversal. Taking the vertical axis as the symmetry axis, we obtain the mirror symmetry signal $f_2(-\tau)$ of $f_2(\tau)$, as shown in ▶ Figure 1.28c.

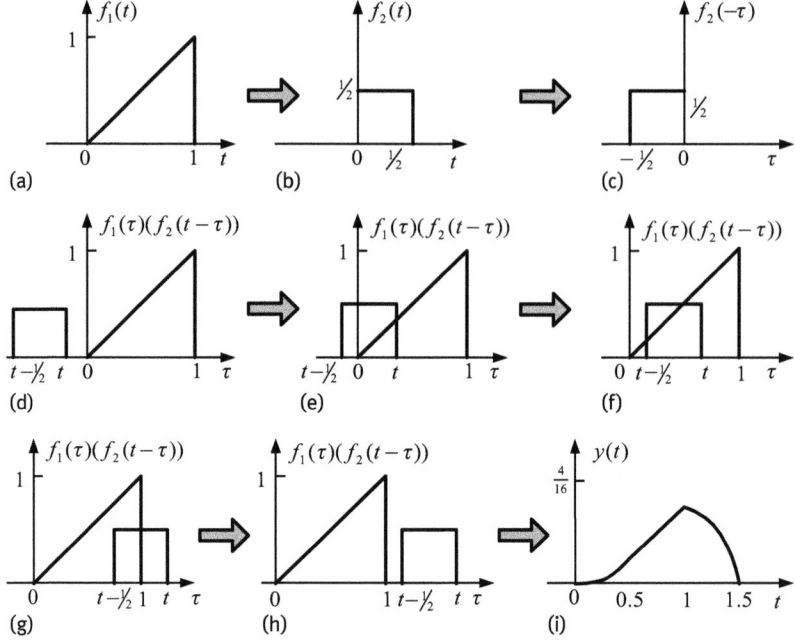

Fig. 1.28: The illustration of convolution.

(3) Shifting. The $f_2(-\tau)$ is shifted by t units to obtain $f_2(t - \tau)$. From ▶ Figure 1.28d to h, we can see that the waveform of $f_2(t - \tau)$ shifts along the τ axis with different values of t.

(4) Multiplication. The product $f_1(\tau) \cdot f_2(t - \tau)$ depends on the values of t.

(5) Integral. The integral to $f_1(\tau) \cdot f_2(t - \tau)$ is calculated and the graphic area corresponding to the product $f_1(\tau) \cdot f_2(t - \tau)$ is computed.

The calculation steps of convolution over different ranges of t are as follows.

(1) For $t \le 0$ shown in ▶ Figure 1.28d, the graphs of $f_1(\tau)$ and $f_2(t - \tau)$ do not overlap and their product is zero; we have

$$y(t) = f_1(t) * f_2(t) = 0 .$$

(2) For $0 < t \le \frac{1}{2}$ shown in ▶ Figure 1.28e, $f_2(t - \tau)$ and $f_1(\tau)$ have a nonzero overlapping portion, which is the area under their product curve. The lower limit is zero and the upper limit is t of the integral, so

$$y(t) = f_1(t) * f_2(t) = \int_0^t \tau \cdot \frac{1}{2} d\tau = \frac{1}{4} t^2 .$$

(3) For $\frac{1}{2} < t \le 1$ shown in ▶ Figure 1.28f, $f_2(t - \tau)$ and $f_1(\tau)$ overlap completely. The lower and upper limits of the integral are, respectively, $t - \frac{1}{2}$ and t, and we have

$$y(t) = f_1(t) * f_2(t) = \int_{t-\frac{1}{2}}^{t} \tau \cdot \frac{1}{2} d\tau = \frac{1}{4}t - \frac{1}{16} \ .$$

(4) For $1 < t \le \frac{3}{2}$ shown in ▶ Figure 1.28g, a portion of $f_2(t - \tau)$ has moved out of the nonzero region of $f_1(\tau)$. The lower and upper limits are, respectively, $t - \frac{1}{2}$ and 1, and we have

$$y(t) = f_1(t) * f_2(t) = \int_{t-\frac{1}{2}}^{1} \tau \cdot \frac{1}{2} d\tau = -\frac{1}{4}t^2 + \frac{1}{4}t + \frac{3}{16} \ .$$

(5) When $t > \frac{3}{2}$, as shown in ▶ Figure 1.28h, $f_2(t - \tau)$ has completely moved out of the nonzero region of $f_1(\tau)$, and their product is zero, that is,

$$y(t) = f_1(t) * f_2(t) = 0 \ .$$

From the above 5 equations, the sectionalized expression of $y(t)$ is

$$y(t) = \begin{cases} \frac{1}{4}t^2 & \left(0 \le t \le \frac{1}{2}\right) \\ \frac{1}{4}t - \frac{1}{16} & \left(\frac{1}{2} < t \le 1\right) \\ -\frac{1}{4}t^2 + \frac{1}{4}t + \frac{3}{16} & \left(1 < t \le \frac{3}{2}\right) \\ 0 & \text{(other)} \end{cases} ,$$

which is pictured in ▶ Figure 1.28i.

As we have seen, convolution can actually be considered as an overlap integral. The above graphic representation also shows that convolution to some functions can be computed directly by using the graphical method to avoid complicated mathematical calculations.

Convolution has the following properties.

(1) A commutative property

$$f_1(t) * f_2(t) = f_2(t) * f_1(t) \ . \tag{1.4-6}$$

Proof.

$$f_1(t) * f_2(t) = \int_{-\infty}^{+\infty} f_1(\tau)f_2(t - \tau)d\tau \overset{\text{Let } t-\tau=x}{=} \int_{-\infty}^{+\infty} f_1(t - x)f_2(x)dx = f_2(t) * f_1(t) \ .$$

$\square$

(2) A distributive property

$$f_1(t) * [f_2(t) + f_3(t)] = f_1(t) * f_2(t) + f_1(t)f_3(t) . \tag{1.4-7}$$

Proof.

$$f_1(t) * [f_2(t) + f_3(t)] = \int_{-\infty}^{+\infty} f_1(\tau)[f_2(t-\tau) + f_3(t-\tau)]d\tau$$

$$= \int_{-\infty}^{+\infty} f_1(\tau)f_2(t-\tau)d\tau + \int_{-\infty}^{+\infty} f_1(\tau)f_3(t-\tau)d\tau$$

$$= f_1(t) * f_2(t) + f_1(t) * f_3(t) .$$

□

(3) An associative property

$$[f_1(t) * f_2(t)] * f_3(t) = f_1(t) * [f_2(t) * f_3(t)] . \tag{1.4-8}$$

Proof.

$$[f_1(t) * f_2(t)] * f_3(t) = \int_{-\infty}^{+\infty} \left[\int_{-\infty}^{+\infty} f_1(\tau)f_2(\eta-\tau)d\tau \right] f_3(t-\eta)d\eta$$

$$\overset{\text{Exchange}}{\underset{\text{sequence}}{=}} \int_{-\infty}^{+\infty} f_1(\tau) \left[\int_{-\infty}^{+\infty} f_2(\eta-\tau)f_3(t-\eta)d\eta \right] d\tau$$

$$\overset{\text{Let } \eta-\tau=x}{=} \int_{-\infty}^{+\infty} f_1(\tau) \left[\int_{-\infty}^{+\infty} f_2(x)f_3(t-\tau-x)dx \right] d\tau$$

Then

$$f_1(t) * [f_2(t) * f_3(t)] = f_1(t) * \left[\int_{-\infty}^{+\infty} f_2(\tau)f_3(t-\tau)d\tau \right]$$

$$= \int_{-\infty}^{+\infty} f_1(\xi) \left[\int_{-\infty}^{+\infty} f_2(\tau)f_3(t-\xi-\tau)d\tau \right] d\xi .$$

By comparing the two equations, the property is proved to be true. □

(4) A differential property
Suppose $y(t) = f_1(t) * f_2(t)$, then

$$y'(t) = f_1(t) * f_2'(t) = f_1'(t) * f_2(t) . \tag{1.4-9}$$

Proof.

$$
\begin{aligned}
y'(t) &= \frac{d}{dt}[f_1(t) * f_2(t)] = \frac{d}{dt}\left[\int_{-\infty}^{+\infty} f_1(\tau)f_2(t-\tau)d\tau\right] \\
&\overset{\substack{\text{Exchange} \\ \text{sequence}}}{=} \int_{-\infty}^{+\infty} f_1(\tau)\left\{\frac{d}{dt}[f_2(t-\tau)]\right\}d\tau = \int_{-\infty}^{+\infty} f_1(\tau)\left\{\frac{d}{d(t-\tau)}[f_2(t-\tau)]\right\}d\tau \\
&= f_1(t) * \frac{df_2(t)}{dt} .
\end{aligned}
$$

□

(5) An integral property

Suppose $y(t) = f_1(t) * f_2(t)$, then

$$
\int_{-\infty}^{t} y(x)dx = f_1(t) * \left[\int_{-\infty}^{t} f_2(x)dx\right] = \left[\int_{-\infty}^{t} f_1(x)dx\right] * f_2(t) . \tag{1.4-10}
$$

Proof.

$$
\begin{aligned}
\int_{-\infty}^{t} y(x)dx &= \int_{-\infty}^{t}\left[\int_{-\infty}^{+\infty} f_1(\tau)f_2(x-\tau)d\tau\right]dx \\
&\overset{\substack{\text{Exchange} \\ \text{sequence}}}{=} \int_{-\infty}^{+\infty} f_1(\tau)\left[\int_{-\infty}^{t} f_2(x-\tau)dx\right]d\tau \overset{x-\tau=\xi}{=} \int_{-\infty}^{+\infty} f_1(\tau)\left[\int_{-\infty}^{t-\tau} f_2(\xi)d\xi\right]d\tau \\
&\overset{\xi=x}{=} \int_{-\infty}^{+\infty} f_1(\tau)\left[\int_{-\infty}^{t-\tau} f_2(x)dx\right]d\tau = f_1(t) * \left[\int_{-\infty}^{t} f_2(x)dx\right] .
\end{aligned}
$$

□

(6) A calculus property

Suppose $y(t) = f_1(t) * f_2(t)$, and if $f_1(t)|_{t=-\infty} = f_2(t)|_{t=-\infty} = 0$, then

$$
y(t) = f_1'(t) * \int_{-\infty}^{t} f_2(t)dt = \int_{-\infty}^{t} f_1(t)dt * f_2'(t) . \tag{1.4-11}
$$

The proof of this property is simple, and the only thing we need to do is to apply the differential property into equation (1.4-10).

(7) Convolution with the impulse signal property

$$
f(t) * \delta(t) = \int_{-\infty}^{+\infty} f(\tau)\delta(t-\tau)d\tau = \int_{-\infty}^{+\infty} f(\tau)\delta(\tau-t)d\tau = f(t) . \tag{1.4-12}
$$

Similarly,

$$f(t) * \delta(t - t_0) = \int_{-\infty}^{+\infty} f(\tau)\delta(t - \tau - t_0)d\tau = \int_{-\infty}^{+\infty} f(\tau)\delta(\tau - t + t_0)d\tau = f(t - t_0). \quad (1.4\text{-}13)$$

Obviously, the convolution of an arbitrary function $f(t)$ and a unit impulse signal that is delayed by t_0 units is just that $f(t)$ is also delayed with t_0 units, and its waveform is unchanged. This phenomenon is called the reappearance characteristic.

Using the properties of convolution differentials and integrals, we also obtain

$$f(t) * \delta'(t) = f'(t), \quad (1.4\text{-}14)$$

$$f(t) * \varepsilon(t) = \int_{-\infty}^{t} f(x)dx. \quad (1.4\text{-}15)$$

We also have the generalized expressions

$$f(t) * \delta^{(k)}(t) = f^{(k)}(t), \quad (1.4\text{-}16)$$

$$f(t) * \delta^{(k)}(t - t_0) = f^{(k)}(t - t_0). \quad (1.4\text{-}17)$$

Example 1.4-5. $f_1(t)$ and $f_2(t)$ are, respectively, depicted in ▶ Figure 1.29a and b; please find the result of $f_1(t) * f_2(t)$.

Solution. According to the calculus feature of convolution, we obtain

$$f_1(t) * f_2(t) = f_1'(t) * \int_{-\infty}^{t} f_2(\tau)d\tau = \int_{-\infty}^{t} f_1(\tau)d\tau * f_2'(t),$$

then

$$f_1(t) * f_2(t) = f_2'(t) * \int_{-\infty}^{t} f_1(\tau)d\tau = [\delta(t) + \delta(t - 1)] * \int_{-\infty}^{t} 2e^{-\tau}\varepsilon(\tau)d\tau$$

$$= 2[\delta(t) + \delta(t - 1)] * \int_{0}^{t} e^{-\tau}d\tau = 2[\delta(t) + \delta(t - 1)] * \left[\left(1 - e^{-t}\right)\varepsilon(t)\right]$$

$$= 2\left(1 - e^{-t}\right)\varepsilon(t) + 2\left[1 - e^{-(t-1)}\right]\varepsilon(t - 1).$$

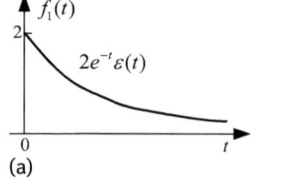

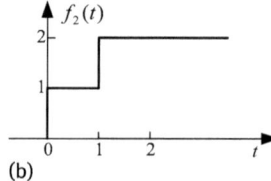

(a)　　　　　　　　　　　　　(b)

Fig. 1.29: E1.4-5.

(8) The time shifting property

If $y(t) = f_1(t) * f_2(t)$, then

$$f_1(t - t_1) * f_2(t - t_2) = y(t - t_1 - t_2) . \qquad (1.4\text{-}18)$$

The proof is omitted here.

Example 1.4-6. Find the convolution $y(t) = f_1(t) * f_2(t)$, where $f_1(t) = \text{sgn}(t - 1)$ and $f_2(t) = e^{-(t+1)}\varepsilon(t + 1)$.

Solution. With time shifting, we have

$$y(t) = f_1(t) * f_2(t) = \text{sgn}(t - 1) * e^{-(t+1)}\varepsilon(t + 1) = \text{sgn}(t) * e^{-t}\varepsilon(t)$$
$$= [2\varepsilon(t) - 1] * e^{-t}\varepsilon(t) = 2\varepsilon(t) * e^{-t}\varepsilon(t) - 1 * e^{-t}\varepsilon(t)$$

and

$$\varepsilon(t) * e^{-t}\varepsilon(t) = \int_{-\infty}^{\infty} \varepsilon(t - \tau)e^{-\tau}\varepsilon(\tau)d\tau = \int_{0}^{t} e^{-\tau}d\tau = - e^{-\tau}\big|_{0}^{t} = \left(1 - e^{-t}\right)\varepsilon(t)$$

$$1 * e^{-t}\varepsilon(t) = \int_{0}^{\infty} e^{-\tau}d\tau = -e^{-\tau}\big|_{0}^{\infty} = 1 ,$$

and then

$$y(t) = f_1(t) * f_2(t) = 2\left(1 - e^{-t}\right)\varepsilon(t) - 1 .$$

For convenience, the main properties of the convolution operation are listed in Table 1.1.

Here, readers may ask: Why do we construct the convolution as such a strange operation? The reason is that it can be used in signal decomposition and plays an important role in system analysis in the time domain.

Tab. 1.1: Main properties of convolution.

No.	Property	Expression
1	Commutative property	$f_1(t) * f_2(t) = f_2(t) * f_1(t)$
2	Distribution property	$f_1(t) * [f_2(t) + f_3(t)] = f_1(t) * f_2(t) + f_1(t)f_3(t)$
3	Associative property	$[f_1(t) * f_2(t)] * f_3(t) = f_1(t) * [f_2(t) * f_3(t)]$
4	Differential property	$y'(t) = f_1(t) * f_2'(t) = f_1'(t) * f_2(t)$
5	Integral property	$\int_{-\infty}^{t} y(x)dx = f_1(t) * \left[\int_{-\infty}^{t} f_2(x)dx\right] = \left[\int_{-\infty}^{t} f_1(x)dx\right] * f_2(t)$
6	Calculus property	$y(t) = f_1'(t) * \int_{-\infty}^{t} f_2(t)dt = \int_{-\infty}^{t} f_1(t)dt * f_2'(t)$
7	Convolution with an impulse signal property	$f(t) * \delta(t) = f(t), \quad f(t) * \delta^{(k)}(t) = f^{(k)}(t)$ $f(t) * \delta(t - t_0) = f(t - t_0), \quad f(t) * \delta^{(k)}(t - t_0) = f^{(k)}(t - t_0)$
8	Time shifting property	$f_1(t - t_1) * f_2(t - t_2) = y(t - t_1 - t_2)$

1.4.9 Plotting

Plotting is the visualization of the mathematical expression of a signal. Plotting methods have been learned by means of the calculus properties of functions in advanced mathematics courses, but in this section, the emphasis will be on ways of mapping with the operation features of signals and the characteristics of signal itself. Herein, most plotting ways relate to sign, step and impulse signals, so readers should understand the definitions and properties of these signals very well. Additionally, readers need to note that the operating ways for a composite function formed by the impulse function are totally different from common composite functions, for example, the derivative operation to this type of composite function. The reason is

$$\delta[f(t)] = \sum_{i=1}^{n} \frac{1}{|f'(t_i)|} \delta(t - t_i),$$ (1.4-19)

where t_i are n different real roots of $f(t) = 0$. For example, if $f(t) = 4t^2 - 1$, we have

$$\delta\left(4t^2 - 1\right) = \frac{1}{4}\delta\left(t + \frac{1}{2}\right) + \frac{1}{4}\delta\left(t - \frac{1}{2}\right).$$

The following examples should be studied carefully to master plotting skills and methods better.

Example 1.4-7. Plot the following signals.
(1) $f_1(t) = \varepsilon(-2t + 3)$
(2) $f_2(t) = e^{-t} \sin 4\pi t[\varepsilon(t - 1) - \varepsilon(t - 2)]$
(3) $f_3(t) = \varepsilon[\cos(\pi t)]$
(4) $f_4(t) = \text{sgn}[\sin(\pi t)]$
(5) $f_5(t) = \delta(t^2 - 4)$
(6) $f_6(t) = \sin[\pi t \, \text{sgn}(t)]$

Solution. Waveforms of these signals are shown in ► Figure 1.30.

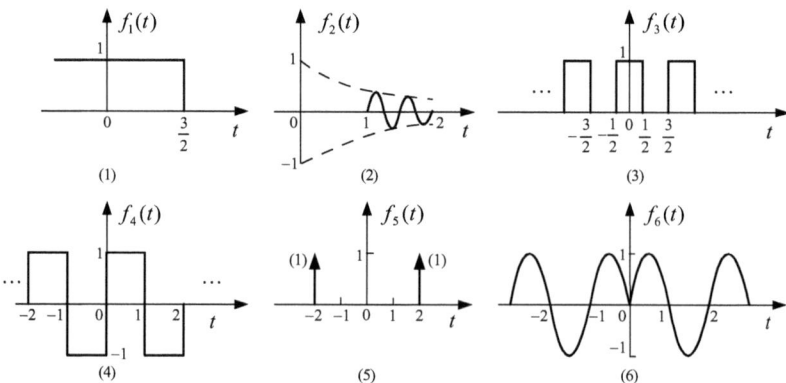

Fig. 1.30: E1.4-7.

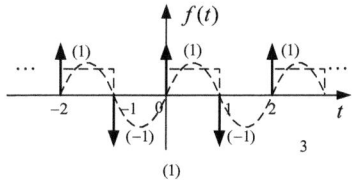

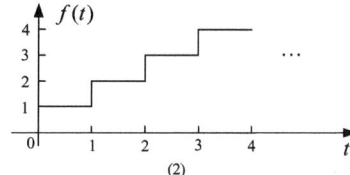

Fig. 1.31: E1.4-8.

Example 1.4-8. Plot the following signals.
(1) $f(t) = \frac{d}{dt}\{\varepsilon[\sin(\pi t)]\}$
(2) $f(t) = \int_{0-}^{t} \delta(\sin(\pi \tau))d\tau$

Solution. The solution for this example can be obtained by (1.4-19), but it is more trouble, so we instead find it by using the definitions and properties of singularity functions.

(1) The independent variable of the composite step signal in this example is a sine function; from the definition of the unit step signal the values of the composite signal are 1 in all the positive sine semicycles, while values are 0 in all the negative semicycles. So, the waveform of the composite signal is a periodic rectangular pulse train, and the derivative waveform of the signal is a series of unit impulse signals located at different time points which alternate between positive and negative values, as shown in ▶ Figure 1.31a.

(2) Referring to the solution of (1), we have

$$f(t) = \int_{0-}^{t} \delta(\sin(\pi \tau))d\tau = \int_{0-}^{t} [\delta(\tau) + \delta(\tau - 1) + \delta(\tau - 2) + \dots]\,d\tau$$

$$= \varepsilon(t) + \varepsilon(t - 1) + \varepsilon(t - 2) + \dots$$

So, the waveform of $f(t)$ is the sum of a series of unit step signals located in different places, as shown in ▶ Figure 1.31b.

1.5 Solved questions

Question 1-1. Calculate the following equations.
(1) $\int_{-\infty}^{\infty} \delta(-at)f(t)dt \quad (a > 0)$;
(2) $\int_{-\infty}^{\infty} \cos 2\pi t \delta(2t - 1)dt$

Solution. (1) From the even, the sampling, and the scaling properties, we have

$$\int_{-\infty}^{\infty} \delta(-at)f(t)dt = \frac{1}{a} \int_{-\infty}^{\infty} \delta(t)f(t)dt = \frac{1}{a}f(0).$$

(2) According to the sampling and scaling properties of the impulse signal, we obtain

$$\int_{-\infty}^{\infty} \cos 2\pi t \delta(2t - 1) dt = \frac{1}{2} \int_{-\infty}^{\infty} \cos 2\pi t \delta \left(t - \frac{1}{2} \right) dt = \frac{1}{2} \cos 2\pi t \Big|_{t=\frac{1}{2}} = -\frac{1}{2}.$$

Question 1-2. Calculate the value of $\delta(t) \cos(2t)$.

Solution. According to the sampling of the impulse signal, we obtain

$$\delta(t) \cos(2t) = \delta(t) \cos(0) = \delta(t).$$

Question 1-3. Calculate $\int_0^t (\tau^2 + 2)\delta(2 - \tau)d\tau$

Solution.

$$\int_0^t \left(\tau^2 + 2 \right) \delta(2 - \tau) d\tau = \int_0^t \left(2^2 + 2 \right) \delta(2 - \tau) d\tau = 6 \int_0^t \delta(2 - \tau) d\tau$$

$$= 6 \int_0^t \delta(\tau - 2) d\tau = 6 \int_{-2}^{t-2} \delta(\eta) d\eta \overset{t-2>0}{=} 6\varepsilon(t).$$

Question 1-4. Given $f(6 - 2t) = 2\delta(t - 3)$, find the value of $\int_{-1}^{\infty} f(t)dt$.

Solution. Let $x = 6 - 2t$, then $f(x) = 2\delta\left(\frac{x}{2}\right)$.
So

$$\int_{-1}^{\infty} f(t)dt = \int_{-1}^{\infty} f(x)dx = \int_{-1}^{\infty} 2\delta\left(\frac{x}{2}\right) dx \overset{y=\frac{x}{2}}{=} \int_{-1/2}^{\infty} 4\delta(y)dy = 4\varepsilon(y) \overset{t=y}{=} 4\varepsilon(t).$$

Question 1-5. Calculate the value of $\int_{-\infty}^{\infty} \left(1 - 2t^2 + \sin \frac{\pi t}{3} \right) \delta(1 - 2t)dt$.

Solution. since the impulse signal is even, from the sampling and the scaling of it, there is

$$\int_{-\infty}^{\infty} \left(1 - 2t^2 + \sin \frac{\pi t}{3} \right) \delta(1 - 2t)dt = \int_{-\infty}^{\infty} \left(1 - 2t^2 + \sin \frac{\pi t}{3} \right) \Big|_{t=\frac{1}{2}} \delta(1 - 2t)dt$$

$$= \int_{-\infty}^{\infty} \delta(1 - 2t)dt = \frac{1}{2} \int_{-\infty}^{\infty} \delta \left(t - \frac{1}{2} \right) dt = \frac{1}{2}.$$

Question 1-6. Calculate the value of $\int_{-\infty}^{\infty} (t^2 + 2) \left[\delta'(t - 1) + \delta(t - 1) \right] dt$.

Solution.

$$\int_{-\infty}^{\infty} \left(t^2 + 2 \right) \left[\delta'(t - 1) + \delta(t - 1) \right] dt = \int_{-\infty}^{\infty} \left(t^2 + 2 \right) \delta'(t - 1)dt + \int_{-\infty}^{\infty} \left(t^2 + 2 \right) \delta(t - 1)dt$$

$$= -2 + 3 = 1.$$

Question 1-7. Calculate the value of $[\varepsilon(t) - \varepsilon(t - 2)]\,\delta(2t - 2)$.

Solution. From sampling and scaling of the impulse signal, we have

$$[\varepsilon(t) - \varepsilon(t - 2)]\,\delta(2t - 2) = [\varepsilon(t) - \varepsilon(t - 2)]\,\delta[2(t - 1)] = [\varepsilon(t) - \varepsilon(t - 2)]\,\frac{1}{2}\delta[(t - 1)]$$

$$= [\varepsilon(1) - \varepsilon(1 - 2)]\,\frac{1}{2}\delta(t - 1) = \frac{1}{2}\delta(t - 1)$$

Question 1-8. $f(t)$ is shown in ▶ Figure Q1-8, plots $0.5\,[f(t) + f(-t)]$ and $0.5\,[f(t) - f(-t)]$.

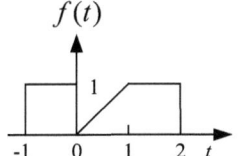

Fig. Q1-8

Solution. Reversing $f(t)$, $f(-t)$ is shown in ▶ Figure Q1-8-1a;
Adding $f(t)$ and $f(-t)$, then $0.5[f(t) + f(-t)]$ is shown in ▶ Figure Q1-8-1b;
Subtracting $f(t)$ and $f(-t)$, then $0.5\,[f(t) - f(-t)]$ is shown in ▶ Figure Q1-8-1c.

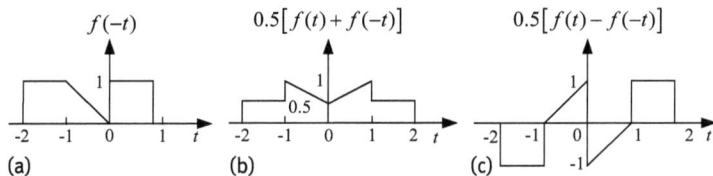

Fig. Q1-8-1

Question 1-9. Signal $f(t)$ is shown in ▶ Figure Q1-9. Plot $f(t + 1)[\varepsilon(t) - \varepsilon(t - 1)]$, $f\left(-\frac{1}{2}t\right)$ and $f(2t - 1)$.

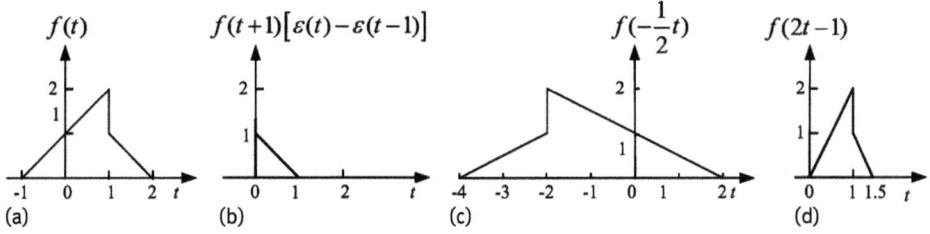

Fig. Q1-9

Solution. From the algorithm and the properties of $\varepsilon(t)$, the above signals can be plotted in ▸ Figure Q1-9a, b and c.

Question 1-10. Calculate $[(1 - t)\delta'(t)] * \varepsilon(t)$.

Solution.

$$[(1 - t)\delta'(t)] * \varepsilon(t) = [\delta'(t) - t\delta'(t)] * \varepsilon(t) = [\delta'(t) - 0\delta'(t) + \delta(t)] * \varepsilon(t)$$
$$= \delta'(t) * \varepsilon(t) + \delta(t) * \varepsilon(t) = \delta(t) + \varepsilon(t) .$$

Question 1-11. The waveforms of $f(t)$ and $h(t)$ are shown in ▸ Figure Q1-11 (1)a and b. Plot $f(t) * h(t)$.

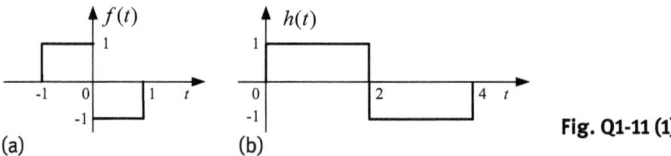

Fig. Q1-11 (1)

(a)   (b)

Solution. Integrating $f(t)$ and differentiating $h(t)$, we obtain the waveforms in ▸ Figure Q1-11 (2)a and b.
    Then

$$f_\Delta(t) = \int_\infty^t f(\tau)d\tau, \quad h'(t) = \delta(t) - 2\delta(t - 2) + \delta(t + 4) .$$

According to the characteristics of convolution, we obtain

$$f(t) * h(t) = f_\Delta(t) * h'(t) = f_\Delta(t) - 2f_\Delta(t - 2) + f_\Delta(t - 4) .$$

Thus, $f(t) * h(t)$ can be plotted as shown in ▸ Figure Q1-11 (3).

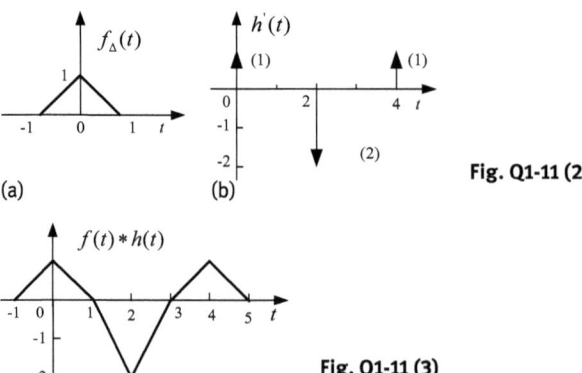

Fig. Q1-11 (2)

(a)   (b)

Fig. Q1-11 (3)

1.6 Learning tips

A signal is an object and a product of system processing, so readers should pay more attention to the following points:
(1) A signal is a function of time.
(2) If the period T tends to infinity, a periodic signal will become a nonperiodic signal.
(3) The complex exponential and impulse signals are the core signals herein.
(4) Usually, a signal can be decomposed into the algebraic sum of limited or unlimited other signals.
(5) Convolution integral operation.
(6) The characteristics of the signal itself can be used to draw its waveform.

1.7 Problems

Problem 1-1. Plot the following composite signals.
(1) $f_1(t) = \varepsilon\,(t^2 - 4)$
(2) $f_2(t) = \text{sgn}\,(t^2 - 1)$
(3) $f_3(t) = \delta\,(t^2 - 9)$
(4) $f_4(t) = \text{sgn}\,[\cos\,(\pi t)]$
(5) $f_5(t) = \delta\,[\cos\,(\pi t)]$
(6) $f_6(t) = \varepsilon\,[\cos\,(2\pi t)]$
(7) $f_7(t) = \sin\,[\pi t\,\text{sgn}(t)]$
(8) $f_8(t) = \frac{d}{dt}\,\{\varepsilon\,[\sin(\pi t)]\}$
(9) $f_9(t) = \frac{d}{dt}\,[e^{-3t}\varepsilon(t)]$
(10) $f_{10}(t) = \int_{0-}^{t} \delta\,(\sin(\pi\tau))d\tau$

Problem 1-2. Plot the following functions.
(1) $f_1(t) = \varepsilon\,(-2t + 3)$,
(2) $f_2(t) = t\varepsilon(t - 1)$
(3) $f_3(t) = (t - 1)\varepsilon(t - 1)$,
(4) $f_4(t) = [\varepsilon(t - 1) - \varepsilon(t - 2)]\,e^{-t}\cos(10\pi t)$
(5) $f_5(t) = [\varepsilon(t) - 2\varepsilon(t - T) + \varepsilon(t - 2T)]\sin\left(\frac{4\pi}{T}t\right)$

Problem 1-3. The waveform of $f(t)$ is shown in ▶ Figure P1-3, try to plot the following functions.
(1) $f(2t)$
(2) $f(t + 3)$
(3) $f(t - 2)$
(4) $f(-4 - t)$
(5) $f(2t - 4)$
(6) $f(4 - 3t)$

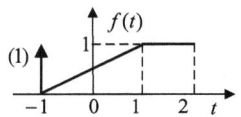

Fig. P1-3

Problem 1-4. What is the numerical value of each of the following integrals?

(1) $\int_{-\infty}^{\infty} f(t - t_0)\delta(t)dt$

(2) $\int_{-\infty}^{\infty} f(t_0 - t)\delta(t)dt$

(3) $\int_{0}^{\infty} \delta(t - 4)\varepsilon(t - 2)dt$

(4) $\int_{-\infty}^{\infty} (e^{-2t} + te^{-t})\delta(2 - t)dt$

(5) $\int_{-\infty}^{\infty} e^{-3t} \sin(\pi t) \left[\delta(t) - \delta\left(t - \frac{1}{3}\right)\right]dt$

(6) $\int_{-\infty}^{\infty} \frac{\sin(2t)}{t}\delta(t)dt$

(7) $\int_{-1}^{1} (t^2 - 3t + 1)\delta(t - 2)dt$

(8) $\int_{-1}^{1} \delta(t^2 - 4)\,dt$

(9) $\int_{-\infty}^{\infty} e^{-\tau}\delta(2\tau)d\tau$

(10) $\int_{-\infty}^{t} e^{-\tau}\delta'(2\tau)d\tau$

(11) $f(t) = 3\delta(t - 3)$ is known, find the value of $\int_{0_-}^{\infty} f(6 - 3t)dt$.

(12) $f(5 - 2t) = 2\delta(t - 3)$ is known, find the value of $\int_{0_-}^{\infty} f(t)dt$.

Problem 1-5. What is the numerical value of each of the following equations?

(1) $\frac{d}{dt}\left[e^{-t}\delta(t)\right]$ (2) $\frac{d}{dt}\left[e^{-2t}\varepsilon(t)\right]$ (3) $\frac{d}{dt}\left[\varepsilon(t) - 2t\varepsilon(t - 1)\right]$

Problem 1-6. Write the expressions of the signals shown in ▶ Figure P1-6.

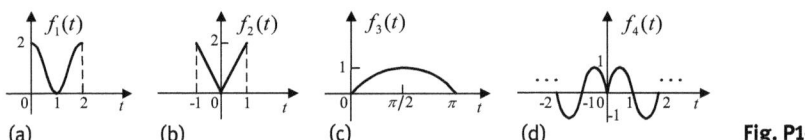

(a)   (b)   (c)   (d)   **Fig. P1-6**

Problem 1-7. The waveform of $f(3 - 2t)$ is shown in ▶ Figure P1-7, plot $f(t)$.

Fig. P1-7

Problem 1-8. Calculate the energy and power of the following signals and classify the signals as energy signals or power signals.

(1) $f_1(t) = \begin{cases} t & 0 \le t \le 1 \\ 2 - t & 1 \le t \le 2 \\ 0 & \text{other} \end{cases}$

(4) $f_4(t) = \begin{cases} 5\cos(\pi t) & -1 \le t \le 1 \\ 0 & \text{other} \end{cases}$

(2) $f_2(t) = \cos(t)\varepsilon(t)$

(5) $f_5(t) = \varepsilon(t) - \varepsilon(t-1)$

(3) $f_3(t) = 5\cos(\pi t) + \sin(5\pi t), \ -\infty < t < \infty$

(6) $f_6(t) = t\varepsilon(t)$

Problem 1-9. Plot the odd and even components of each signal shown in ▸ Figure P1-9.

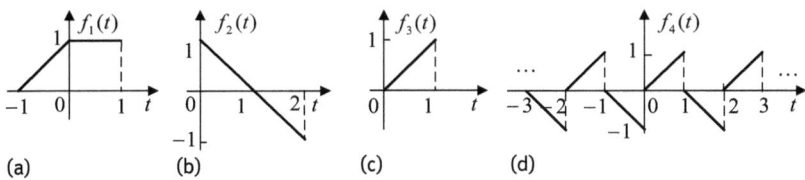

Fig. P1-9

Problem 1-10. Signals are plotted in ▸ Figure P1-10, find each expression of $\int_{-\infty}^{t} f_i(\tau)d\tau$ and sketch it.

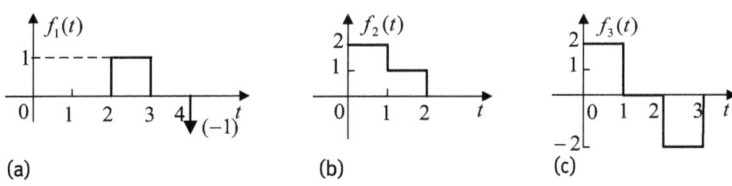

Fig. P1-10

Problem 1-11. Known is $f_1(t) = e^{-2t}$ $(-\infty < t < \infty)$, $f_2(t) = e^{-t}\varepsilon(t)$, find $f_1(t) * f_2(t)$.

Problem 1-12. Known are
(1) $f_1(t) * t\varepsilon(t) = (t + e^{-t} - 1)\varepsilon(t)$
(2) $f_2(t) * [e^{-t}\varepsilon(t)] = (1 - e^{-t})\varepsilon(t) - [1 - e^{-(t-1)}]\varepsilon(t-1)$
Find $f_1(t)$ and $f_2(t)$.

Problem 1-13. Find the values of $f_1(t) * f_2(t)$ for each group of signals shown in ▶ Figure P1-13 and sketch them.

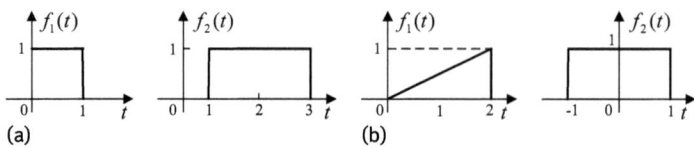

(a) (b)

Fig. P1-13

Problem 1-14. Find these convolutions.

(1) $f_1(t) = e^{-t}\varepsilon(t) * \varepsilon(t)$

(2) $f_2(t) = \sin(2\pi t)[\varepsilon(t) - \varepsilon(t-1)] * \varepsilon(t)$

2 Systems

Questions: Many tasks in productive practice can be considered as being performed by a system. Then what is a system? How can we analyze a system?

Solution: Develop a system concept from practice → seek similarities among systems → categorize systems → find out the basic analysis methods for a system.

Results: 20 types of systems, such as linear and nonlinear, time variant and time invariant, continuous and discrete systems, and so on. There are six basic operations, such as addition, subtraction, multiplication, time delay, differentiation and integration. There are four basic analysis methods, for example, the external ways in the time and transform domains, the state space ways in the time and transform domains.

2.1 The concept of a system

1. The definition of system

As stated in the Preface, there are many kinds of systems in our lives; we can find some common features of these systems through analysis on them, such as: having a certain scale; consisting of various things, equipment, facilities, and related staff; performing special tasks, etc. As a result, a system can be said to be

a set of some elements that are mutually independent and relate to each other, and can achieve some specific functions together.

This definition is applicable in all physical and nonphysical systems.

The systems discussed herein are mainly physical systems, for example, circuits or electric networks and related models composed of resistors R, inductors L and capacitors C. Therefore, the term system in this book can be defined as

the general designation for circuits, devices or algorithms which can transmit, transform or process signals.

Obviously, a system can be abstracted into a functional module or an entirety with cause-and-effect relationships, which can produce one or more outputs (effects or results) from one or more inputs (causes or reasons), as is sketched in ▶ Figure 2.1, where the transforming relationship between the output $y(t)$ and the input $f(t)$ or the purpose of a system can be denoted by the equation

$$y(t) = T[f(t)] \, , \tag{2.1-1}$$

where the notation $T[\bullet]$ represents a transformation to an element in brackets [ ]. If the variable t is replaced by n, equation (2.1-1) also fits the description of discrete sys-

https://doi.org/10.1515/9783110419535-002

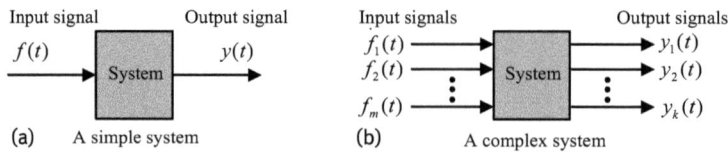

Fig. 2.1: Schematic diagram of the system.

tems. If $T[\bullet]$ can be further expressed by a specific mathematical equation, the explicit equation should be called the mathematical model of the system or, simply, the model.

Note: The systems mentioned subsequently will be the models of systems in theory, not the physical systems themselves, if there is no special explanation.

2. Contents and purposes of system analysis
The main contents of system analysis can be described equivalently by words as follows:
(1) After a signal has passed through a system, what will it look like?
(2) After an excitation has acted in a system, what response will be produced by the system?

To sum up, system analysis aims to discuss the characteristics of transmission, transformation or treatment of a signal by a known system, or is a process to analyze the relationship between the excitation and the response of a given system. The main purposes of system analysis are the following:
(1) To provide theoretical support and a method of guidance for the design, building up or using a system that is described by the differential or difference equation such as a real communication system or a control system.
(2) To lay the foundation for the study of the principles of communication, automatic control and other professional courses.

It should be explained that this course only focuses on electronic systems, but the analysis methods can be generalized to other analogous systems, because many physical systems are similar to electronic systems. So called analogous systems are systems that have the same mathematical models, for example, an LC circuit and a mechanism simple pendulum are analogous systems.

2.2 Excitation, response and system state

Before analyzing a system, we need to know the basic concepts such as excitation, response and state.

The *excitation* is a kind of external energy or impact acting on a system, it is also known as the input signal.

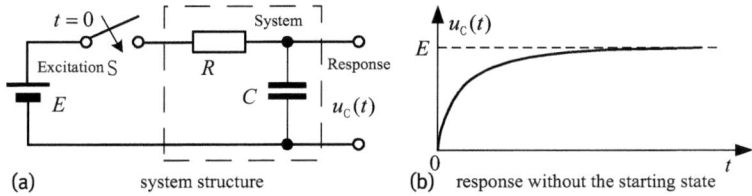

Fig. 2.2: An RC charging circuit and its response with zero starting state.

The *response* is a kind of reaction or result from a system to an excitation, it is also known as the output signal.

The RC circuit shown as ▶ Figure 2.2 is a typical electric system. If the power E is set as an excitation, then the voltage $u_C(t)$ on the capacitor C is the response produced by E after the switch S is closed. If at $t = 0$ the switch S is closed, according to the three-factor method in circuits analysis, there is

$$y(t) = y(\infty) + [y(0_+) - y(\infty)]\, e^{-\frac{1}{\tau}t}, \qquad (2.2\text{-}1)$$

then the voltage $u_C(t)$, that is, the response of the system is

$$u_C(t) = u_C(\infty) + [u_C(0_+) - u_C(\infty)]\, e^{-\frac{1}{\tau}t} = E + (0 - E)e^{-\frac{1}{\tau}t} = E\left(1 - e^{-\frac{1}{\tau}t}\right), \qquad (2.2\text{-}2)$$

where, $u_C(\infty)$ is the final value of $u_C(t)$ and $u_C(0_+)$ is the initial value, which is the instantaneous value after switch S is closed at time $t = 0$; $\tau = RC$ is called the time constant whose size can reflect the transient process length of $u_C(t)$; the greater the time constant, the longer the transient process, and the waveform changes more slowly. The waveform of $u_C(t)$ is shown in ▶ Figure 2.2b. It can be seen that the system response is generated entirely by the excitation existing after time $t = 0$.

Note: The above conclusion is reached under the premise that $u_C(t)$ is zero before switch S is closed, i.e. $u_C(0_-) = 0$. If $u_C(t)$ is not zero before switch S is carried out, for example, $u_C(0_-) = E_1$, then what will be the performance of the system response?

In ▶ Figure 2.3a, before switch S is closed the circuit has been in a stable state, i.e. $u_C(0_-) = E_1$. At time $t = 0$, S is pulled to position 2, according to the Law of Switching in circuits theory, the voltage across a capacitor cannot jump (the current through an inductor cannot jump either). Then $u_C(0_+) = u_C(0_-) = E_1$, again using the three-factor method, and we have

$$
\begin{aligned}
u_C(t) &= u_C(\infty) + [u_C(0_+) - u_C(\infty)]\, e^{-\frac{1}{\tau}t} = E + (E_1 - E)\, e^{-\frac{1}{\tau}t} \\
&= E_1 e^{-\frac{1}{\tau}t} + E\left(1 - e^{-\frac{1}{\tau}t}\right).
\end{aligned} \qquad (2.2\text{-}3)
$$

Obviously, the response $u_C(t)$ consists of two parts, the first is $E_1 e^{-\frac{1}{\tau}t}$ generated by the stored voltage (or energy) E_1 of the capacitor before time $t = 0$; the second is $E(1-e^{-\frac{1}{\tau}t})$ caused by the external excitation E after time $t = 0$, as shown in ▶ Figure 2.3b.

It is necessary to point out that the stored voltage on the capacitor before $t = 0$ can be called the starting state of the circuit or the system. The first part in the response is

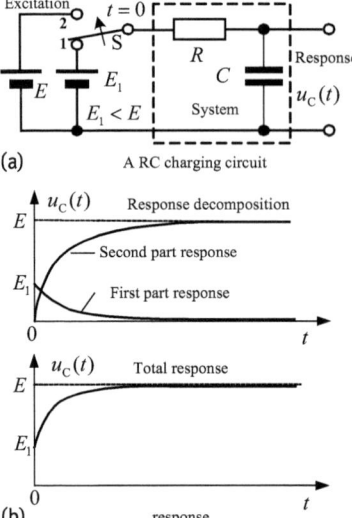

(a)   A RC charging circuit

(b)   response

Fig. 2.3: An RC charging circuit and its response with starting state.

generated by the starting state and is unrelated to the excitation existing after $t = 0$. So, the state that can cause the response is also a kind of excitation in concept. To differentiate from the input signal, the state is also called the inner excitation, whereas the input signals we often mention all refer to the external excitations of a system.

Note: Although E_1 is an external excitation in form, it does not work after $t = 0$, and its role has been converted to the storage energy in the capacitor, so the external excitation is only E in the example.

This way, the state of the system will be introduced as a new concept. The state is considered as a kind of characteristic of things. The changing of the state means the development or the changing of things, thus, the state can be considered as a basis for dividing up the development stages of things. So, the state of a system is a set of the fewest data that needs to be known and can be used to determine the response at any time after $t = t_0$ with the excitation appearing for $t \geq t_0$. In general, this set of data represents the energy storage conditions of every energy storage component before the excitations are applied to the system. In other words,

under the condition of which all external excitations are given, a set of necessary and sufficient data that must be known in order to determine the future response of system is called the state of system.

Due to the influence of switching to a circuit, the state values of system may change instantaneously at moment $t = t_0$. In order to distinguish values before and after the state instantaneously changes, we use the notations t_{0_-} and t_{0_+} to denote, respectively, the instants at which before and after the excitation acts to the system at $t = t_0$.

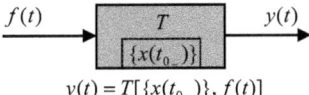

$$y(t) = T[\{x(t_{0_-})\}, f(t)]$$

Fig. 2.4: Relation between response and excitations.

Then the state at the instant before the excitation is accessed is named the starting state, its values are denoted as $x_1(t_{0_-})$, $x_2(t_{0_-})$, ..., $x_n(t_{0_-})$. Obviously, this set of data records all the related historical information of the system. The state at the instant after the excitation is accessed is called the initial state, and its values are denoted as $x_1(t_{0_+})$, $x_2(t_{0_+})$, ..., $x_n(t_{0_+})$.

From the above concept of the state of the system, we can see that system response $y(t)$ for $t \geq t_0$ is a function of the starting state and the input $f(t)$, which is connected to the system at time $t = t_0$, and is expressed as

$$y(t) = T[x_1(t_{0_-}), x_2(t_{0_-}), \dots, x_n(t_{0_-}), f(t)] \quad t \geq t_0 . \tag{2.2-4}$$

For convenience, the starting state values $x_1(t_{0_-})$, $x_2(t_{0_-})$, ..., $x_n(t_{0_-})$ at $t = t_0$ can be represented by the symbol $\{x(t_{0_-})\}$; then (2.2-4) can be expressed as

$$y(t) = T[\{x(t_{0_-})\}, f(t)] \quad t \geq t_0 , \tag{2.2-5}$$

as plotted in ▸ Figure 2.4.

To sum up, the response $y(t)$ of system for $t \geq t_0$ is codetermined by the starting state $\{x(t_{0_-})\}$ and the excitation $f(t)$ existing in a range $[t_0, t]$. This conclusion is also applicable in multi-input multi-output systems. The state of a system at a given moment can give us all the information of the system at that moment. Hence, the analysis of a system must focus on all behaviors of the system in the past, at present and in the future.

2.3 Classification of systems

According to the different characteristics of a system, systems herein are divided mainly into 20 types, including linear and nonlinear systems.

2.3.1 Simple and complex systems

A system with single input and single output signals is known as a simple system (SISO system), while a system with multi-input and multi-output signals is known as a complex system (MIMO system), these are sketched in ▸ Figure 2.1. The simple system is the main topic that will be discussed in this book.

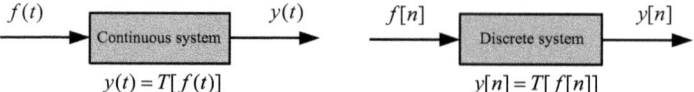

Fig. 2.5: Continuous system and discrete system.

2.3.2 Continuous-time and discrete-time systems

Like the classification of signals, systems are also classified into continuous-time and discrete-time systems.

A system with continuous-time input and output is a continuous-time system or, simply, a CT system, while a system with discrete-time input and output is a discrete-time system or, simply, a DT system or a digital system.

The excitation and response in a continuous system are represented by $f(t)$ and $y(t)$, while the input and the output for a discrete system are represented by $f[n]$ and $y[n]$, as is shown in ▶ Figure 2.5. Practically, there may be a hybrid system with both continuous and discrete signals.

2.3.3 Linear and nonlinear systems

A system that consists of some linear devices such as linear resistors, inductors and capacitor can be called the linear system. A system containing some nonlinear devices such as diodes is a nonlinear system. Both are plotted in ▶ Figure 2.6.

Then, the question is how to judge the linearity of a system whose internal structure and components are unknown. Firstly, we must know a lot about the linearity concept.

The linearity contains homogeneity and additivity.

(1) Homogeneity. If an input (excitation) increases k times, the output (response) will also increase k times (k is any constant). That is,

if

$$f(t) \rightarrow y(t) \,,$$

then

$$kf(t) \rightarrow ky(t) \,, \tag{2.3-1}$$

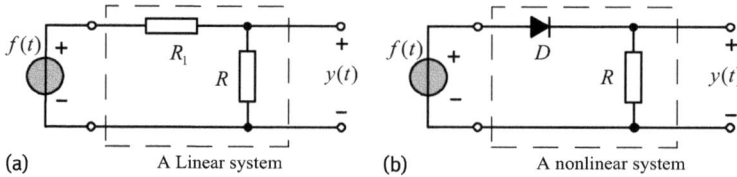

(a)   A Linear system   (b)   A nonlinear system

Fig. 2.6: Examples of linear and nonlinear system.

where the symbol "→" represents a kind of transform or process to a signal; it will appear frequently later.

Note: If the coefficient $k = 0$,

a conclusion of which zero input must result in zero output for a linear system will be obtained, i.e. the "zero-in/zero-out" property.

(2) Additivity. This means that if several excitations act on a system at the same time, the total response of the system can be represented as an algebraic sum of responses generated by each excitation acting alone separately (others are zero). This characteristic can be expressed as:

if

$$f_1(t) \rightarrow y_1(t) \quad \text{and} \quad f_2(t) \rightarrow y_2(t),$$

then

$$f_1(t) + f_2(t) \rightarrow y_1(t) + y_2(t). \tag{2.3-2}$$

The above two criteria can be combined to yield the superposition property or the linearity,

if

$$f_1(t) \rightarrow y_1(t) \quad \text{and} \quad f_2(t) \rightarrow y_2(t),$$

then

$$k_1 f_1(t) + k_2 f_2(t) \rightarrow k_1 y_1(t) + k_2 y_2(t). \tag{2.3-3}$$

The linearity is shown in ▶ Figure 2.7.

So far, the basic definition of the linear system can be given by

a system whose relationship between the response and the excitation meets linearity or superposition property is linear.

This concept of the linear system has been widely accepted and has appeared in many textbooks and monographs. Moreover, linearity is also regarded as a basis for estimat-

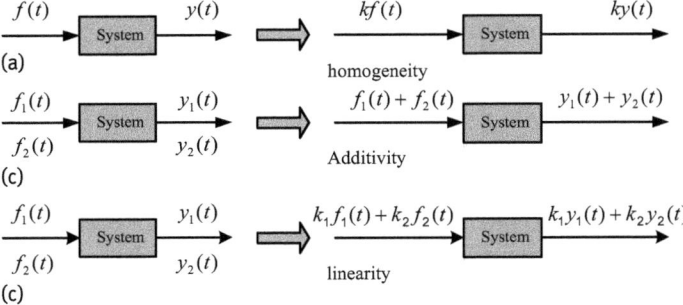

Fig. 2.7: Schematic diagram of linear characteristic.

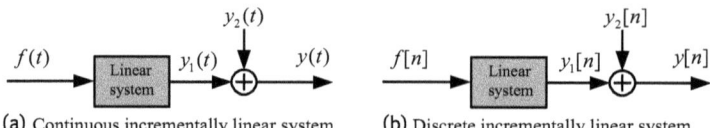

(a) Continuous incrementally linear system   (b) Discrete incrementally linear system

Fig. 2.8: Incrementally linear systems.

ing if a system is linear and is called the linearity judgment condition. The conclusion of "zero-in/zero-out" can be directly used to test whether a system is nonlinear. Obviously, the condition can work well with the external measurements of a system but does not need to understand the internal structure of a system.

A deficiency of the condition can be found by careful analysis; the basic definition of linear systems just applies to systems whose response is only produced by external excitation. However, this is not suitable for linear systems whose complete response may also contain a part produced by the starting state of the system. So, under this condition, systems that have a response component from the starting state should be judged as nonlinear systems and are further defined as the incrementally linear systems displayed in ▶ Figure 2.8. The complete response can be considered as a superposition of a response that satisfies the linearity condition and an increment response in the systems. For example, $y(t) = af(t) + b$ can be written as $y(t) = y_1(t) + y_2(t)$, where $y_1(t) = af(t)$ is the response that holds the linear condition, and $y_2(t) = b$ is an increment response. It can be known that the linear systems meeting the basic definition or the linearity condition only belong to a subset of incrementally linear systems or are a special case in incrementally linear systems when the increment is zero.

American scholar B.P. Lathi put forward another definition for linear systems in his work *Signals, Systems, and Controls* published in 1974:

A system is linear if and only if it can satisfy the response decomposability, the zero-state linearity and the zero-input linearity.

For convenience, this definition is called the Lathi definition in this book.

The so called response decomposability, zero-state linearity and zero-input linearity are illustrated separately as follows:

(1) Response decomposability. From Section 2.2, it is known that the response $y(t)$ of a system depends not only on the excitation $f(t)$ but also on the starting state $\{x(t_{0-})\}$. This fact also considers that the response of a system is generated by two different excitations of $f(t)$ and $\{x(t_{0-})\}$. Thus, the complete response $y(t)$ should be the sum of two parts; one is the response $y_x(t)$ which is only caused by the starting state $\{x(t_{0-})\}$ when the input $f(t)$ is zero; the other is the response $y_f(t)$ which only results from the input signal $f(t)$ when the starting state $\{x(t_{0-})\}$ is zero. Usually, $y_x(t)$ is called the zero-input component or zero-input response, and $y_f(t)$ is called the zero-state component or zero-state response. Thus, the complete response $y(t)$ can be expressed as

$$y(t) = y_x(t) + y_f(t) .$$ (2.3-4)

A living example can deepen the understanding of equation (2.3-4) for us: You hurt your left foot yesterday when you carelessly kicked a stone, so your left foot is still very sore today. This existing soreness can be compared to the zero-input response of your nervous system now. Unfortunately, your left foot was also hit by a falling cup today, and this new pain can be considered to the zero-state response. Your pain is the result of one disaster after another; the old pain caused by the stone and new pain from the cup together form the complete response of your nervous system.

Equation (2.3-4) is just response decomposability, that is

the response decomposability is that the complete response of a system can be decomposed into two components, a zero-input response and a zero-state response.

The response decomposability allows us to obtain the complete response with a relative simple method, that is, we can calculate, respectively, two components caused by two different excitations and then superimpose them to form the complete response. In equation (2.2-3), the first term $E_1 e^{-\frac{1}{\tau}t}$ is the zero-input component, and the second $E(1 - e^{-\frac{1}{\tau}t})$ is the zero-state component.

(2) Zero-state linearity. When the starting state $\{x(t_{0_-})\}$ is zero, the zero-state response $y_f(t)$ must satisfy the linear condition to various inputs.
If

$$f_1(t) \to y_{f1}(t) \quad \text{and} \quad f_2(t) \to y_{f2}(t) ,$$

then

$$k_1 f_1(t) + k_2 f_2(t) \to k_1 y_{f1}(t) + k_2 y_{f2}(t) .$$ (2.3-5)

(3) Zero-input linearity. When the input $f(t)$ is zero, the zero-input response $y_x(t)$ must satisfy the linear condition to various starting-state values.
If

$$x_1(t_{0_-}) \to y_{x1}(t) \quad \text{and} \quad x_2(t_{0_-}) \to y_{x2}(t) ,$$

then

$$k_1 x_1(t_{0_-}) + k_2 x_2(t_{0_-}) \to k_1 y_{x1}(t) + k_2 y_{x2}(t) .$$ (2.3-6)

Obviously, if the zero-state response and zero-input response are all nonzero, not only the complete response $y(t)$ and the excitation $f(t)$ do not hold the linear condition, but the complete response $y(t)$ and the starting state $\{x(t_{0_-})\}$ do not either. Under the Lathi definition, $y_1(t)$ and $y_2(t)$ in ▶ Figure 2.7 can be regarded as $y_f(t)$ and $y_x(t)$ in equation (2.3-4), so the incrementally linear system is linear. Clearly, a linear system that only satisfies the basic definition is actually a Lathi system whose zero-input response is zero. The Lathi definition is generalized by the basic definition and is more comprehensive than the basic one. Thus, linear systems in the book are all considered as Lathi systems or incrementally linear systems in the basic definition, if there is no special explanation.

There is an important property in application of linear systems, that is, the response and the excitation must be of the same frequency. In other words, the response will not contain different components from the excitation in frequency. This property is the theoretical basis for which alternating stable circuits can be analyzed by the phasor method from circuit analysis courses. So far, a nonlinear system can be easily judged from the linearity concept. *A system must be nonlinear, if it is not linear.*

Based on the above characteristics, the analysis method for a linear system can be greatly simplified. The two components in the complete response are easily calculated by using the decomposability (the zero-input response can be obtained if the input signal is set to be zero, the zero-state response can be obtained if the starting-state values are set to be zero). Then, if the excitation $f(t)$ can be decomposed into the algebraic sum of many simple functions (signals)

$$f(t) = a_1 f_1(t) + a_2 f_2(t) + a_3 f_3(t) + \dots , \tag{2.3-7}$$

so, according to zero-state linearity, the zero-state response can be expressed as

$$y_f(t) = a_1 y_{f1}(t) + a_2 y_{f2}(t) + a_3 y_{f3}(t) + \dots y_{fi}(t) + \dots , \tag{2.3-8}$$

where, $y_{fi}(t)$ is the zero-state response of system to the input $f_i(t)$.

The following examples show the usages and the characteristics of linear systems.

Example 2.3-1. If the starting-state values of a linear system are $x_1(0_-) = 1$ and $x_2(0_-) = 2$ or, simply, $\{1, 2\}$, the zero-input response is $2 + 3e^{-2t}$. Please give the zero-input response when the starting-state values increase five times compared to the original ones.

Solution. Because the starting state is $\{5, 10\}$, according to zero-input linearity, the zero-input response can expressed as $5(2 + 3e^{-2t})$.

Example 2.3-2. Suppose that the zero-input response is $2 - 3e^{-2t}$ when the starting state is $\{1, 2\}$ for a linear system, and the zero-input response is $5 + 2e^{-2t}$ when the starting state is $\{4, 1\}$. Find the zero-input response for the starting state is $\{5, 3\}$.

Solution. Obviously, the starting state $\{5, 3\}$ is the superposition of $\{1, 2\}$ and $\{4, 1\}$, so the corresponding zero-input response is also the superposition of the zero-input response of $2 - 3e^{-2t}$ and $5 + 2e^{-2t}$, that is, $7 - e^{-2t}$.

Example 2.3-3. Please judge whether or not the following systems are linear.

(1) $y(t) = t \cdot f^2(t)$

(2) $y(t) = t \cdot f(t)$

(3) $y(t) = x(0_-) + f^2(t)$

(4) $y(t) = x^2(0_-) + \int_0^t f(\tau)d\tau$

(5) $y(t) = 5x(0_-)f(t)$

(6) $y(t) = 3f(t) + 6$

Solution. Suppose $f_1 \rightarrow y_1; f_2 \rightarrow y_2$, then

(1) because $af_1 + bf_2 \rightarrow t(af_1 + bf_2)^2 = t(af_1)^2 + t(bf_2)^2 + t2abf_1f_2 \neq ay_1 + by_2$, the system does not satisfy the linear condition and is nonlinear.

(2) Because $af_1 + bf_2 \rightarrow t(af_1 + bf_2) = atf_1 + btf_2 = ay_1 + by_2$, the system satisfies the linear condition and is a linear system.

(3) The system can satisfy the decomposability and zero-input linearity, but not zero-state linearity, so it is a nonlinear system.

(4) The system can satisfy decomposability and zero-state linearity, but not zero-input linearity, so it is a nonlinear system.

(5) The system cannot satisfy decomposability, so it is a nonlinear system.

(6) The system can satisfy decomposability, zero-input linearity and zero-state linearity, so it is a linear system under the Lathi definition.

2.3.4 Time-variant and time-invariant systems

If the zero-state response is $y_f(t)$ to the excitation $f(t)$ for a system, and the response can become $y_f(t - t_0)$ when the excitation is $f(t - t_0)$, or, if a time shift in the input only results in the same time shift in the output, this system is said to be a time-invariant system or, simply, a TI system. This can be expressed as

if

$$f(t) \rightarrow y_f(t) ,$$

then

$$f(t - t_0) \rightarrow y_f(t - t_0) . \tag{2.3-9}$$

The time invariance of the system is illustrated in ▶ Figure 2.9.

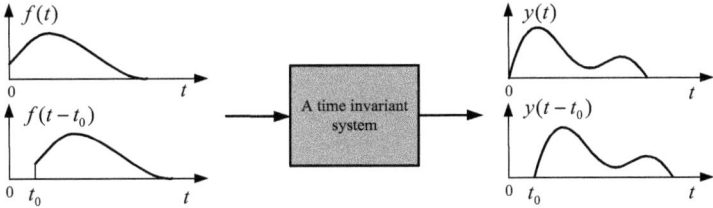

Fig. 2.9: Time invariant property.

Time invariance means that all parameters in the mathematical model of a system do not change with time or are constant. A TI system is also called a fixed system.

Most systems in real life are TI systems or can be reasonably approximated to this kind of system. Otherwise, a system without this property is a time-variant system.

If a system not only meets the linear condition, but also meets the time invariant condition, the system is called a linear time-invariant system or, simply, an LTI system.

The differential and integral properties of an LTI system can be shown as follows: if

$$f(t) \rightarrow y(t) ,$$

then

$$\frac{df(t)}{dt} \rightarrow \frac{dy(t)}{dt} \tag{2.3-10}$$

and

$$\int_{-\infty}^{t} f(\tau)d\tau \rightarrow \int_{-\infty}^{t} y(\tau)d\tau . \tag{2.3-11}$$

RLC networks, which are often studied, and circuits that include some active devices (such as electronic tubes and transistors) are all TI systems. Note that here $y(t)$ refers to the zero-state response $y_f(t)$.

Example 2.3-4. Judge whether or not a system described by $y_f(t) = tf(t)$ is time-invariant.

Solution. The expression of response with time shifting t_0 is

$$y_f(t - t_0) = (t - t_0)f(t - t_0) .$$

The response corresponding to the excitation with time shifting t_0 is

$$T[f(t - t_0)] = tf(t - t_0) .$$

Obviously, $T[f(t - t_0)] \neq y_f(t - t_0)$, so the system is time variant.

The decomposability, the superposition and time invariance of the response can provide many methods for linear system analysis; for example, the convolution integral, the convolution sum and the analysis methods in transform domains are all based on these properties. So,

the linearity and time invariance are the essence of system analysis.

Because there are none of the above properties, the analysis of nonlinear systems is very difficult, not only are the analysis methods not simple, but also the general solution of the system cannot be obtained. Luckily, many nonlinear systems can be approximated to a linear system working in a limited range, which is the shortcut for the analysis of nonlinear systems. So, the analysis methods for linear systems are also important for the analysis of nonlinear systems.

2.3.5 Causal and noncausal systems

A system is causal if its output at a certain moment is only determined by the inputs at that moment and before, and is unrelated to the input in the future. Or we can say the output of a causal system cannot precede its input, or a response will not be produced by the system without an excitation; put simply, the result follows the cause.

The causal system is also the achievable system, and its property is

$$f(t) = 0, \quad t < t_0 \quad \rightarrow \quad y(t) = 0, \quad t < t_0 . \tag{2.3-12}$$

Note: We generally suppose $t_0 = 0$. Thus, equation (2.3-12) can be regarded as the judgment condition for a causal system. A system that cannot satisfy the causal condition is a noncausal system. A noncausal system cannot be achieved physically.

Example 2.3-5. Test the causality of the following systems.
(1) $y(t) = f(t) - f(t - 3)$ 　　　　　　(3) $y(t) = f(t - 1) + f(t + 2)$
(2) $y(t - 1) = f(t) + f(t - 1)$ 　　　　　(4) $y(t) = f(3t)$

Solution. From the definition of causal system, if $t_0 = 0$, and put it into all expressions, there are
(1) $y(0) = f(0) - f(-3)$
　　The excitation values $f(0)$ and $f(-3)$ are both before the response $y(0)$, so the system is causal.
(2) $y(-1) = f(0) + f(-1)$
　　The response value $y(-1)$ arises before $t_0 = 0$ and $f(0)$, so the system is noncausal.
(3) $y(0) = f(-1) + f(2)$
　　The $f(2)$ arises after the response $y(0)$, which means the response $y(0)$ at $t_0 = 0$ relates to not only $f(-1)$ before it but also $f(2)$ after it, so the system is noncausal.
(4) If $t = 1$, there is $y(1) = f(3)$, the response leads the excitation, so the system is noncausal.

Example 2.3-6. Judge the linearity, time invariance and causality of the following systems.
(1) $y'(t) + a_0 y(t) = b_0 f(t) + b_1 f'(t)$ 　　　(3) $y(t) = \int_{-\infty}^{5t} f(t) dt$
(2) $y(t) = 2f(t)\varepsilon(t)$

Solution. (1) The system is a linear, time-invariant and causal system.
(2) The system satisfies linearity and causality. However,

$$y(t - t_0) = 2f(t - t_0)\varepsilon(t - t_0) \neq 2f(t - t_0)\varepsilon(t) ,$$

so, the system is linear, causal and time variant.
(3) The system satisfies the linear condition.
　　If $f(t) = e^t$, then $y(t) = \int_{-\infty}^{5t} f(t) dt = \int_{-\infty}^{5t} e^t dt = e^{5t}$.
　　If $f(t - 1) = e^{t-1}$, then $y(t) = \int_{-\infty}^{5t} e^{t-1} dt = e^{5t-1} = e^{5(t-\frac{1}{5})}$.

Thus, the excitation delay is 1, but the response delay is only 0.5; the response appears before the excitation, so the system cannot satisfy the causality. Additionally, because $y(t-1) = e^{5(t-1)}$, there is $T[f(t-1)] = e^{5(t-\frac{1}{2})} \neq y(t-1) = e^{5(t-1)}$, the system does not satisfy time invariance, so it is linear, noncausal and time variant.

2.3.6 Dynamic and static systems

If the response $y(t_0)$ of a system at time t_0 relates to not only the excitation value $f(t_0)$ at t_0, but also values of excitation before time t_0, this system is called a dynamic system or a memory system. If the response $y(t_0)$ at time t_0 only depends on the excitation $f(t_0)$ at t_0, the system is called a static, memoryless, real time or instantaneous system. For example, a system that only consists of resistors is a real-time system, because a resistor cannot store energy. A system that includes storage elements (such as capacitors, inductors and magnetic cores) or memory circuits (such as registers) is a dynamic system. The instances of the two systems are shown in ▶ Figure 2.10.

Note: Although a sinusoidal steady state circuit in a circuits analysis course contains dynamic devices in the form of a circuit, it is not a dynamic system because the excitation is a sinusoidal signal and the steady-state response is only determined by the excitation at present. Therefore, if we want to test whether a system is dynamic, dynamic elements must be included in it, the types of excitation and the focus on response must also considered.

Usually, the mathematical model of a static system in the time domain is an algebraic equation, but the mathematical model of a dynamic system is a differential or difference equation.

2.3.7 Open-loop and closed-loop systems

A system that only allows the signal to be transmitted forward (from the input end to the output end) is called an open-loop system or a no feedback system. A system that also includes a back branch from the output to the input as well as a forward branch is called a closed-loop system or a feedback system. These are plotted in ▶ Figure 2.11.

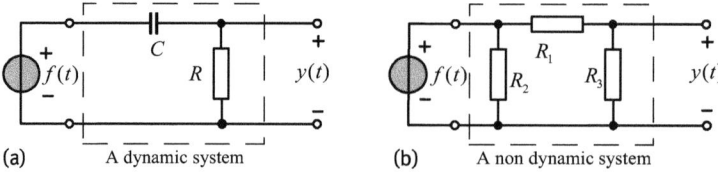

(a)   A dynamic system   (b)   A non dynamic system

Fig. 2.10: Dynamic system and nondynamic systems.

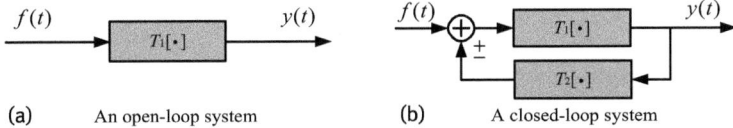

(a)   An open-loop system   (b)   A closed-loop system

Fig. 2.11: Open-loop and closed-loop systems.

From the topology or physical structure, the diagram of an open-loop system is like a straight line, and that of a closed-loop system is actually like a loop because the feedback branch exists.

According to the different effects of a feedback signal in a feedback system, closed-loop systems can be divided into positive and negative feedback systems. Systems whose feedback signal enhances the effect of the input signal are positive feedback systems (the operation symbol "+" is used in the graph), while systems whose feedback signal weakens the effect of the input signal are negative feedback systems (the symbol "−" is used in the graph).

2.3.8 Stable and unstable systems

For any system which has no energy storage at the starting moment, if a bounded input produces a bounded output (BIBO: bound input/bound output), then the system is called a stable system; it can be expressed as:

$$|f(t)| < \infty \rightarrow |y(t)| < \infty . \tag{2.3-13}$$

If the input is bounded but the output is unbounded (infinite), then the system is known as an unstable system. Note that the $y(t)$ here still refers to the zero-state response $y_f(t)$. A positive feedback system is usually unstable, but a negative feedback system is stable.

2.3.9 Lumped and distributed parameter systems

Systems composed of lumped parameter elements are called lumped parameter systems.

The lumped parameter element refers to one whose characteristics can be represented intensely by one parameter. Generally speaking, electronic elements whose physical sizes are much smaller than their working wavelengths are lumped parameter elements, such as common resistors, inductors, capacitors, transistors and transformers, etc.

The main characteristics of a lumped parameter system are as follows:

(1) The effects from an excitation applied to any point of the system instantaneously spreads to the whole system. In other words, once an excitation is applied on the

input port of the system, the response of the system will immediately be produced on the output port.

(2) There is only one independent variable in a lumped parameter system, and it can be time, frequency or other physical parameters, so a lumped parameter system can be described by an ordinary differential equation.

The systems composed of distributed parameter elements are called distributed parameter systems. The distributed parameter element refers to one whose characteristics cannot be intensively represented by a parameter. That is, electronic elements whose physical sizes are approximately same as their working wavelengths are distributed parameter elements, for example, transmission lines, antennas, ducts of waves and mechanical shafts, etc. The characteristics of a distributed parameter element cannot be reflected by only one parameter, but by multiparameters at the same time. For example, in a section of a transmission line because the resistance, inductance and capacitance exist on every point along the line, their three effects will appear in the section at same time together, we must use three parameters such as R, L and C to represent the features of the transmission line.

The main characteristics of a distributed parameter system are as follows:

(1) The effects from an excitation applied on any point of the system will spread to the entire system after a period of time. In other words, although an excitation is applied on the input port, the response cannot be produced immediately on the output port.

(2) The variables involved in a distributed parameter system include not only time but also space. This means the system variables are more than one, so the system needs to be described by the partial differential equation.

In fact, all actual systems are distributed parameter systems, but some of them can be approximated as lumped parameter systems. This approximation is reasonable only if the element sizes of the system are much smaller than the input signal wavelength. Usually, all the low frequency circuits are the lumped parameter systems. Only microwave circuits where the working frequency is greater than 300 MHz are likely to be treated as distributed parameter systems. All systems discussed herein are the lumped parameter systems.

2.3.10 Invertible and nonreversible systems

If different excitations can result in different responses, the system is an invertible system, otherwise it is a nonreversible system. For example, $y(t) = 3f(t)$ is an invertible system, but $y(t) = 3f^2(t)$ is not invertible because $+f(t)$ and $-f(t)$ will result in the same $y(t)$.

The invertible system has an important property: If a system is invertible, there must be an inverse system corresponding to it. The response of the inverse system is

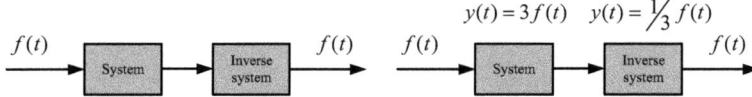

Fig. 2.12: The feature of invertible system and an example.

the excitation of the system when the system and its inverse system are connected in cascade form, namely, the input of the whole system is equal to output. For example, a system $y(t) = \frac{1}{3}f(t)$ is the inverse system of a system $y(t) = 3f(t)$; they are plotted in ▶ Figure 2.12. Usually, if the response is equal to the excitation, the system can be called an identity system.

The invertibility of a system is an important concept; for example, the coder must be an invertible system in a communication system.

From the above content, we know that a real system can have several properties, such as linearity, time invariance and causality, etc. Among various systems, the LTI causal system is the most basic and important, and it is also the basis for the analysis of other systems, so we will mainly discuss LTI causal systems or, simply LTI systems, in this book.

Note: The above concepts or definitions of various continuous systems hold in discrete time in the book if there are no further instructions.

2.4 Models of LTI systems

2.4.1 Mathematical models

Because the structures and functions of different LTI systems are quite different from each other, it is necessary to seek a general analysis method in order to analyze the various LTI systems effectively. Based on their structures and functions, the various physical systems can be abstracted as a unified model – the mathematical expression, which is the basis of various analytical methods. Therefore, we have the following definition:

The mathematical expression that can fully reflect the characteristics of a system is called the mathematical model of the system or, simply, the model. The process of looking for this mathematical expression is called modeling.

It can be proved that the mathematical model for nth-order continuous LTI systems is an nth-order linear constant coefficient differential equation like

$$a_n \frac{\mathrm{d}^n y(t)}{\mathrm{d}t^n} + \cdots + a_i \frac{\mathrm{d}^i y(t)}{\mathrm{d}t^i} + \cdots + a_1 \frac{\mathrm{d}y(t)}{\mathrm{d}t} + a_0 y(t)$$

$$= b_m \frac{\mathrm{d}^m f(t)}{\mathrm{d}t^m} + \cdots + b_i \frac{\mathrm{d}^i f(t)}{\mathrm{d}t^i} + \cdots + b_1 \frac{\mathrm{d}f(t)}{\mathrm{d}t} + b_0 f(t) \quad (2.4\text{-}1a)$$

or

$$\sum_{i=0}^{n} a_i \frac{d^i y(t)}{dt^i} = \sum_{j=0}^{m} b_j \frac{d^j f(t)}{dt^j} \ . \tag{2.4-1b}$$

The mathematical model of Nth-order discrete LTI systems is an Nth-order linear constant coefficient difference equation like

$$a_N y[n] + a_{N-1} y[n-1] + \cdots + a_i y[n-i] + \cdots + a_0 y[n-N]$$
$$= b_M f[n] + b_{M-1} f[n-1] + \cdots + b_i f[n-i] + \cdots + b_0 f[n-M] \tag{2.4-2a}$$

or

$$\sum_{k=0}^{N} a_{N-k} y[n-k] = \sum_{r=0}^{M} b_{M-r} f[n-r] \ . \tag{2.4-2b}$$

Note: If equation (2.4-1) satisfies the slacking (at rest) condition

$$f(t) = 0 \ , \ t < t_0 \quad \rightarrow \quad y(t) = 0 \ , \ t < t_0 \ ; \tag{2.4-3}$$

if equation (2.4-2) satisfies the slacking condition

$$f[n] = 0 \ , \ n < n_0 \quad \rightarrow \quad y[n] = 0 \ , \ n < n_0 \ , \tag{2.4-4}$$

then the system described by equation (2.4-1) or equation (2.4-2) is a continuous linear or discrete linear system, which can only satisfy the basic definition of a linear system. Because a system with a nonslacking starting condition can produce the zero-input response caused by the starting state, namely, the increment response, so the system will become an incrementally linear system. Additionally, it can be proved that the system with a slacking condition and modeled by equation (2.4-1) or equation (2.4-2) is also time invariant.

By this observation, the slacking condition equation (2.4-3) is the same as the causal condition equation (2.3-12), which means that a system described by a linear differential equation under the slacking condition is both an LTI system and a causal system. This conclusion fits well for discrete systems.

For the convenience of analysis, the t_0 in equation (2.4-3) and the n_0 in equation (2.4-4) are set to be $t_0 = 0$ and $n_0 = 0$. The a_i, b_j in equation (2.4-1) and the a_{N-k}, b_{M-r} in equation (2.4-2) are all real constants. Values of m, n, M and N depend on the related parameters of the system, usually $m \le n$ and $M \le N$. If the coefficients a_i, b_j, a_{N-k} and b_{M-r} are functions of time t or n, the system will be a linear time variant one.

Note: The starting slacking condition of an equation is equal to that the starting condition of a system is zero. Because the starting condition is determined by the starting state, zero starting means zero-state, and zero-state means zero stored energy.

2.4.2 Mathematical modeling

It is depends largely on the comprehensiveness and accuracy of the mathematical description of the system whether we can completely and accurately analyze a system, so

Tab. 2.1: Component constraint.

Component	Symbol	Constraint relationship
Resistor	i_R R $+ \; u_R \; -$	$i_R(t) = \frac{u_R(t)}{R}$
Capacitor	i_C C $+ \quad u_C \quad -$	$i_C(t) = C\frac{du_C(t)}{dt}$
Inductor	i_L L $+ \quad u_L \quad -$	$u_L(t) = L\frac{di_L(t)}{dt}$

modeling is a premise and a key to system analysis. For an electronic system, modeling is used to seek the equation that can reflect the motion behavior of a system based
on some related circuit principles (constraint conditions) and the system structure.

The essential basis to establish the mathematical model of an electric system is
two constraint conditions which exist in all circuits:

(1) The constraint condition for a component is a relation expression to describe the
characteristics of a component and is usually the relationship between voltage
and current on the component. It is expressed as the V-A characteristic. Table 2.1
shows the constraint relations for common components.

(2) The constraint conditions for a network are constraint relations among voltages
and currents in a circuit and are determined by the network structure. They are expressed by Kirchhoff's voltage law (KVL) and Kirchhoff's current law (KCL). Thus,
we can say that

*the V-A characteristics of components and Kirchhoff's laws of circuits are the basic
theories of modeling of electric systems.*

Kirchhoff's current law (the first law) states that for any node in a circuit, the algebraic sum of all branch currents on this node is zero at any time. That is,

$$\sum i(t) = 0 \quad \text{or} \quad \sum i_{in}(t) = \sum i_{out}(t) . \tag{2.4-5}$$

Kirchhoff's voltage law (the second law) states that for any loop in a circuit, the
algebraic sum of all branch voltages along the loop is zero at any time. That is,

$$\sum u(t) = 0 \quad \text{or} \quad \sum u_{up}(t) = \sum u_{down}(t) . \tag{2.4-6}$$

Details of modeling of an electronic system will be shown by the following examples.

Example 2.4-1. ▶ Figure 2.13 shows a third-order electronic system, where $u_S(t)$ is the
excitation, $u(t)$ is the response, $L_1 = L_2 = 1\,\text{H}$ and $R_1 = R_2 = 1\,\Omega$, $C = 2\,\text{F}$. Try to give
the differential equation of this system.

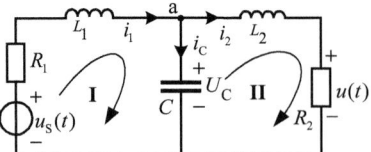

Fig. 2.13: E2.4-1.

Solution. Meshes I, II and their current reference directions are shown in ▸ Figure 2.13. Using KCL to write the current equation on node a, we have

$$i_C = i_1 - i_2 .$$

Using KVL to write equations of meshes I, II separately

$$R_1 i_1 + L_1 \frac{di_1}{dt} + u_C = u_S \quad \text{and} \quad R_2 i_2 + L_2 \frac{di_2}{dt} = u_C .$$

Putting the component parameters into them, we have

$$i_1 + \frac{di_1}{dt} + u_C = u_S \quad \text{and} \quad i_2 + \frac{di_2}{dt} = u_C .$$

From the above two expressions, we obtain

$$i_1 + \frac{di_1}{dt} + i_2 + \frac{di_2}{dt} = u_S . \tag{2.4-7}$$

Then we can write the relationships between the voltage and current of elements, and putting the component parameters into them, we have

$$i_2 = \frac{u}{R_2} = u , \tag{2.4-8}$$

$$u_C = L_2 \frac{di_2}{dt} + u = \frac{du}{dt} + u ,$$

$$i_C = C \frac{du_C}{dt} = 2 \frac{d^2 u}{dt^2} + 2 \frac{du}{dt} = i_1 - i_2 .$$

Therefore,

$$i_1 = 2 \frac{d^2 u}{dt^2} + 2 \frac{du}{dt} + i_2 . \tag{2.4-9}$$

Putting equations (2.4-8) and (2.4-9) into equation (2.4-7), the differential equation of the system is

$$\frac{d^3 u}{dt^3} + 2 \frac{d^2 u}{dt^2} + 2 \frac{du}{dt} + u = \frac{1}{2} u_S . \tag{2.4-10}$$

Note: The order number of a differential or a difference equation is also the order of the system and also the number of independent energy storage devices in the system.

To sum up, the modeling steps for an electronic system are the following:

(1) From the V-A characteristic of each device, write the V-A relations of all devices in a system.

(2) Put these relations into the KCL and KVL equations in the system to obtain several algebra and differential equations.

(3) All these equations can be rearranged and simplified by means of elimination into a differential equation that only contains excitation and response, which is the mathematical model of the system.

Herein, the mathematical model of an electronic system is exactly a constant coefficient linear differential or difference equation.

2.4.3 Block models

From the picture books of our childhood, we know that the knowledge represented by graphics is much easier to understand and to remember. In other words, the graphical representation of knowledge is the highest level of teaching. Thus the question is, can the formula in equation (2.4-1) or equation (2.4-2) be represented by graphics? Can a differential or a difference equation be diagramed?

By observing equation (2.4-1) carefully, we find that the mathematical model of a system is an expression whose excitation and response are connected by addition, multiplication and differential operations. Naturally, we think that if these operations can be shown by graphs, then $f(t)$ and $y(t)$ can be connected by them, and thereby the graphical model of a system can be obtained. For this case, operations such as addition, multiplication, differentiation, delay and integration all are abstracted as a unit or a module and further represented by block diagrams, which are shown in ▶ Figure 2.14. Note that a differentiator is an inverse system of an integrator.

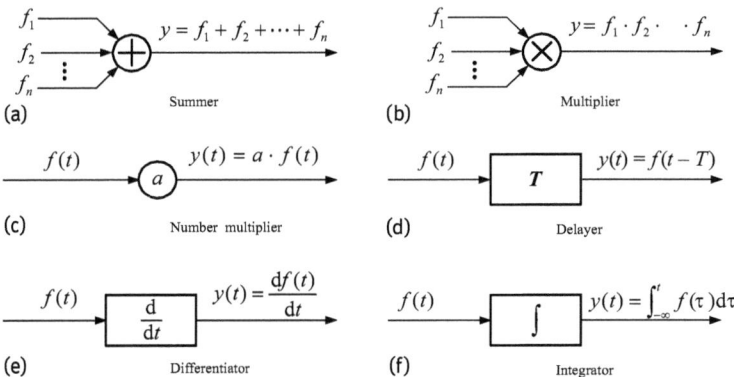

Fig. 2.14: Basic block diagrams of system operations.

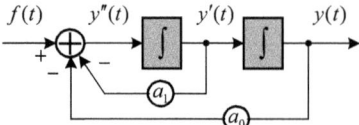

Fig. 2.15: E2.4-2.

According to the linear condition, the above operators can be proved as linear systems. Thus, using these graphical components for basic operators, we can translate a mathematical model that is an abstract and obscure differential equation into a visual and comprehensible graphical model. It is known as the block diagram simulation method and uses block diagrams to express a system; the similar flow diagram simulation method will be introduced in Chapter 7.

Example 2.4-2. $\frac{d^2y(t)}{dt^2} + a_1\frac{dy(t)}{dt} + a_0y(t) = f(t)$ is a two-order system model; try to simulate this system with basic block diagrams.

Solution. The basic principle of solving a differential equation is to extract the original function using the integral operation. For the above equation, obviously, twice as many integral operations are necessary. Thus, we transform the original expression into

$$\frac{d^2y(t)}{dt^2} = -a_1\frac{dy(t)}{dt} - a_0y(t) + f(t) \ .$$

It can be seen that the right side of this equation can be considered as the sum of three terms, that is, the output of an adder is a second-order function term $\frac{d^2y(t)}{dt^2}$, which can become the original function after it has been integrated twice. Then the function is multiplied separately by corresponding coefficients and is fed back to the adder, and the problem is solved. The system simulation is plotted in ▶ Figure 2.15.

Example 2.4-3. $\frac{d^2y(t)}{dt^2} + a_1\frac{dy(t)}{dt} + a_0y(t) = b_1\frac{df(t)}{dt} + b_0f(t)$ is a second-order linear model; simulate the system by basic operators.

Solution. Because $b_1\frac{df(t)}{dt}$ appears on the right side, the solution in Example 2.4-2 cannot be borrowed directly. Thus, an assistant signal $m(t)$ is introduced to connect $f(t)$ and $y(t)$ to form two submodels, and the solution in Example 2.4-2 can be used to solve this problem.

The ways to design the submodels are:
(1) change $y(t)$ into $m(t)$ on the left side; only $f(t)$ is kept on the right side, so the first submodel is obtained.

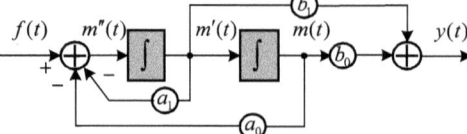

Fig. 2.16: E2.4-3.

(2) Only keep $y(t)$ on left side and change $f(t)$ into $m(t)$ on the right side to obtain the second submodel. We have

$$\frac{\mathrm{d}^2 m(t)}{\mathrm{d}t^2} + a_1 \frac{\mathrm{d}m(t)}{\mathrm{d}t} + a_0 m(t) = f(t) , \tag{1}$$

$$y(t) = b_1 \frac{\mathrm{d}m(t)}{\mathrm{d}t} + b_0 m(t) . \tag{2}$$

Simulating the above two equations, we obtain ▶ Figure 2.16.

The solution here also fits well for situations where there are higher orders of derivatives of the excitation on the right side. In addition, this method is only used to explain the simulation principle, but Mason's formula in Chapter 7 is the common simulation method in real applications.

The above examples illustrate that a system can be represented by both the mathematical model and the block diagram model. Note that the block diagram model is not a new system model that is different from the mathematical model, but just the graphical expression of the mathematical model, and thus, the mathematical model is the basis. A system can be modeled with different of its mathematical model.

2.5 Analysis methods for LTI systems

As stated in above, LTI system models can reflect directly the relation between excitation and response, namely, the analysis results of the models can only reflect the external properties or the I/O properties of a system and are unrelated to the internal parameters. This analysis method is called the external analysis method or the I/O or ports analysis method. The external method is the main method used for analysis of a system herein.

In real applications, people often want to know what influence an internal change of a system has on the response, so state space analysis method can be used.

The state space analysis method can connect excitations, responses and internal state variables by a set of state equations and a set of output equations, so that the change laws of the responses are revealed with changes of the internal parameters in a system.

The state equations and the output equations are all mathematical models of a system, so, whether the external analysis method or the state space analysis method is used, the basic principle of both is to solve mathematical models or to solve equations. Moreover, the solutions to an equation can be classified into time domain and transform domain methods. So, the ways to analyze an LTI system mainly consist of four analysis methods: the external time domain, transform domains, the state space time domain, and transform domains.

In conclusion, there are four analysis methods for a LTI system, and their principles are all based on building the model of the system and finding its solution. In other words, the essence of system analysis is just to solve equations.

2.6 Solved questions

Question 2-1. For an LTI system with the same initial condition, the complete response is $y_1(t) = [2e^{-3t} + \sin(2t)]\,\varepsilon(t)$ when the excitation is $e(t)$, while the complete response is $y_2(t) = [e^{-3t} + 2\sin(2t)]\,\varepsilon(t)$ when the excitation is $2e(t)$. Then
(1) When the initial condition is invariant, find the complete response $y_3(t)$ when the excitation is $e(t - t_0)$, and t_0 is a real constant and greater than 0.
(2) If the initial condition increase by one, find the complete response $y_4(t)$ when the excitation is $0.5e(t)$.

Solution. (1) Letting the zero-input and the zero-state responses be $y_x(t)$ and $y_f(t)$, we have

$$y_1(t) = y_x(t) + y_f(t) = \left[2e^{-3t} + \sin(2t)\right]\varepsilon(t)\,,$$
$$y_2(t) = y_x(t) + 2y_f(t) = \left[e^{-3t} + 2\sin(2t)\right]\varepsilon(t)\,.$$

Then
$$y_x(t) = 3e^{-3t}\varepsilon(t) \quad\text{and}\quad y_f(t) = \left[-e^{-3t} + \sin(2t)\right]\varepsilon(t)\,.$$

When the excitation is $e(t - t_0)$, the complete response is
$$y_3(t) = 3e^{-3t}\varepsilon(t) + \left[-e^{-3(t-t_0)} + \sin(2t - 2t_0)\right]\varepsilon(t - t_0)\,.$$

(2) From the results of (1), we know that $y_x(t)$ and $y_f(t)$, so, when the initial condition increases by one and the excitation is $0.5e(t)$, the complete response is

$$y_4(t) = 2y_x(t) + 0.5y_f(t) = 2 \times 3e^{-3t}\varepsilon(t) + 0.5 \times \left[-e^{-3t} + \sin(2t)\right]\varepsilon(t)$$
$$= \left[5.5e^{-3t} + 0.5\sin(2t)\right]\varepsilon(t)\,.$$

Question 2-2. The initial conditions of an LTI system are $x_1(0)$ and $x_2(0)$.
(1) When $x_1(0) = 1$, $x_2(0) = 0$, the zero-input response is $y_{x1}(t) = (e^{-t} + e^{-2t})\varepsilon(t)$.
(2) When $x_1(0) = 0$, $x_2(0) = 1$, the zero-input response is $y_{x2}(t) = -(e^{-t} - e^{-2t})\varepsilon(t)$.
If the complete response is $(2 + e^{-t})\varepsilon(t)$ when the excitation is $f(t)$, $x_1(0) = 1$ and $x_2(0) = -1$, find the complete response when the excitation is $2f(t)$, $x_1(0) = -1$ and $x_2(0) = -2$.

Solution. From the known conditions (1) and (2), we have

$$y_{x1}(t) = T\{f(t) = 0, x_1(0) = 1,\ x_2(0) = 0\} = \left(e^{-t} + e^{-2t}\right)\varepsilon(t)\,,$$
$$y_{x2}(t) = T\{f(t) = 0,\ x_1(0) = 0, x_2(0) = 1\} = -\left(e^{-t} - e^{-2t}\right)\varepsilon(t)\,.$$

Because
$$y_f(t) = T\{f(t), x_1(0) = 0, x_2(0) = 0\},$$
from the known condition, we obtain
$$y_1(t) = T\{f(t), x_1(0) = 1, x_2(0) = -1\} = y_f(t) + y_{x1}(t) - y_{x2}(t)$$
$$= y_f(t) + 2e^{-t}\varepsilon(t) = \left(2 + e^{-t}\right)\varepsilon(t).$$

Thus, the zero-state response is
$$y_f(t) = (2 - e^{-t})\varepsilon(t).$$

Hence, for the excitation $2f(t)$, $x_1(0) = -1$, $x_2(0) = -2$, the complete response $y(t)$ is
$$y(t) = T\{2f(t), x_1(0) = -1, x_2(0) = -2\} = 2y_f(t) - y_{x1}(t) - 2y_{x2}(t)$$
$$= \left(4 - e^{-t} - 3e^{-2t}\right)\varepsilon(t).$$

Question 2-3. Known the zero-state response of a CT system is $y_f(t) = f(4t)$ when the excitation is $f(t)$. Judge whether the system is time variant and linear. Prove your conclusions.

Solution. According to the given conditions, we have
$$f_1(t) \to y_{f1}(t) = f_1(4t) \quad \text{and} \quad f_2(t) \to y_{f2}(t) = f_2(4t),$$
then
$$af_1(t) + bf_2(t) \to ay_{f1}(t) + by_{f2}(t) = af_1(4t) + bf_2(4t).$$
Thus, the system satisfies linearity. According to the given conditions, we also know that
$$f(t - t_0) \to f(4t - t_0).$$
From the time invariance property, this can be written as
$$f(t - t_0) \to y_f(t - t_0) = f(4t - 4t_0).$$

Obviously, the system cannot satisfy the time invariant condition, so it is linear, time variant.

Question 2-4. Judge whether or not the system described by $y(t) = (t + 5)\cos\left(\frac{1}{x(t)}\right)$ is causal and time variant. Prove your conclusions.

Solution. The output $y(t)$ is only related to the current output, so the system is causal.
Letting
$$x(t) \to y(t) = (t + 5)\cos\left[\frac{1}{x(t)}\right],$$
we have
$$x(t - \tau) \to (t + 5)\cos\left[\frac{1}{x(t - \tau)}\right] \neq y(t - \tau).$$
So, the system is time variant.

Question 2-5. Judge whether or not the system described by $y(t) = \int_{-\infty}^{2t-1} f(\tau)d\tau$ is linear, time variant. Prove your conclusions.

Solution. The known conditions can be expressed as

$$f(t) \rightarrow y(t) = \int_{-\infty}^{2t-1} f(\tau)d\tau .$$

Let

$$f_1(t) \rightarrow y_1(t) = \int_{-\infty}^{2t-1} f_1(\tau)d\tau, f_2(t) \rightarrow y_2(t) = \int_{-\infty}^{2t-1} f_2(\tau)d\tau .$$

Because

$$\alpha f_1(t) + \beta f_2(t) \rightarrow \int_{-\infty}^{2t-1} [\alpha f_1(\tau) + \beta f_2(\tau)]\,d\tau = \alpha y_1(t) + \beta y_2(t) ,$$

the system is linear. Because

$$f(t - t_0) \rightarrow \int_{-\infty}^{2t-1} f(\tau - t_0)d\tau = \int_{-\infty}^{2t-1-t_0} f(\eta)d\eta \neq y(t - t_0) ,$$

the system is time variant.

2.7 Learning tips

The system is the main object of research herein, and readers should pay attention to the following points:
(1) Theoretically, a system is a kind of signal converter and is also a type of algorithm, the characteristics of which can be described by a mathematical model. In fact, systems mainly refers to circuits herein. Therefore, the basic knowledge from circuit analysis.
(2) The state and the response of a system are separate reactions of the system to the storage energy situation and the excitations.
(3) The LTI system is the basis for analysis of other systems.
(4) The system can be described by models, so analyzing a system is actually analyzing its models.
(5) Each system model is a bridge to connect excitation and response, which is established by means of the characteristics and the structure of system itself.
(6) The system mathematical model can be simulated by operating block diagrams, or the mathematical model and the block diagram model of a system are equivalent.

2.8 Problems

Problem 2-1. If the starting state of a system is $x(0_-)$, the excitation is $f(t)$ and the response is $y(t)$, try to judge whether or not the following systems are linear.

(1) $y(t) = x^2(0_-) + f^2(t)$

(2) $y(t) = x(0_-) \log f(t)$

(3) $y(t) = x(0_-) \sin(t) + tf(t)$

(4) $y(t) = x^2(0_-) + \int_0^t f(\tau)d\tau$

(5) $y'(t) = \log[x(0_-)] + f^2(t) + ty(t)$

Problem 2-2. If the starting state of a system is $x(0_-)$, the excitation is $f(t)$ and the response is $y(t)$, try to judge whether or not the following systems are time invariant.

(1) $y(t) = f(t) + f(t - t_0)$

(2) $y(t) = x(0_-) + 3tf^2(t)$

(3) $y(t) = f(t) + tx(0_-)$

(4) $y''(t) = y'(t)y(t) + x_1(0_-) + x_2(0_-) + \lg[f(t)]$

Problem 2-3. Test whether or not the following systems are causal.

(1) $y(t) = \cos(t) \cdot f(t)$

(2) $y(t) = f(-t)$

(3) $y(t) = f(t - 1) - f(1 - t)$

(4) $y(t) = f(t) \cdot f(t - b)$

(5) $y(t) = 2f(t) \cdot \varepsilon(t)$

Problem 2-4. Judge whether or not the following systems are linear, time invariant and causal.

(1) $y(t) = x(0_-) \sin(t) + at^2 f(t)$

(2) $y(t) = f(t + 10) + f^2(t)$

(3) $y(t) = (t + 1)f(t)$

(4) $y'(t) + 10y(t) = f(t)$

(5) $y'(t) + y(t) = f(t + 10)$

(6) $y'(t) + t^2 y(t) = f(t)$

Problem 2-5. When the input of a system is $\delta(t - \tau)$, the output is $h(t) = \varepsilon(t - \tau) - \varepsilon(t - 3\tau)$, judge whether or not the system is time invariant and causal.

Problem 2-6. The starting state of an LTI causal system is zero, when the excitation is $f_1(t) = \varepsilon(t)$, the response is $y_1(t) = (3e^{-t} + 4e^{-2t})\varepsilon(t)$, find the response when the excitation is instead the signal as shown in the ▶ Figure P2-6.

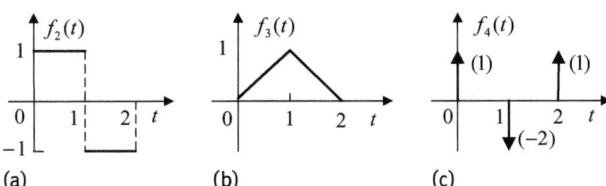

(a)　(b)　(c)　**Fig. P2-6**

Problem 2-7. An LTI system has a starting state, and the complete response is $y_1(t) = 3e^{-2t} + \sin(4t)$　$t > 0$ when the excitation is $f(t)$. If the starting state is constant and the excitation is $2f(t)$, the complete response is $y_2(t) = 4e^{-2t} + 2\sin(4t)$ $t > 0$. Find the complete response when the excitation is $3f(t)$ and the starting state is the same.

Problem 2-8. There is an LTI system with starting conditions $x_1(0_-)$ and $x_2(0_-)$, the excitation is $f(t)$, and the response is $y(t)$. It is known that
(1) when $f(t) = 0$, $x_1(0_-) = 5$, $x_2(0_-) = 2$, we have $y(t) = e^{-t}(7t + 5)$, $t > 0$;
(2) when $f(t) = 0$, $x_1(0_-) = 1$, $x_2(0_-) = 4$, we have $y(t) = e^{-t}(5t + 1)$, $t > 0$;
(3) when $f(t) = \begin{cases} 0, t < 0 \\ 1, t > 0 \end{cases}$, $x_1(0_-) = 1$, $x_2(0_-) = 1$, we have $y(t) = e^{-t}(t + 1)$, $t > 0$.

Find the zero-state response of the system when $f(t) = \begin{cases} 0, t < 0 \\ 3, t > 0 \end{cases}$.

Problem 2-9. Try to write the differential equation of $u_C(t)$ in ▶ Figure P2-9.

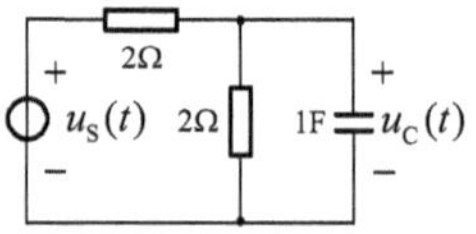

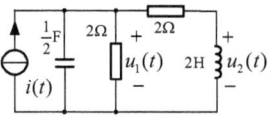

Fig. P2-9

Problem 2-10. An excitation is $f(t) = \sin(2t)\varepsilon(t)$, voltages across capacitors are zero at initial moment, find the expression of the output signal $u_C(t)$ in ▶ Figure P2-10.

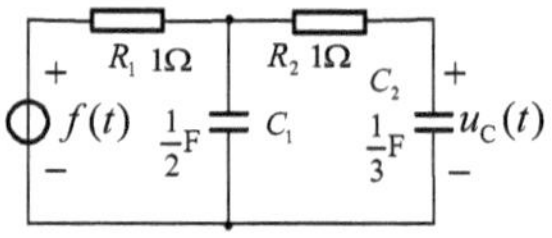

Fig. P2-10

Problem 2-11. The circuit is shown in ▶ Figure P2-11, try to write out the differential equations between $u_1(t)$, $u_2(t)$ and $i(t)$.

Fig. P2-11

Problem 2-12. Plot block diagram models of the following systems.
(1) $y''(t) + 7y'(t) + 12y(t) = f(t)$
(2) $y'''(t) + 4y''(t) + 10y'(t) + 3y(t) = f''(t) + 10f(t)$

3 Analysis of continuous-time systems in the time domain

Questions: In order to analyze the relationship between the excitation and response of a continuous LTI system, we need to solve the system model, that is, the differential equation in the time domain.

Solution: (1) Use classical methods of advanced mathematics.

(2) Introduce the operator to simplify the solution process and to provide support for solving from impulse response.

(3) Consider a basic signal (impulse signal or step signal) as the input → Obtain the solution (impulse response or step response) → Find the relationship between the basic signal and other signals → Use linearity to obtain system response when other signals serve as the excitation.

Results: Response decomposition, impulse response, step response and transport operator.

The system analysis process is generally divided into three stages, as shown in ▶ Figure 3.1.

(1) Establish the system model. Write the mathematical expression that can show the relationship between the input and the output signals of a system. Usually, the model of a continuous system is a differential equation, whereas the model of a discrete system is a difference equation.

(2) Solve the system model. Analyze and solve the model using appropriate mathematical methods; in short, solve the linear differential or difference equation.

(3) Analyze the results. Give the physical interpretations for the responses (equation solutions) obtained in the *time* or *frequency* domain, enhance the understanding level for the transfer or processing of a system to signals, and obtain the desired conclusions from them.

Herein, only LTI systems will be discussed. The mathematical model for these systems is an nth-order constant coefficient linear differential or difference equation. Therefore,

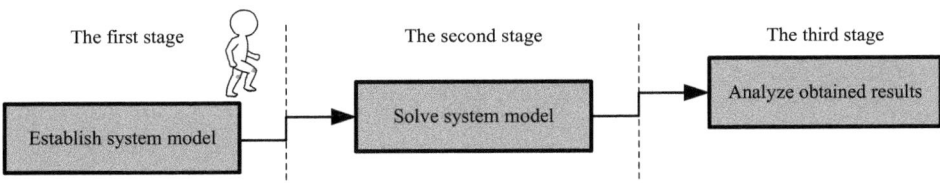

Fig. 3.1: The process of system analysis.

https://doi.org/10.1515/9783110419535-003

the system analysis is to establish and solve an nth-order constant coefficient linear dif-
ferential or difference equation, which is the model of a system, then to analyze the re-
sults.

For a continuous system, with respect to how to solve the differential equation, we
will successively introduce three basic methods in the *time* domain, *frequency* domain
and *complex frequency* domain. For a discrete system, with respect to how to solve the
difference equation, we will also introduce the corresponding two methods in the time
domain and the *z* domain.

After having studied this material, readers should know that these analysis methods are all based on two cornerstones: the decomposability of a signal and the linearity and time-invariance of a system. In this chapter, we will present external analysis
methods of a continuous system in the time domain, that is, we will discuss the variation of the response of a system with time, or the time characteristics of a system, or
solving methods of system equations in the time domain.

3.1 Analysis methods with differential equations

3.1.1 The classical analysis method

Usually, "differential equation" refers to an equality containing derivative functions
of unknown functions. The maximum order in derivative terms is also the order of the
differential equation.

We studied the classical analysis method to solve differential equations in an advanced mathematics course and will present a brief review here. As we know, the
mathematical model of an LTI system is a constant coefficient linear differential equation, and its complete solution $y(t)$ is composed of the homogeneous solution $y_c(t)$
and the particular solution $y_p(t)$ in two parts, that is,

$$y(t) = y_c(t) + y_p(t) . \tag{3.1-1}$$

First, find the homogeneous solution. For a constant coefficient differential equation

$$\sum_{i=0}^{n} a_i y^{(i)}(t) = \sum_{j=0}^{m} b_j f^{(j)}(t) , \tag{3.1-2}$$

it can be also written as

$$a_n y^{(n)}(t) + a_{n-1} y^{(n-1)}(t) + \cdots + a_1 y'(t) + a_0 y(t)$$
$$= b_m f^{(m)}(t) + b_{m-1} f^{(m-1)}(t) + \cdots + b_1 f'(t) + b_0 f(t)$$

When its right side equals zero, the solution is homogeneous. This means that the
solution satisfies a homogeneous equation

$$a_n y^{(n)}(t) + a_{n-1} y^{(n-1)}(t) + \cdots + a_1 y'(t) + a_0 y(t) = 0 . \tag{3.1-3}$$

Usually, the homogeneous solution is combined with several functions in the form $Ce^{\lambda t}$.

Substituting the term $Ce^{\lambda t}$ into equation (3.1-3), we have

$$a_n C\lambda^n e^{\lambda t} + a_{n-1} C\lambda^{n-1} e^{\lambda t} + \cdots + a_1 C\lambda e^{\lambda t} + a_0 Ce^{\lambda t} = 0 \, ,$$

if $C \neq 0$, so,

$$a_n \lambda^n + a_{n-1}\lambda^{n-1} + \cdots + a_1 \lambda + a_0 = 0 \, . \tag{3.1-4}$$

Equation (3.1-4) is called the characteristic equation of the differential equation (3.1-2); n roots $\lambda_1, \lambda_2, \ldots, \lambda_n$ of the characteristic equation are said to be the characteristic roots, the natural frequencies or inherent frequencies of the differential equation.

If there are no repeated roots in (3.1-4), the homogeneous solution should be

$$y_c(t) = \sum_{i=1}^{n} c_i e^{\lambda_i t} = c_n e^{\lambda_n t} + c_{n-1} e^{\lambda_{n-1} t} + \cdots + c_1 e^{\lambda_1 t} \, . \tag{3.1-5}$$

If there are repeated roots, the form of the homogeneous solution should be different. Suppose λ_1 is an r-repeated root, that is, $\lambda_1 = \lambda_2 = \cdots = \lambda_r$, then $n - r$ rest roots are simple, so the homogeneous solution becomes

$$
\begin{aligned}
y_c(t) &= \sum_{i=1}^{r} c_i t^{r-i} e^{\lambda_i t} + \sum_{j=r+1}^{n} c_j e^{\lambda_j t} \\
&= c_1 t^{r-1} e^{\lambda_1 t} + c_2 t^{r-2} e^{\lambda_2 t} + \cdots + c_r e^{\lambda_r t} + c_{r+1} e^{\lambda_{r+1} t} + \cdots + c_n e^{\lambda_n t} \, .
\end{aligned}
\tag{3.1-6}
$$

Second, find the particular solution. The form of the particular solution is related to the form of excitation, several typical excitation signals $f(t)$ and their corresponding particular solutions $y_p(t)$ are listed in Table 3.1.

After each particular solution $y_p(t)$ in Table 3.1 is substituted into the original differential equation, according to the criterion that the corresponding coefficients on both sides of an equation are the same, the unknown coefficients p and B in $y_p(t)$ can be determined.

Tab. 3.1: Typical excitation signals and their particular solutions.

No.	Excitation $f(t)$	Particular solution $y_p(t)$
1	t^m	$p_m t^m + p_{m-1} t^{m-1} + \cdots + p_0$
2	$e^{\alpha t}$	$p e^{\alpha t}$ (α is not the characteristic root)
		$\sum_{i=0}^{r} p_i t^i e^{\alpha t}$ (α is the r-repeated roots)
3	$\cos \beta t$	$p_1 \cos \beta t + p_2 \sin \beta t$
4	$\sin \beta t$	$p_1 \cos \beta t + p_2 \sin \beta t$
5	A (constant)	B (constant)

Finally, n unknown coefficients c_i ($i = 1, 2, \ldots n$) in the homogeneous solution need to be determined. Now we only need to substitute n values of initial condition of the system into the total solution to obtain the coefficients. What is the initial condition?

Similarly to the concepts of initial state and starting state in Chapter 2, because of the effect and the influence of the switching of circuit, the response and its all-order derivatives may change instantaneously at time $t_0 = 0$. To distinguish between their values before and after the step occurs, we use "0_-" to represent the moment before the excitation is accessed or the switch is activated (the starting moment), and we use "0_+" to represent the moment after the excitation is accessed or the switch activated (the initial moment). Hence, we can say that the starting condition of a system refers to the values of system response and its all-order derivatives at time 0_-, which can be expressed as $\{y^{(i)}(0_-), i = 0, 1, \ldots n - 1\}$. While the initial condition of the system is the values of system response and its all-order derivatives at time 0_+, which can be expressed as $\{y^{(i)}(0_+), i = 0, 1, \ldots n - 1\}$.

Usually, the response of system is the output of the system after the system is driven by an excitation, that is, the response exists over a range $0_+ \le t < +\infty$. Therefore, we can work out the unknown coefficients in the homogeneous solution using the initial condition of the system.

Note: The starting condition is different from the starting state, and the initial condition is also different from the initial state. The state of a system refers to the situations or data of the energy storage of the system. Because energy storage elements in the electric system are generally inductors and capacitors, the energy storage situations can be reflected by the currents through inductors and the voltages across capacitors. Therefore, in an electronic system, the starting state refers to the current values through inductors and the voltage values across capacitors at the starting moment. For example, with $i_L(0_-)$ and $u_C(0_-)$. The initial state refers to the current values through inductors and the voltage values across capacitors at the initial moment. For example, $i_L(0_+)$ and $u_C(0_+)$.

The responses of a system are not always capacitor voltages and inductor currents but may be the terminal voltage across or the current through a resistor or other parameters, hence, the values of response and the all-order derivatives at the starting or initial moment $\{y^{(i)}(0_-), i = 0, 1, \ldots n - 1\}$ or $\{y^{(i)}(0_+), i = 0, 1, \ldots n - 1\}$ cannot be directly expressed by the starting or the initial state. However, according to the concepts of initial condition and initial state, we know that the initial state should be included by the initial condition, and the initial state is definitely the initial condition, whereas the initial condition is not always the initial state but can be calculated from the initial state. Similarly, the starting state is included by the starting condition and is certainly the starting condition, but not vice versa; the starting condition can be calculated based on the starting state. Therefore, in an electronic system, the initial condition or starting condition usually refers to including inductor currents and capacitor voltages and values of other circuit parameters or variables (may include their nth-order derivatives) caused by them at the initial or starting moment.

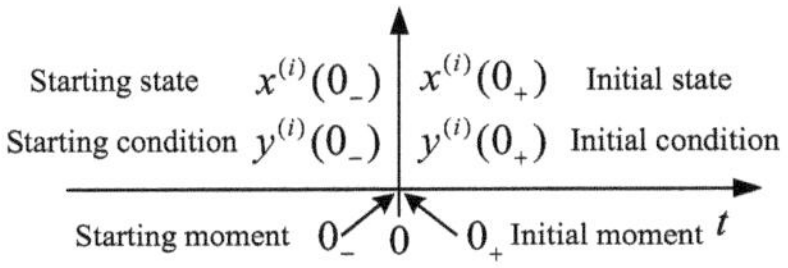

Starting state $x^{(i)}(0_-)$ | $x^{(i)}(0_+)$ Initial state
Starting condition $y^{(i)}(0_-)$ | $y^{(i)}(0_+)$ Initial condition
Starting moment 0_- 0 0_+ Initial moment t

Fig. 3.2: Starting/initial moment and state/condition.

Because the two conditions can definitely be calculated from the two states, we often equate the condition and the state. Moreover, since the value of the state generally cannot jump, this means the inductor current and the capacitor voltage cannot change suddenly, so the starting state is equal to the initial state, i.e. $i_L(0_-) = i_L(0_+)$ and $u_C(0_-) = u_C(0_+)$. However, we must note that the starting condition is not always equal to the initial condition. ▶ Figure 3.2 shows the starting and initial moments and the corresponding states and conditions.

Example 3.1-1. An LTI system model is

$$\frac{d^2}{dt^2}y(t) + 3\frac{d}{dt}y(t) + 2y(t) = \frac{d}{dt}f(t) + 2f(t) .$$

If an excitation $f(t) = t^2$ and the initial condition $y(0_+) = 1$, $y'(0_+) = 1$, work out the complete response of the system.

Solution. The homogeneous equation of the system is

$$\frac{d^2}{dt^2}y(t) + 3\frac{d}{dt}y(t) + 2y(t) = 0 .$$

Its characteristic equation is
$$\lambda^2 + 3\lambda + 2 = 0 .$$

The characteristic roots are
$$\lambda_1 = -1, \quad \lambda_2 = -2 ,$$

so, the homogeneous solution is

$$y_c(t) = c_1 e^{-t} + c_2 e^{-2t} .$$

Consider $f(t) = t^2$; the particular solution can be set as

$$y_p(t) = p_2 t^2 + p_1 t + p_0 .$$

Substituting the above expression and $f(t) = t^2$ into the system model, we have

$$2p_2 t^2 + (2p_1 + 6p_2) t + (2p_0 + 3p_1 + 2p_2) = 2t^2 + 2t ,$$

According to the balance equation rule, then

$$\begin{cases} 2p_2 = 2 \\ 2p_1 + 6p_2 = 2 \\ 2p_0 + 3p_1 + 2p_2 = 0 \end{cases} ,$$

and we obtain

$$p_2 = 1, \quad p_1 = -2, \quad p_0 = 2 .$$

Thus, the particular solution

$$y_p(t) = t^2 - 2t + 2 .$$

The complete solution

$$y(t) = c_1 e^{-t} + c_2 e^{-2t} + t^2 - 2t + 2 .$$

Substituting the initial condition $y(0_+) = 1$; and $y'(0_+) = 1$ into the above expression,

$$\begin{cases} c_1 + c_2 + 2 = 1 \\ -c_1 - 2c_2 - 2 = 1 \end{cases} ,$$

thus,

$$c_1 = 1, \quad c_2 = -2 .$$

The complete response is

$$y(t) = e^{-t} - 2e^{-2t} + t^2 - 2t + 2 \quad t \geq 0 .$$

3.1.2 Response decomposition analysis method

We know that the complete solution of a linear constant coefficient differential equation can be decomposed into homogeneous and particular solutions. According to different standards, the complete response of a system can be also decomposed into other kinds of responses, such as zero-input and zero-state responses, transient and steady state responses, or natural and forced responses. Thus, besides the classical methods, other ways can be provided to solve the system model, for example, the zero-input/zero-state response method in the following, which has also been called the response decomposition method or the modern analysis method.

From Chapter 2, the complete response of an LTI system can be decomposed into zero-input response and zero-state response, that is,

$$y(t) = y_x(t) + y_f(t) , \tag{3.1-7}$$

where

$$y_x(t) = T[x_1(0_-), x_2(0_-), \ldots x_n(0_-), 0] = T[\{x(0_-)\}, 0] , \tag{3.1-8}$$

$$y_f(t) = T[0, f_1(t), f_2(t), \ldots, f_n(t)] = T[0, \{f(t)\}] . \tag{3.1-9}$$

Because the starting state $\{x(0_-)\}$ can be equivalent to n excitation sources, the complete response can be considered as the combined effects of the external excitation source $f(t)$ and the internal equivalent excitation sources $\{x(0_-)\}$. Accordingly, a

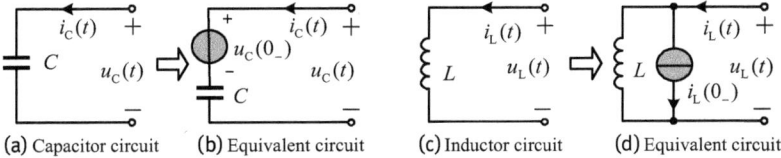

(a) Capacitor circuit   (b) Equivalent circuit   (c) Inductor circuit   (d) Equivalent circuit

Fig. 3.3: Capacitor and inductor circuits and their equivalent circuits.

method to analyze a system, that is, the response decomposition method has been put forward.

There are two steps in the response decomposition analysis method. First, find the zero-input response and the zero-state response of a system; second, add them together to obtain the complete response.

To use this method, we should first research the voltage response $u_C(t)$ on a capacitor and the current response $i_L(t)$ through an inductor over the range $[-\infty, t]$ for $t \geq 0$.

For the capacitor C in ▶ Figure 3.3a, suppose the starting voltage is $u_C(0_-)$. Then the response $u_C(t)$ can be written as

$$u_C(t) = \frac{1}{C} \int_{-\infty}^{t} i_C(\tau)d\tau = \frac{1}{C} \int_{-\infty}^{0_-} i_C(\tau)d\tau + \frac{1}{C} \int_{0_-}^{0_+} i_C(\tau)d\tau + \frac{1}{C} \int_{0_+}^{t} i_C(\tau)d\tau$$

$$= u_C(0_-) + 0 + \frac{1}{C} \int_{0_+}^{t} i_C(\tau)d\tau = u_C(0_-) + \frac{1}{C} \int_{0_+}^{t} i_C(\tau)d\tau \, (t \geq 0) \, .$$

Over the range $[-\infty, t]$ a capacitor with the nonzero starting voltage $u_C(0_-)$ can be considered as a series form of a capacitor with zero starting voltage and a voltage source $u_C(0_-)$, which is shown in ▶ Figure 3.3b.

Similarly, the analysis results of an inductor current response can be obtained. For the inductor L with the starting current $i_L(0_-)$ shown in ▶ Figure 3.3c, $i_L(t)$ can be written as

$$i_L(t) = \frac{1}{L} \int_{-\infty}^{t} u_L(\tau)d\tau = \frac{1}{L} \int_{-\infty}^{0_-} u_L(\tau)d\tau + \frac{1}{L} \int_{0_-}^{0_+} u_L(\tau)d\tau + \frac{1}{L} \int_{0_+}^{t} u_L(\tau)d\tau$$

$$= i_L(0_-) + 0 + \frac{1}{L} \int_{0_+}^{t} u_L(\tau)d\tau = i_L(0_-) + \frac{1}{L} \int_{0_+}^{t} u_L(\tau)d\tau \, (t \geq 0) \, .$$

It can be seen that over the range $[-\infty, t]$, an inductor L with the nonzero starting current $i_L(0_-)$ can be considered as the parallel form of an inductor with zero starting current and a current source $i_L(0_-)$, which is shown in ▶ Figure 3.3d.

Obviously, both $u_C(0_-)$ and $i_L(0_-)$ belong to the zero-input response, whereas $\frac{1}{C} \int_{0_+}^{t} i_C(\tau)d\tau$ and $\frac{1}{L} \int_{0_+}^{t} u_L(\tau)d\tau$ are both the zero-state response.

Based on the above response decomposition concepts of two energy storage elements, a method has been developed to find the zero-input and the zero-state responses of an LTI system.

1. Find the zero-input response $y_x(t)$

Assume that the right side of the equation (3.1-2) is zero

$$\sum_{i=0}^{n} a_i y^{(i)}(t) = 0 , \tag{3.1-10}$$

then the zero-input response is only caused by the starting state of the system for $t \geq 0$ under $f(t)$ and its all-order derivative values are zero, and it meets equation (3.1-10) and has the same form as the homogeneous solution. In other words, its form is just like equation (3.1-5) or equation (3.1-6).

Assuming that the characteristic roots are dissimilar simple roots, the zero-input response is

$$y_x(t) = \sum_{i=1}^{n} c_{x_i} e^{\lambda_i t} \quad t \geq 0 , \tag{3.1-11}$$

where coefficients c_{x_i} ($i = 1, 2, \ldots, n$) are determined by the initial condition $\left\{ y_x^{(k)}(0_+) \right\}$ of the zero-input response.

According to the linearity concept of the LTI system, the initial condition $y^{(k)}(0_+)$ of the complete response (actually the initial condition of the system) is the sum of the initial condition $y_x^{(k)}(0_+)$ of the zero-input response and the initial condition $y_f^{(k)}(0_+)$ of the zero-state response, that is, $y^{(k)}(0_+) = y_x^{(k)}(0_+) + y_f^{(k)}(0_+)$. Under the zero-input condition, the equation is considered to exist over the range $-\infty < t < \infty$; thus the starting condition $y^{(k)}(0_-)$ of the complete response is actually equal to the initial condition $y_x^{(k)}(0_+)$ of the zero-input response, and at this moment, $y^{(k)}(0_-) = y_x^{(k)}(0_+) = y_x^{(k)}(0_-)$. This shows that the coefficient c_{x_i} can be determined by the starting condition $\left\{ y^{(k)}(0_-) \right\}$ of the complete response. Note that $y^{(k)}(0_+)$ cannot usually be used to determine c_{x_i}, because it may include the effects of excitations.

Example 3.1-2. The homogeneous equation of an LTI system is $y''(t) + 2y'(t) + 2y(t) = 0$, and the system starting state values are $y(0_-) = 0$, $y'(0_-) = 2$. Try to find the zero-input response of the system.

Solution. The characteristic equation of the system is

$$\lambda^2 + 2\lambda + 2 = 0 ,$$

its characteristic roots are

$$\lambda_1 = -1 + j$$
$$\lambda_2 = -1 - j ,$$

and the zero-input response is

$$y_x(t) = c_1 e^{(-1+j)t} + c_2 e^{(-1-j)t} \ .$$

Using Euler's relations, the homogeneous solution can be transformed into trigonometric form

$$
\begin{aligned}
y_x(t) &= c_1 e^{(-1+j)t} + c_2 e^{(-1-j)t} \\
&= e^{-t}(c_1 \cos t + j c_1 \sin t + c_2 \cos t - j c_2 \sin t) \\
&= e^{-t}[(c_1 + c_2) \cos t + j(c_1 - c_2) \sin t] \\
&= e^{-t}(A_1 \cos t + A_2 \sin t) \ .
\end{aligned}
$$

From the starting state values $y(0_-) = 0$ and $y'(0_-) = 2$, we have

$$
\begin{cases}
A_1 = 0 \\
A_2 = 2 \ ,
\end{cases}
$$

so, the zero-input response is

$$y_x(t) = 2e^{-t} \sin t \quad t \geq 0 \ .$$

2. Find the zero-state response $y_f(t)$

The zero state refers to the system not having any energy storage before the excitation is applied. Therefore, the system response for $t \geq 0$ can only be caused by the excitation applied when $t \geq 0$. At this moment, the system model is a nonhomogeneous differential equation. Obviously, the zero-state response should satisfy this equation, that is,

$$\sum_{i=0}^{n} a_i y_f^{(i)}(t) = \sum_{j=0}^{m} b_j f^{(j)}(t) \tag{3.1-12}$$

In order to solve equation (3.1-12), we must also know a set of data $\{y_f^{(k)}(0_+)\}$, which is the initial condition of the zero-state response. Note that $\{y_f^{(k)}(0_+)\}$ relate to the excitation $f(t)$ applied to the system at $t = 0$, so the zero state may not mean $\{y_f^{(k)}(0_+)\} = 0$. In fact, under the zero-state condition, we have $y_f^{(k)}(0_+) = y^{(k)}(0_+) - y^{(k)}(0_-) = y^{(k)}(0_+)$, and therefore, $y_f^{(k)}(0_+)$ is also called the step change value of the system response.

Supposing that the characteristic roots $\lambda_1, \lambda_2, \ldots, \lambda_n$ of the system expressed by equation (3.1-12) are simple and dissimilar, the homogeneous solution of the zero-state response is

$$y_{f_c}(t) = \sum_{i=1}^{n} c_{f_i} e^{\lambda_i t} \ . \tag{3.1-13}$$

The method for the solution of the zero-state response is the same the classical method, so

$$y_f(t) = y_{f_c}(t) + y_p(t) = \sum_{i=1}^{n} c_{f_i} e^{\lambda_i t} + y_p(t) \ . \tag{3.1-14}$$

The coefficients c_{f_i} in Equations (3.1-13) and (3.1-14) can be determined by the data $\{y_f^{(k)}(0_+)\}$ via equation (3.1-14). In real system analysis, the starting state values at the moment 0_- are always known, so $\{y_f^{(k)}(0_+)\}$ should be determined by the values. Two methods are usually employed to solve this problem; one is the impulse function balance method, and the other one is the model analysis method based on some physical concepts like the Law of Switching.

The fundamental principles of the impulse function balance method are the following:

(1) The differential equation to describe a system must be tenable over a range $(-\infty, \infty)$. However, because the derivative of a function at a step point is usually inexistent, the differential equation formed by this function does not hold over the range $(-\infty, \infty)$. The impulse function introduced in Chapter 1 can solve the problem of the derivative existing at a step point, that is, $\delta(t) = \frac{d\varepsilon(t)}{dt}$, so that a differential equation with the step change can be tenable over the interval $(-\infty, \infty)$.

(2) If an impulse component in the excitation leads to the emergence of the impulse function and its all-order derivatives on the right side of the differential equation, the left side of the equation should also have the corresponding impulse signal and its all-order derivatives to make the equations tenable. Obviously, the balancing refers to generating the corresponding function terms on the left side of the equation to terms on the right side. However, the production of these function terms means that for some terms in $y^{(k)}(t)$ a step change must occur at $t = 0$.

In short, two key points of the method are:

(1) The balance refers to that $\delta(t)$ and its all-order derivative terms which appear on both sides of a differential equation are the same order and have the same number of terms.

(2) If there is no impulse function on the right side of the differential equation, a step change will not occur in $y(t)$ at $t = 0$.

Herein, we will mainly discuss the circuit model analysis method based on the Law of Switching. In order to determine the initial condition, we need to use the internal energy storage continuity features of the system, which are the electric charge continuity across a capacitor and the flux linkage continuity through an inductor. Suppose that a capacitor current $i_C(t)$ and an inductor voltage $u_L(t)$ are bounded over an interval $0_- \leq t \leq 0_+$, then

$$u_C(0_+) - u_C(0_-) = \frac{1}{C} \int_{0_-}^{0_+} i_C(t)dt = 0$$

$$i_L(0_-) - i_L(0_+) = \frac{1}{L} \int_{0_-}^{0_+} u_L(t)dt = 0 \, ,$$

Namely,

$$u_C(0_+) = u_C(0_-) , \tag{3.1-15}$$

$$i_L(0_+) = i_L(0_-) . \tag{3.1-16}$$

Equations (3.1-15) and (3.1-16) show that if the capacitor current values are limited, the voltage $u_C(t)$ is continuous at $t = 0$, and if the inductor voltage values are limited, the current $i_L(t)$ is continuous at $t = 0$. In short, **neither the capacitor voltage nor the inductor current can change instantaneously**. This conclusion is called the Law of Switching, which we have learned in the circuits analysis course.

The following example will illustrate how can determine the zero-input and zero-state responses using the Law of Switching.

Example 3.1-3. If the circuit pictured in ▶ Figure 3.4a is in steady state, and the switch S is closed quickly at $t = 0$, find the zero-input response $u_{C_x}(t)$ and zero-state response $u_{C_f}(t)$ of $u_C(t)$ for $t \geq 0_+$.

Solution. When the switch S is closed, the equivalent circuit of the system is as shown in ▶ Figure 3.4b. Accordingly, the differential equations of the equivalent circuit can be written by the V-A relations of components and KCL, and

$$i_C(t) = C\frac{du_C(t)}{dt} = 0.2\frac{du_C(t)}{dt} ,$$

$$i_R(t) = \frac{u_C(t)}{R} = u_C(t) ,$$

$$i_L(t) = i_C(t) + i_R(t) = 0.2\frac{du_C(t)}{dt} + u_C(t) ,$$

$$u_L(t) = L\frac{di_L(t)}{dt} = 1.25\frac{d}{dt}\left[0.2\frac{du_C(t)}{dt} + u_C(t)\right] = 0.25\frac{d^2u_C(t)}{dt^2} + 1.25\frac{du_C(t)}{dt} .$$

According to KVL, we obtain

$$u_L(t) + u_C(t) = u_S(t) = 2 ,$$

namely,

$$0.25\frac{d^2u_C(t)}{dt^2} + 1.25\frac{du_C(t)}{dt} + u_C(t) = 2 .$$

After rearranging the expression above, we obtain

$$\frac{d^2u_C(t)}{dt^2} + 5\frac{du_C(t)}{dt} + 4u_C(t) = 8 .$$

The characteristic equation can be written as

$$\lambda^2 + 5\lambda + 4 = 0 .$$

Thus, the characteristic roots are

$$\lambda_1 = -1, \lambda_2 = -4 .$$

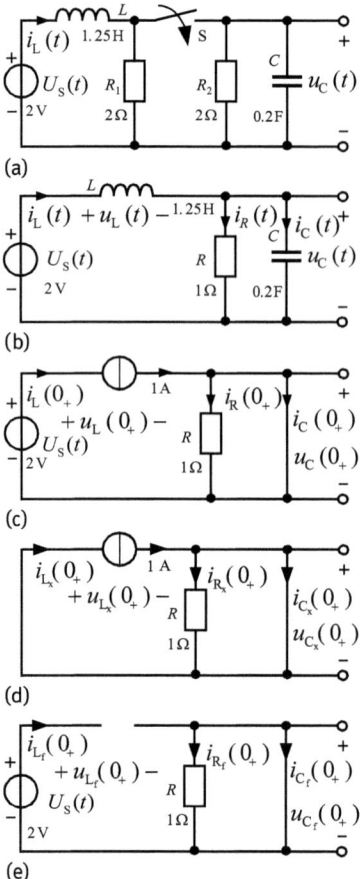

(a)

(b)

(c)

(d)

(e)

Fig. 3.4: E3.1-4.

The zero-input response is assumed as

$$u_{C_x}(t) = c_{x_1} e^{-t} + c_{x_2} e^{-4t} .$$ (3.1-17)

The zero-state response of homogeneous solution is assumed as

$$u_{C_{fc}}(t) = c_{f_1} e^{-t} + c_{f_2} e^{-4t} .$$

It is easy to obtain the solution of zero-state response

$$u_{C_{fp}}(t) = 2 .$$

Then the zero-state response is

$$u_{C_f}(t) = c_{f_1} e^{-t} + c_{f_2} e^{-4t} + 2 .$$ (3.1-18)

To obtain coefficients c_{x_1}, c_{x_2}, c_{f_1}, c_{f_2}, we need to find $u_{C_x}(0_+)$, $u'_{C_x}(0_+)$ and $u_{C_f}(0_+)$, $u'_{C_f}(0_+)$ using the Law of Switching. ▶ Figure 3.4a shows that the circuit is powered by a constant excitation and is stable when $t = 0_-$; the inductor is a short circuit, and the capacitor as an open circuit, so,

$$i_L(0_-) = \frac{u_S}{R_1} = 1\,\text{A}, \quad u_C(0_-) = 0\,\text{V}.$$

According to the Law of Switching, at $t = 0_+$, we have

$$i_L(0_+) = i_L(0_-) = 1\,\text{A}, \quad u_C(0_+) = u_C(0_-) = 0\,\text{V}.$$

The equivalent circuit is shown in ▶ Figure 3.4c when $t = 0_+$. Therefore, we can draw the equivalent circuits of zero input and zero state as in ▶ Figure 3.4d and e.

From ▶ Figure 3.4d, we have $u_{C_x}(0_+) = 0\,\text{V}$ and $i_{C_x}(0_+) = 1\,\text{A}$. Because $i_{C_x}(0_+) = Cu'_{C_x}(0_+)$, $u'_{C_x}(0_+) = \frac{1}{C}i_{C_x}(0_+) = 5\,\text{V/s}$. Thus, the initial conditions of the zero-input response are

$$\begin{cases} u_{C_x}(0_+) = 0\,\text{V} \\ u'_{C_x}(0_+) = 5\,\text{V/s} \end{cases},$$

Putting the initial conditions into (3.1-17), we obtain $c_{x_1} = 5/3$, $c_{x_2} = -5/3$. Thus, the zero-input response is

$$u_{C_x}(t) = \frac{5}{3}e^{-t} - \frac{5}{3}e^{-4t} \quad t \geq 0.$$

We find $u_{C_f}(0_+) = 0\,\text{V}$, $i_{C_f}(0_+) = 0\,\text{A}$ from ▶ Figure 3.4e. Because $i_{C_f}(0_+) = Cu'_{C_f}(0_+)$, $u'_{C_f}(0_+) = \frac{1}{C}i_{C_f}(0_+) = 0\,\text{V/s}$. Thus, the initial conditions of the zero-state response are

$$u_{C_f}(0_+) = 0\,\text{V}$$
$$u'_{C_f}(0_+) = 0\,\text{V/s}$$

Putting the initial conditions into (3.1-18), we have $c_{f_1} = -8/3$, $c_{f_2} = 2/3$. Thus, the zero-state response is

$$u_{C_f}(t) = -\frac{8}{3}e^{-t} + \frac{2}{3}e^{-4t} + 2\,\text{V} \quad t \geq 0.$$

3. Find the complete response $y(t)$

From the zero-input response and zero-state response, the complete response can be

$$y(t) = y_x(t) + y_f(t) = \underbrace{\sum_{i=1}^{n} c_{x_i}e^{\lambda_i t}}_{\text{zero input response}} + \underbrace{\sum_{i=1}^{n} c_{f_i}e^{\lambda_i t}}_{\text{zero state response}} + y_p(t) \tag{3.1-19}$$

$$\text{natural response} \qquad \text{forced response}$$

$$= \underbrace{\sum_{i=1}^{n} c_i e^{\lambda_i t}}_{\text{natural response}} + \underbrace{y_p(t)}_{\text{forced response}} \tag{3.1-20}$$

According to the two expressions above and features of the response, the complete response can be expressed as follows:
(1) Complete response = zero-input response + zero-state response.
(2) Complete response = natural response + forced response.
(3) Complete response = transient response + steady state response.

Usually, the exponential functions in the response that fit the form $ae^{\lambda t}$ can be called natural type terms, so the component composed of all natural type terms in the response can be called the natural response or the free response. The form of natural response is determined by the characteristic roots of the system. Aside from the natural response, those terms whose form is determined by an excitation are called the forced response. Therefore, the homogeneous solution is the natural response and the particular solution is the forced response in the classical method.

It can be seen from Equations (3.1-19) and (3.1-20) that:
(1) The natural response can be decomposed into two parts; one being caused by the starting state, and another being generated by the excitation.
(2) Both the natural response and zero-input response are solutions that can satisfy the homogeneous equation, but their coefficients are different. The c_{x_i} is only determined by the starting state of the system, but c_i is dependent on the starting state and the excitation when $t = 0$.
(3) The natural response includes the whole zero-input response and a part of the zero-state response, which is the homogeneous solution in the zero-state response. For a stable system, the zero-input response must be a part of the natural response.

According to the variation characteristic of a response with time, the complete response can be divided into transient response and steady state response. The part last decreases to zero with time increase is called the transient response, and the part that becomes a constant or an oscillation term with time increase is called the steady state response (this often includes step signals and periodic signals). Usually, transient response may include the transient components in the forced response.

For a steady system, the characteristic roots that are also known as the free frequencies are all negative, so all the natural terms will tend to zero with time increase, and the natural response must be the transient response. For an unstable system, if a root is a plural or a positive value, the natural term composed of this root will not tend to zero with time increase, so the natural response will not be transient.

To enable a better understanding of the above, ▶ Figure 3.5 shows the relationships between the complete response and each sub response.

Note that the zero-state response is generated by a zero-state system, which is accessed with an excitation at $t = 0$, so we always multiply the zero-state response $y_f(t)$ by $\varepsilon(t)$ or mark a time start $t \geq 0$ behind $y_f(t)$, in order to show that $y_f(t)$ exists from $t = 0$. The mark $t \geq 0$ can be considered as $t \geq 0_+$ here.

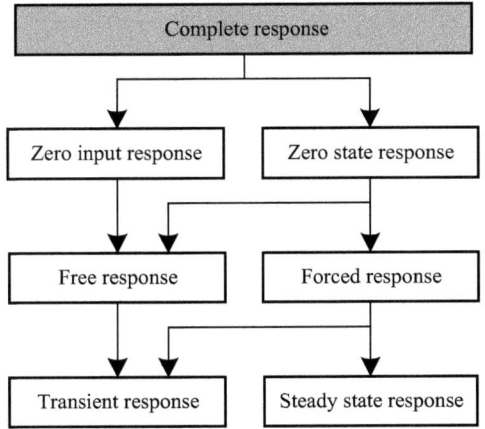

Fig. 3.5: Complete response and its composition diagram.

Example 3.1-4. The input-output equation of a system is $y'(t) + 3y(t) = 3\varepsilon(t)$, and the starting state is $y(0_-) = \frac{3}{2}$. Find the natural response, the forced response, the zero-input response and the zero-state response.

Solution. Firstly, find the natural response and the forced response. The characteristic equation is

$$\lambda + 3 = 0 \, ,$$

and the characteristic root is

$$\lambda = -3 \, ,$$

so, the homogeneous solution is

$$y_c(t) = c_1 e^{-3t} \, .$$

Suppose the particular solution is $y_p(t) = A$ and substitute it into the system equation; then $A = 1$ can be found. Thus,

$$y_p(t) = 1 \, ,$$

and the complete solution is

$$y(t) = c_1 e^{-3t} + 1 \, . \tag{3.1-21}$$

Because $y(t)$ is continuous at $t = 0$ from the impulse signal balance, we obtain

$$y(0_-) = y(0_+) = \frac{3}{2} \, .$$

Substituting the initial condition into (3.1-21), we have

$$c_1 = \frac{1}{2} \, ,$$

and, therefore, the natural response and the forced response are, respectively,

$$y_c(t) = \frac{1}{2} e^{-3t}, \quad y_p(t) = 1 \, .$$

Supposing the zero-input response and the zero-state response are, respectively,

$$y_x(t) = c_2 e^{-3t}, \quad y_f(t) = c_3 e^{-3t} + 1 ,$$

and because $y_x(0_+) = y(0_-) = \frac{3}{2}$, $y_f(0_+) = y_f(0_-) = 0$ are known, we obtain

$$c_2 = \frac{3}{2}, \quad c_3 = -1 .$$

Thus, the zero-input response and the zero-state response are,

$$y_x(t) = \frac{3}{2} e^{-3t} \quad t \geq 0 ,$$
$$y_f(t) = 1 - e^{-3t} \quad t \geq 0 .$$

3.2 Impulse and step responses

The ultimate objective of system analysis is to find the response of a system to any excitation, and to achieve it we expect to seek a simple and general approach. Fortunately, the impulse response and the step response, which are the zero-state responses produced by a system to impulse and step signals, can help us to make this possible.

3.2.1 Impulse response

The zero-state response generated by a system to a unit impulse signal $\delta(t)$ is called the impulse response and is denoted as $h(t)$.

If $y(t)$ and $f(t)$ in the system model represented by Equations (2.4-7) or (3.1-2) are, respectively, replaced by $h(t)$ and $\delta(t)$, then

$$\sum_{i=0}^{n} a_i h^{(i)}(t) = \sum_{j=0}^{m} b_j \delta^{(j)}(t) . \tag{3.2-1}$$

Because values of $\delta(t)$ are zero for $t > 0$, $h(t)$ and the homogeneous solution of the differential equation must have the same form. If the equation has n different characteristic roots such as $\lambda_1, \lambda_2, \ldots, \lambda_n$, the form of the impulse response should be

$$h(t) = \sum_{i=1}^{n} c_i e^{\lambda_i t} \varepsilon(t) . \tag{3.2-2}$$

Substituting equation (3.2-2) into equation (3.2-1), it can be seen that the impulse function and its all-order derivatives are on the right side of the equation, and the highest-order derivative is $\delta^{(m)}(t)$. Obviously, to balance all the corresponding singular functions on both sides of equation (3.2-1), the left side should also include the $\delta^{(m)}(t), \delta^{(m-1)}(t), \ldots, \delta(t)$ terms. Because the highest-order term on the left side of the equation is $h^{(n)}(t)$, it should include the $\delta^{(m)}(t)$ term, at least. We can see that the form of $h(t)$ relates to n and m. By comparing the highest-order m and n for both sides (the corresponding excitation and response, respectively) of the equation, $h(t)$ can be expressed in the following three forms:

(1) $n > m$

If $h(t)$ includes $\delta(t)$ term, the $h^{(n)}(t)$ on the left side of equation (3.2-1) will contain $\delta^{(n)}(t)$. However, the highest-order derivative term is $\delta^{(m)}(t)$ on the right side of the equation here, and the singular functions on both sides of the equation do not match, so it not possible to obtain the coefficients of all the terms. Therefore, $h(t)$ cannot contain the $\delta(t)$ term and can be expressed as

$$h(t) = \sum_{i=1}^{n} c_i e^{\lambda_i t} \varepsilon(t) \ . \tag{3.2-3}$$

(2) $n = m$

If $h(t)$ includes the derivative terms of $\delta(t)$, the derivative order of $\delta(t)$ on the left side will be higher than that on the right side. Thus, here $h(t)$ must include the impulse function $\delta(t)$ itself rather than its all-order derivatives at this moment, that is

$$h(t) = \sum_{i=1}^{n} c_i e^{\lambda_i t} \varepsilon(t) + B\delta(t) \ . \tag{3.2-4}$$

(3) $n < m$

Here, $h(t)$ contains not only the impulse function term but also its derivative terms. Supposing $n + 1 = m$, then

$$h(t) = \sum_{i=1}^{n} c_i e^{\lambda_i t} \varepsilon(t) + A\delta'(t) + B\delta(t) \ . \tag{3.2-5}$$

We also structure this form of $h(t)$ in order to substitute it into equation (3.2-1) and make the corresponding singular functions match each other on both sides of the equation.

The undetermined coefficients, such as A, B and c_i in Equations (3.2-3), (3.2-4) and (3.2-5), should be substituted into equation (3.2-1) and be determined by the principle that the coefficients of each corresponding singular function must be equal to each other on both sides.

Example 3.2-1. The differential equation of a second-order continuous system is

$$\frac{d^2}{dt^2}y(t) + 4\frac{d}{dt}y(t) + 3y(t) = \frac{d}{dt}f(t) + 2f(t)$$

Find the impulse response $h(t)$ of this system.

Solution. Considering $f(t) = \delta(t)$ and the zero-state condition, we know that $y(t) = h(t)$, so,

$$h''(t) + 4h'(t) + 3h(t) = \delta'(t) + 2\delta(t) \ , \tag{3.2-6}$$

the characteristic equation is

$$\lambda^2 + 4\lambda + 3 = 0 ,$$

and the characteristic roots are

$$\lambda_1 = -1, \quad \lambda_2 = -3 .$$

Because $n > m$, the impulse response is

$$h(t) = \left(c_1 e^{-t} + c_2 e^{-3t}\right) \varepsilon(t) . \tag{3.2-7}$$

Taking the derivative of equation (3.2-7), we get

$$h'(t) = (c_1 + c_2)\, \delta(t) - \left(c_1 e^{-t} + 3c_2 e^{-3t}\right) \varepsilon(t) , \tag{3.2-8}$$

$$h''(t) = (c_1 + c_2)\, \delta'(t) - (c_1 + 3c_2)\, \delta(t) + \left(c_1 e^{-t} + 9c_2 e^{-3t}\right) \varepsilon(t) . \tag{3.2-9}$$

Substituting Equations (3.2-7), (3.2-8) and (3.2-9) into equation (3.2-6), we obtain

$$(c_1 + c_2)\, \delta'(t) + (3c_1 + c_2)\, \delta(t) = \delta'(t) + 2\delta(t) .$$

Comparing the coefficients on both sides of the above equation, we obtain

$$\begin{cases} c_1 + c_2 = 1 \\ 3c_1 + c_2 = 2 \end{cases} ,$$

and then we find

$$c_1 = \frac{1}{2}, c_2 = \frac{1}{2} .$$

The impulse response is

$$h(t) = \frac{1}{2}\left(e^{-t} + e^{-3t}\right) \varepsilon(t) .$$

3.2.2 Step response

The zero-state response generated by a system to a unit step signal $\varepsilon(t)$ is called the step response and is denoted as $g(t)$.

If $y(t)$ and $f(t)$ in the system model are, respectively, replaced by $g(t)$ and $\varepsilon(t)$, we obtain

$$\sum_{i=0}^{n} a_i g^{(i)}(t) = \sum_{j=0}^{m} b_j \varepsilon^{(j)}(t) . \tag{3.2-10}$$

Since $\delta(t)$ is the derivative of $\varepsilon(t)$, the right side of equation (3.2-10) is a constant when $t \geq 0_+$, and then

$$\sum_{i=0}^{n} a_i g^{(i)}(t) = b_0 \quad t \geq 0_+ . \tag{3.2-11}$$

Obviously, the step response $g(t)$ should have the same form as the nonhomogeneous solution, which is composed of the homogeneous solution $g_c(t)$ and the particular solution $g_p(t)$. However, the form of the solution should be a constant, which can be assumed as $B\varepsilon(t)$, and after it is substituted into equation (3.2-11) we obtain $B = \frac{b_0}{a_0}$, so the solution is

$$g_p(t) = \frac{b_0}{a_0}\varepsilon(t) . \tag{3.2-12}$$

If n characteristic roots $\lambda_1, \lambda_2, \ldots, \lambda_n$ are different, then the homogeneous solution is

$$g_c(t) = \sum_{i=1}^{n} c_i e^{\lambda_i t}\varepsilon(t) . \tag{3.2-13}$$

Adding equation (3.2-12) to equation (3.2-13), we find the step response

$$g(t) = \sum_{i=1}^{n} c_i e^{\lambda_i t}\varepsilon(t) + \frac{b_0}{a_0}\varepsilon(t) . \tag{3.2-14}$$

Similarly to the impulse response, at the moment $t = 0$, the step response may also contain the impulse signal and its all-order derivatives. When $n \geq m$, $g(t)$ will not contain the impulse function term because the highest-order of derivatives for $\delta(t)$ is $m - 1$ on the right side of equation (3.2-10). If $g(t)$ contains the impulse function term, then the highest-order of derivatives for $\delta(t)$ is n on the left side of the equation, and the coefficients on both sides of the singular functions obviously do not match. So, the step response $g(t)$ must be

$$g(t) = \sum_{i=1}^{n} c_i e^{\lambda_i t}\varepsilon(t) + \frac{b_0}{a_0}\varepsilon(t) . \tag{3.2-15}$$

When $n < m$, $g(t)$ will contain the impulse function $\delta(t)$ and its derivatives, but the derivative order will depend on the result of $m-n$. If $n+1 = m$, then it will only contain $\delta(t)$; if $n + 2 = m$, it will contain $\delta(t)$ and $\delta'(t)$; and so on. The step response $g(t)$ will be given in the following, when $n + 1 = m$,

$$g(t) = A\delta(t) + \sum_{i=1}^{n} c_i e^{\lambda_i t}\varepsilon(t) + \frac{b_0}{a_0}\varepsilon(t) . \tag{3.2-16}$$

All the undetermined coefficients A and c_i in equation (3.2-16) can be substituted into the original equation (3.2-10) and be obtained by the method that the coefficients of each singular function on both sides should be equal.

Because we only discuss LTI systems, and the step and the impulse signals satisfy a calculus relationship, the step and the impulse responses can also satisfy this relationship, that is,
if

$$\varepsilon(t) \xrightarrow{\text{System transformation}} g(t) ,$$

$$\delta(t) \xrightarrow{\text{System transformation}} h(t) ,$$

because

$$\delta(t) = \frac{d\varepsilon(t)}{dt}$$

or

$$\varepsilon(t) = \int_{-\infty}^{t} \delta(\tau)d\tau ,$$

there are

$$h(t) = \frac{dg(t)}{dt} \tag{3.2-17}$$

and

$$g(t) - g(-\infty) = \int_{-\infty}^{t} h(\tau)d\tau , \tag{3.2-18}$$

For a causal system, because $g(-\infty) = 0$, equation (3.2-18) can be written as

$$g(t) = \int_{-\infty}^{t} h(\tau)d\tau . \tag{3.2-19}$$

Equation (3.2-17) or equation (3.2-19) shows an important characteristic of an LTI causal system, which can reveal the relationship between the above two important responses of a system and plays an important role in the analysis of an LTI causal system.

Example 3.2-2. The input-output differential equation of an LTI system is

$$y''(t) + 5y'(t) + 6y(t) = 3f'(t) + f(t)$$

Find the step response and the impulse response of this system.

Solution. Considering $f(t) = \varepsilon(t)$ and the zero-state condition, the equation can be rewritten as

$$g''(t) + 5g'(t) + 6g(t) = 3\varepsilon'(t) + \varepsilon(t) . \tag{3.2-20}$$

We can write the characteristic equation as

$$\lambda^2 + 5\lambda + 6 = 0 ,$$

and obtain the characteristic roots

$$\lambda_1 = -2, \quad \lambda_2 = -3 .$$

Then, the homogeneous solution part in the step response is

$$g_c(t) = \left(c_1 e^{-2t} + c_2 e^{-3t}\right) \varepsilon(t) ,$$

and the solution part in the step response is

$$g_p(t) = \frac{b_0}{a_0} \varepsilon(t) = \frac{1}{6}\varepsilon(t) .$$

Because $n > m$, the step response is

$$g(t) = \left(c_1 e^{-2t} + c_2 e^{-3t}\right) \varepsilon(t) + \frac{1}{6}\varepsilon(t) .$$

The first-order derivative is

$$g'(t) = \left(c_1 + c_2 + \frac{1}{6}\right) \delta(t) - 2c_1 e^{-2t} - 3c_2 e^{-3t} ,$$

and the second-order derivative is

$$g''(t) = \left(c_1 + c_2 + \frac{1}{6}\right) \delta'(t) - (2c_1 + 3c_2) \delta(t) + 4c_1 e^{-2t}\varepsilon(t) + 9c_2 e^{-3t}\varepsilon(t) .$$

Substituting above three equations into equation (3.2-20), we obtain

$$\left(c_1 + c_2 + \frac{1}{6}\right) \delta'(t) + \left(3c_1 + 2c_2 + \frac{5}{6}\right) \delta(t) = 3\delta(t) .$$

Comparing the coefficients on both sides of the equation, we have

$$\begin{cases} c_1 + c_2 + \frac{1}{6} = 0 \\ 3c_1 + 2c_2 + \frac{5}{6} = 3 \end{cases} ,$$

and we obtain

$$c_1 = \frac{5}{2}, \quad c_2 = -\frac{8}{3} .$$

The step response is

$$g(t) = \left(\frac{5}{2}e^{-2t} - \frac{8}{3}e^{-3t}\right) \varepsilon(t) + \frac{1}{6}\varepsilon(t) .$$

Taking the derivative of $g(t)$, we obtain $h(t)$ as

$$h(t) = \left(8e^{-3t} - 5e^{-2t}\right) \varepsilon(t) .$$

Example 3.2-3. The $i(t)$ is a response of the circuit shown in ▶ Figure 3.6; find the step and impulse responses.

Solution. According to KCL and KVL laws,

$$i_1(t) = i(t) + i_C(t) ,$$

$$2i(t) = 2i_C(t) + u_C(t) ,$$

$$u_C(t) = \frac{1}{C} \int_{-\infty}^{t} i_C(\tau)d\tau ,$$

$$2i_1(t) + 2i(t) = u_S(t) .$$

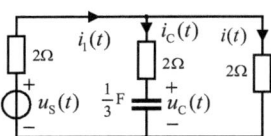

Fig. 3.6: E3.2-3.

The above four equations are considered as a set of equations, so $i_1(t)$, $u_C(t)$, $i_C(t)$ can be eliminated. All element values are substituted into the equation set, and the differential equation can be obtained,

$$i'(t) + i(t) = \frac{1}{6}u'_S(t) + \frac{1}{4}u_S(t) .$$

Letting $u_S(t) = \varepsilon(t)$ and the system state be zero,

$$g'(t) + g(t) = \frac{1}{6}\delta(t) + \frac{1}{4}\varepsilon(t) .$$

According to the impulse function balance method, we obtain $g(0_+) = \frac{1}{6}$. To solve the differential equation, the step response can be expressed as

$$g(t) = \left(ce^{-t} + \frac{1}{4}\right)\varepsilon(t) .$$

After substituting the initial condition into the equation, $c = -\frac{1}{12}$ can be found. The step response is

$$g(t) = \left(-\frac{1}{12}e^{-t} + \frac{1}{4}\right)\varepsilon(t) .$$

Taking the derivative of the step response, we get

$$h(t) = \frac{1}{12}e^{-t}\varepsilon(t) + \frac{1}{6}\delta(t) .$$

3.3 The operator analysis method

3.3.1 Differential and transfer operators

To facilitate solving the differential equations in the time domain, we will introduce the differential operator as a new concept. The differential operator is a simplified symbol for the differential operation, which can be represented by lower case p, that is,

$$p = \frac{d}{dt} \tag{3.3-1}$$

or

$$\frac{1}{p} = \int_{-\infty}^{t} (\ )dt . \tag{3.3-2}$$

Then

$$px = \frac{dx}{dt} , \quad p^n x = \frac{d^n x}{dt^n} , \quad \frac{1}{p}x = \int_{-\infty}^{t} xdt .$$

At the same time,

$$p\frac{1}{p}x = \frac{d}{dt}\int_{-\infty}^{t} xdt = x \tag{3.3-3}$$

Equation (3.3-3) indicates that the function will remain the same if it is first integrated and then differentiated. In equation (3.3-3), two characters of p can be reduced as variables. Note: If a function is first operated by the differential and then by the integral, we have

$$\frac{1}{p}px = \int_{-\infty}^{t} \frac{dx}{d\tau} d\tau = x(t) - x(-\infty),$$

so,

$$\frac{1}{p}px \neq x,$$

which means that two p cannot be reduced as two variables here, unless $x(-\infty) = 0$.

Using the differential operator, a linear constant coefficient differential equation as an LTI system model,

$$a_n \frac{d^n y(t)}{dt^n} + a_{n-1} \frac{d^{n-1} y(t)}{dt^{n-1}} + \cdots + a_1 \frac{dy(t)}{dt} + a_0 y(t)$$
$$= b_m \frac{d^m f(t)}{dt^m} + \cdots + b_1 \frac{df(t)}{dt} + b_0 f(t),$$

can be simplified as

$$\left(a_n p^n + a_{n-1} p^{n-1} + \cdots + a_1 p + a_0\right) y(t)$$
$$= \left(b_m p^m + b_{m-1} p^{m-1} + \cdots + b_1 p + b_0\right) f(t). \quad (3.3\text{-}4)$$

Obviously, because of the introduction of the differential operator, solving to a differential equation can be changed to solving to an operator equation or a pseudo-algebraic equation. It has the same form and some properties of the algebraic equation but cannot be equivalent to the algebraic equation, so it is called the pseudo-algebraic equation. As a result, we should learn some operational properties related to the operator.

Property 1: A positive power polynomial of p can be operated by expansion and factorization like an algebraic polynomial.

Property 2: The order/sequence of two polynomials before a signal can be exchanged, for example,

$$(p + 1)\left(p^2 + 2p + 3\right) f(t) = \left(p^2 + 2p + 3\right) (p + 1)f(t).$$

Property 3: The common factors that include p on both sides of an operator equation cannot be eliminated casually. For example, the equation $py(t) = pf(t)$ is usually shown as $y(t) = f(t) + c$ instead of $y(t) = f(t)$, where c is a constant.

Property 4: The order/sequence of multiplication and division operations for a signal cannot be randomly changed.

Conclusion: The variables in a differential equation set consisting of operators can be reduced with Cramer's rule like an algebraic equation set; this is the main purpose of introducing the differential operator. In other words, the introduction of the differential operator can simplify a differential equation (or set) whose solution process is relatively complex into an algebraic equation (set) whose solution process is relatively easy.

Equation (3.3-4) can be rearranged as

$$y(t) = \frac{b_m p^m + b_{m-1} p^{m-1} + \cdots + b_1 p + b_0}{a_n p^n + a_{n-1} p^{n-1} + \cdots + a_1 p + a_0} f(t) = \frac{N(p)}{D(p)} f(t) , \qquad (3.3\text{-}5)$$

where

$$N(p) = b_m p^m + b_{m-1} p^{m-1} + \cdots + b_1 p + b_0 , \qquad (3.3\text{-}6)$$

$$D(p) = a_n p^n + a_{n-1} p^{n-1} + \cdots + a_1 p + a_0 , \qquad (3.3\text{-}7)$$

and $D(p)$ is the characteristic polynomial of a differential equation (or system); $D(p) = 0$ is the characteristic equation.

The transfer operator can be extracted from the above content as another important concept. *The ratio of the response $y(t)$ to the excitation $f(t)$ in an operator equation (model) is defined as a transfer operator of the system, which is represented as $H(p)$,* that is

$$H(p) = \frac{y(t)}{f(t)} = \frac{N(p)}{D(p)} . \qquad (3.3\text{-}8)$$

The importance/purpose of constructing $H(p)$ is to extract the content that is unrelated to the input and the output of system model but can show the system characteristics, and facilitates the analysis of the system. This result is exactly the concrete reflection of the linearity of the system.

From equation (3.3-8), we have

$$y(t) = H(p)f(t) . \qquad (3.3\text{-}9)$$

The input $f(t)$ seems to be transferred from the input end to the output end of the system using $H(p)$ as the medium, which is the reason for $H(p)$ being named as the transfer operator. So, once we know the transfer operator $H(p)$ of a system, the system model can be determined, and the system response $y(t)$ to any input $f(t)$ can also be obtained by $H(p)$. In addition, the structure of $H(p)$ is the same as that of the system function $H(j\omega)$ or $H(s)$ to be introduced later, where $H(s) = H(p)|_{p=s}$ or $H(j\omega) = H(p)|_{p=j\omega}$. So, the introduction of $H(p)$ provides a new method to find the system function (transfer function) also.

Example 3.3-1. Find the transfer operator of the system in Example 2.4-2.

Solution. The mathematical model of the system in Example 2.4-2 is

$$\frac{d^3 u}{dt^3} + 2\frac{d^2 u}{dt^2} + 2\frac{du}{dt} + u = \frac{1}{2} u_s .$$

The operator is substituted into the above equation, and then the operator equation is found as

$$(p^3 + 2p^2 + 2p + 1)u = \frac{1}{2}u_s .$$

Then, the transfer operator is

$$H(p) = \frac{1}{2(p^3 + 2p^2 + 2p + 1)} .$$

Example 3.3-2. For the system shown in ▶ Figure 3.7, $i_2(t)$ and $f(t)$ are, respectively, the response and the excitation. Find the system model and the transfer operator.

Solution. According to KVL, we can write a set of loop equations

$$\begin{cases} i_1 + 3\frac{di_1}{dt} - \frac{di_2}{dt} = f \\ 3i_2 + \frac{di_2}{dt} - \frac{di_1}{dt} = 0 \end{cases} ,$$

and transfer it to a set of operator equations

$$\begin{cases} (3p + 1)i_1 - pi_2 = f \\ -pi_1 + (p + 3)i_2 = 0 \end{cases} .$$

Then, using Cramer's rule to eliminate the variables in the operator equations

$$i_2 = \frac{\begin{vmatrix} 3p+1 & f \\ -p & 0 \end{vmatrix}}{\begin{vmatrix} 3p+1 & -p \\ -p & p+3 \end{vmatrix}} = \frac{p}{2p^2 + 10p + 3}f$$

and

$$\left(2p^2 + 10p + 3\right)i_2 = pf .$$

We can write the system model as

$$2\frac{d^2 i_2}{dt^2} + 10\frac{di_2}{dt} + 3i_2 = \frac{df}{dt} ,$$

and then the transfer operator is

$$H(p) = \frac{p}{2p^2 + 10p + 3} .$$

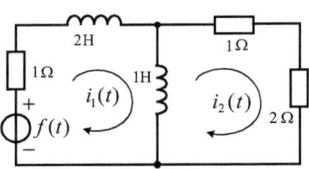

Fig. 3.7: E3.3-2.

Example 3.3-3. Knowing the transfer operator and the initial conditions of a system, find the corresponding zero-input response,

$$H(p) = \frac{1}{p^2 + 2p + 5}, \quad y_x(0_+) = 1, \quad y_x'(0_+) = 1.$$

Solution. The characteristic equation is

$$\lambda^2 + 2\lambda + 5 = 0,$$

and the characteristic roots are

$$\lambda_1 = -1 + 2j, \quad \lambda_2 = -1 - 2j,$$

so, the zero-input response is

$$y_x(t) = e^{-t} [A_1 \cos(2t) + A_2 \sin(2t)],$$

and

$$y_x'(t) = -e^{-t} [A_1 \cos(2t) + A_2 \sin(2t)] + e^{-t} [-2A_1 \sin(2t) + 2A_2 \cos(2t)].$$

Substituting the initial values $y_x(0_+) = 1$ and $y_x'(0_+) = 1$ into the above equations, we have

$$A_1 = A_2 = 1.$$

So, the zero-input response is

$$y_x(t) = e^{-t} [\cos(2t) + \sin(2t)] = \sqrt{2} e^{-t} \sin\left(2t + \frac{\pi}{4}\right), \quad t \geq 0.$$

From the above examples, we can summarize two common methods to find the transfer operator:
(1) Using the system model namely the differential equation (or simulation block diagram).
(2) Using the circuit constraint conditions.

Another purpose of introducing the transfer operator is to solve the zero-state response. The different transfer operators correspond to different impulse responses, so we can obtain a lot of relevant formulas through mathematical deduction. The form of a transfer operator is the same as that of a transfer function (system function), so the methods and formulas for solving the zero-state response are also similar to those of for the transfer function, and the details will not be given here.

3.3.2 Determining impulse response by the transfer operator

The transfer operator defined by equation (3.3-8) can be processed in partial fractions

$$H(p) = \sum_{i=0}^{q} K_i p^i + \sum_{j=1}^{l} \frac{K_j}{(p - \lambda_j)^{r_j}} \tag{3.3-10}$$

According to the different situations of equation (3.3-10), the corresponding different impulse responses $h(t)$ can be obtained by

(1)
$$H(p) = \frac{K}{p - \lambda} \quad \rightarrow \quad h(t) = Ke^{\lambda t}\varepsilon(t) . \tag{3.3-11}$$

(2)
$$H(p) = \frac{K}{(p - \lambda)^2} \quad \rightarrow \quad h(t) = Kte^{\lambda t}\varepsilon(t) . \tag{3.3-12}$$

(3)
$$H(p) = \frac{K}{(p - \lambda)^r} \quad \rightarrow \quad h(t) = \frac{K}{(r - 1)!}t^{r-1}e^{\lambda t}\varepsilon(t) . \tag{3.3-13}$$

(4)
$$H(p) = Kp^n \quad \rightarrow \quad h(t) = K\delta^{(n)}(t) . \tag{3.3-14}$$

This way, we can obtain general steps to find the impulse response using the transfer operator.

Step 1: Find the transfer operator $H(p)$.

Step 2: Decompose $H(p)$ into partial fraction form, such as equation (3.3-10).

Step 3: Determine each $h_i(t)$ corresponding to each fraction, according to the formulas in Equations (3.3-11) to (3.3-14).

Step 4: Adding all $h_i(t)$ together, the impulse response $h(t)$ of the system can be obtained.

Example 3.3-4. Find $h(t)$ of the system described by the following equation:

$$y^{(3)} + 5y^{(2)} + 8y^{(1)} + 4y = f^{(3)} + 6f^{(2)} + 10f^{(1)} + 6f$$

Solution. The operator equation of the system is

$$(p^3 + 5p^2 + 8p + 4)y = (p^3 + 6p^2 + 10p + 6)f .$$

Based on the long division method, we have

$$H(p) = \frac{p^3 + 6p^2 + 10p + 6}{p^3 + 5p^2 + 8p + 4} = 1 + \frac{1}{p + 1} - \frac{2}{(p + 2)^2} .$$

According to Equations (3.3-11) to (3.3-14), we have

$$h(t) = \delta(t) + e^{-t}\varepsilon(t) - 2te^{-2t}\varepsilon(t) = \delta(t) + (e^{-t} - 2te^{-2t})\varepsilon(t) .$$

3.4 The convolution analysis method

What is the purpose of introducing the impulse response? It is only the zero-state response of a system to a specific signal, namely, the impulse signal, and its solution process is also more complex, so what is the meaning of solving the zero-state responses of a system to a nonperiodic signal? In fact, the purpose is exactly to find the zero-state responses of a system to a nonperiodic signal, and there may be easier methods to solve the impulse response itself. Next, we will give the convolution analysis method for solving zero-state response.

The $f_1(t)$ and $p(t)$ shown in ▶ Figure 3.8 can be related by

$$f_1(t) = A \cdot \Delta \cdot p(t). \tag{3.4-1}$$

Now we will research the relation between any signal $f(t)$ and $p(t)$; $\hat{f}(t)$ can be considered as an approximate signal of $f(t)$ in ▶ Figure 3.9 and is composed of the sum of some rectangular pulses as shown in ▶ Figure 3.9; each pulse can be expressed by using $p(t)$.

The pulse 0 has height $f(0)$ and duration time $-\frac{\Delta}{2} \sim \frac{\Delta}{2}$ and can be expressed as $f(0)\Delta p(t)$.

The pulse 1 has height $f(\Delta)$ and duration time $\frac{\Delta}{2} \sim \frac{3\Delta}{2}$ and can be expressed as $f(\Delta)\Delta p(t - \Delta)$. The pulse k has height $f(k\Delta)$ and duration time $\frac{(2k-1)\Delta}{2} \sim \frac{(2k+1)\Delta}{2}$ and

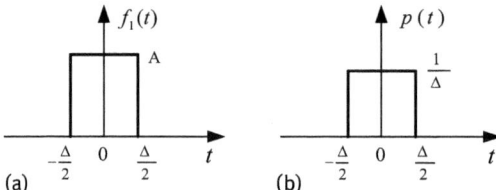

(a)　　　　　　　　　　　(b)

Fig. 3.8: $f_1(t)$ and $p(t)$.

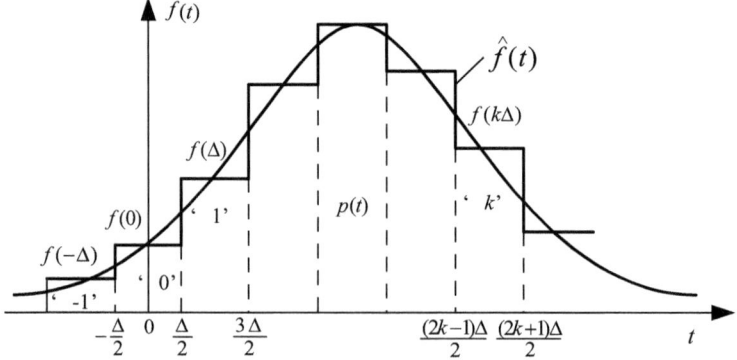

Fig. 3.9: The relationship between arbitrary $f(t)$ and $p(t)$.

can be expressed as $f(k\Delta)\Delta p(t - k\Delta)$. Hence,

$$f(t) \cong \hat{f}(t) = \sum_{k=-\infty}^{+\infty} f(k\Delta)\Delta p(t - k\Delta) \quad (k \text{ is an integer}).$$

If the zero-state response is written as $x(t)$ for the excitation $p(t)$, then from the linearity and time invariant properties, the zero-state response $y_f(t)$ of an LTI system to any excitation $f(t)$ can be approximated as

$$y_f(t) \cong \sum_{k=-\infty}^{+\infty} f(k\Delta)\Delta x(t - k\Delta) \quad (k \text{ is an integer}) .$$

When $\Delta \to 0$, Δ can be written as $d\tau$, $k\Delta$ as τ, $p(t-k\Delta)$ as $\delta(t-\tau)$. Here, the summation notation can be written as the integral symbol

$$f(t) = \lim_{\Delta \to 0} \sum_{k=-\infty}^{+\infty} f(k\Delta)\Delta p(t - k\Delta) = \int_{-\infty}^{+\infty} f(\tau)\delta(t - \tau)d\tau , \qquad (3.4\text{-}2)$$

$$y_f(t) = \lim_{\Delta \to 0} \sum_{k=-\infty}^{+\infty} f(k\Delta)\Delta x(t - k\Delta) = \int_{-\infty}^{+\infty} f(\tau)h(t - \tau)d\tau . \qquad (3.4\text{-}3)$$

Equation (3.4-2) shows that any signal $f(t)$ can be decomposed into a continuous sum of countless impulse signals located at different times and with different weights. Moreover, equation (3.4-3) shows that the zero-state response $y_f(t)$ of a system can also be also decomposed into a continuous sum of countless impulse responses. Obviously, according to the convolution definition, the zero-state response $y_f(t)$ is equal to the convolution integral of the excitation $f(t)$ and the impulse response $h(t)$; as a result, equation (3.4-3) can be written as

$$y_f(t) = h(t) * f(t) . \qquad (3.4\text{-}4)$$

The above deduction process is described in ▶ Figure 3.10.

Moreover, $f(t)$ can be decomposed into the continuous sum of impulse signals and it can also be decomposed into the continuous sum of step signals, that is,

$$f(t) = \lim_{\Delta \to 0} \sum_{k=-\infty}^{\infty} \frac{f(k\Delta) - f((k - 1)\Delta)}{\Delta} \cdot \varepsilon(t - k\Delta) \cdot \Delta . \qquad (3.4\text{-}5)$$

When $\Delta \to 0$, equation (3.4-5) can be reduced to

$$f(t) = \int_{-\infty}^{\infty} \frac{df(\tau)}{d\tau} \varepsilon(t - \tau)d\tau . \qquad (3.4\text{-}6)$$

Equation (3.4-6) shows that any signal $f(t)$ can be expressed as a continuous sum or an integral of infinite step signals. Equation (3.4-6) is also called *Duhamel's integral*.

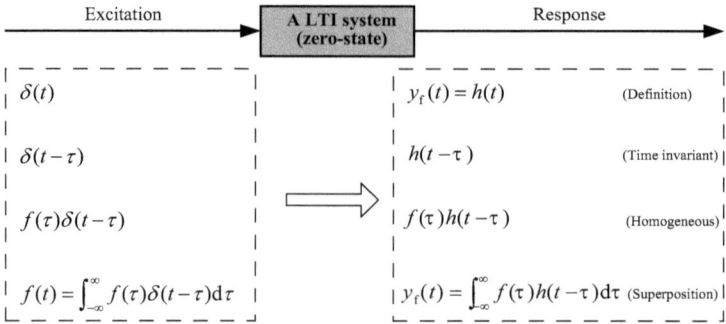

Fig. 3.10: The derivation process of the relationship between zero-state response and impulse response.

From the relation between the impulse signal and the step signal, the zero-state response $y_f(t)$ of a system can be written as

$$y_f(t) = f(t) * h(t) = f'(t) * \int_{-\infty}^{t} h(\tau)d\tau = f'(t) * g(t) .$$

When the unit step response of a system is known, the zero-state response can be directly calculated by the following equation:

$$y_f(t) = g(t) * f'(t) . \tag{3.4-7}$$

Equation (3.4-7) can also be deduced by a process similar to that shown in ▶ Figure 3.10.

In conclusion, equation (3.4-4) is the greatest contribution to system analysis from signal decomposability, system linearity and time-invariance, and it can provide a new way to find the zero-state response of a system. Moreover, this also is the main purpose of the introduction of the convolution operation.

Example 3.4-1. The impulse response of an LTI system is $h(t) = e^{\alpha t}\varepsilon(t)$, and the excitation is $f(t) = \varepsilon(t - 1)$. Find the zero-state response of this system.

Solution.

$$y_f(t) = h(t) * f(t) = \int_{-\infty}^{+\infty} f(\tau)h(t - \tau)d\tau = \int_{-\infty}^{+\infty} \varepsilon(\tau - 1)e^{\alpha(t-\tau)}\varepsilon(t - \tau)d\tau = \int_{1}^{t} e^{\alpha(t-\tau)}d\tau$$

$$= e^{\alpha t}\int_{1}^{t} e^{-\alpha\tau}d\tau = e^{\alpha t}\left(-\frac{1}{\alpha}\right)e^{-\alpha\tau}\Big|_{1}^{t} = \frac{1}{\alpha}e^{\alpha t}\left(e^{-\alpha} - e^{-\alpha t}\right)$$

$$= \frac{1}{\alpha}\left[e^{\alpha(t-1)} - 1\right], \quad t \geq 0 .$$

Example 3.4-2. The waveforms of the excitation $f(t)$ and the impulse response $h(t)$ of a system are, respectively, shown in ▶ Figure 3.11a and b. Find the zero-state response $y_f(t)$ and plot it.

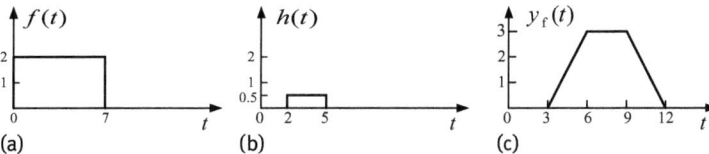

Fig. 3.11: E3.4-2.

Solution. The excitation and the impulse response can, respectively, be written as

$$f(t) = 2\varepsilon(t - 1) - 2\varepsilon(t - 7),$$

$$h(t) = \frac{1}{2}\varepsilon(t - 2) - \frac{1}{2}\varepsilon(t - 5).$$

Thus,

$$f'(t) = 2\delta(t - 1) - 2\delta(t - 7),$$

$$\int_{-\infty}^{t} h(\tau)d\tau = \int_{-\infty}^{t} \frac{1}{2}\left[\varepsilon(\tau - 2) - \varepsilon(\tau - 5)\right]d\tau$$

$$= \frac{1}{2}(t - 2)\varepsilon(t - 2) - \frac{1}{2}(t - 5)\varepsilon(t - 5).$$

Then

$$y_f(t) = f(t) * h(t) = f'(t) * \int_{-\infty}^{t} h(\tau)d\tau$$

$$= [2\delta(t - 1) - 2\delta(t - 7)] * \left[\frac{1}{2}(t - 2)\varepsilon(t - 2) - \frac{1}{2}(t - 5)\varepsilon(t - 5)\right]$$

$$= (t - 3)\varepsilon(t - 3) - (t - 6)\varepsilon(t - 6) - (t - 9)\varepsilon(t - 9) + (t - 12)\varepsilon(t - 12)$$

This can be expressed by a piecewise function

$$y_f(t) = \begin{cases} 0 & (t \le 3) \\ t - 3 & (3 < t \le 6) \\ 3 & (6 < t \le 9) \\ 12 - t & (9 < t \le 12) \\ 0 & (t > 12) \end{cases},$$

and is pictured in ▶ Figure 3.11c.

3.5 Judgment of dynamics, reversibility and causality

As mentioned before, the impulse response can describe or represent an LTI system, so the properties of an LTI system can be also described or judged by the impulse response.

3.5.1 Judgment of dynamics

If the impulse response $h(t)$ of a continuous system satisfies

$$h(t) = K\delta(t) , \tag{3.5-1}$$

then the response $y(t)$ and the excitation $f(t)$ satisfy

$$y(t) = Kf(t) . \tag{3.5-2}$$

Obviously, the system is a static/memoryless system. Thus, equation (3.5-1) can be considered as the judgment condition for whether a system is dynamic or static. If the coefficient $K > 1$, the system is an ideal amplifier and it is an ideal attenuator if $0 < K < 1$.

Example 3.5-1. Prove the condition $h(t) = K\delta(t)$ to make an LTI system a memoryless system.

Proof. If the excitation of an LTI system is $f(t)$, the response produced by $f(t)$ is $y(t)$, so when $t = t_0$, we have

$$y(t_0) = f(t) * h(t) \big|_{t=t_0} = \int_{-\infty}^{+\infty} f(\tau)h(t_0 - \tau)d\tau . \tag{3.5-3}$$

According to the sampling property of the impulse signal, we have

$$\int_{-\infty}^{\infty} f(t)\delta(t - t_0)dt = f(t_0) . \tag{3.5-4}$$

As an even signal, $\delta(t)$ is helpful for changing the above formula into

$$\int_{-\infty}^{\infty} f(t)\delta(t_0 - t)dt = f(t_0) . \tag{3.5-5}$$

Comparing equation (3.5-3) with equation (3.5-5), if

$$h(t) = K\delta(t) , \tag{3.5-6}$$

where K is positive or negative constant, then we have

$$y(t_0) = \int_{-\infty}^{+\infty} f(\tau)K\delta(t_0 - \tau)d\tau = Kf(t_0) .$$

Thus, the present response is only determined by the present excitation for the system, namely the system is memoryless. □

3.5.2 Judgment of reversibility

If the impulse responses of two systems are $h(t)$ and $h_i(t)$, separately, then when

$$h(t) * h_i(t) = \delta(t) , \qquad (3.5\text{-}7)$$

the system with $h(t)$ as the impulse response is a reversible system or an original system, and the system with $h_i(t)$ as the impulse response is the inverse system of the original system.

3.5.3 Judgment of causality

If the impulse response of a continuous system $h(t)$ satisfies

$$h(t) = 0 \quad t < 0 , \qquad (3.5\text{-}8)$$

the system is causal. In other words, a system is causal if its impulse response is a causal signal.

3.6 Solved questions

Question 3-1. The unit step response of an LTI system is $g(t) = e^{-3t}\varepsilon(t)$. Then the impulse response of the system is $h(t) =$ _____ .

Solution.

$$h(t) = \frac{dg(t)}{dt} = \delta(t) - 3e^{-3t}\varepsilon(t) .$$

Question 3-2. The unit step response of an LTI system is $g(t) = (3e^{-2t} - 1)\varepsilon(t)$. Using time domain analysis ways find:
(1) The impulse response $h(t)$ of the system.
(2) The zero-state response $y_{f1}(t)$ of the system to an input $f_1(t) = t\varepsilon(t)$.
(3) The zero-state response $y_{f2}(t)$ of the system to an input $f_2(t) = t[\varepsilon(t) - \varepsilon(t - 1)]$.

Solution. (1) According to the relation between $h(t)$ and $g(t)$, we have

$$h(t) = \frac{dg(t)}{dt} = 2\delta(t) - 3e^{-2t}\varepsilon(t) .$$

(2) From the relation between $r(t) = t\varepsilon(t)$ and $\varepsilon(t)$, and the linearity, we obtain

$$y_{f1}(t) = \int_0^t g(t)dt = \left(-\frac{3}{2}e^{-2t} - t\right)\Big|_0^t = \left(\frac{3}{2} - \frac{3}{2}e^{-2t} - t\right)\varepsilon(t) .$$

(3) Because $f_2(t) = t[\varepsilon(t) - \varepsilon(t - 1)] = t\varepsilon(t) - (t - 1)\varepsilon(t - 1) - \varepsilon(t - 1)$, and based on system linearity and time invariance, we obtain

$$y_{f2}(t) = \left(\frac{3}{2} - \frac{3}{2}e^{-2t} - t\right)\varepsilon(t) - \left(\frac{3}{2} + \frac{3}{2}e^{-2(t-1)} - t\right)\varepsilon(t - 1) .$$

Question 3-3. The step response of an LTI system is $g(t) = e^{-t}\varepsilon(t)$. Find the zero-state response $y_f(t)$ when the input is $f(t) = 3e^{2t}$.

Solution. Because

$$h(t) = \frac{dg(t)}{dt} = \delta(t) - e^{-t}\varepsilon(t) ,$$

$$y_f(t) = f(t) * h(t) = 3e^{2t} * \left[\delta(t) - e^{-t}\varepsilon(t)\right] = 3e^{2t} - 3e^{2t} * e^{-t}\varepsilon(t) ,$$

$$3e^{2t} * e^{-t}\varepsilon(t) = \int_0^\infty 3e^{2(t-\tau)}e^{-\tau}d\tau = 3e^{2t}\int_0^\infty e^{-3\tau}d\tau = 3e^{2t}\left.\frac{e^{-3\tau}}{-3}\right|_0^\infty = e^{2t} ,$$

we have

$$y_f(t) = 3e^{2t} - e^{2t} = 2e^{2t}\varepsilon(t) .$$

Question 3-4. The impulse response of a system is $h(t) = e^{-t}\varepsilon(t)$, and the excitation is $f(t) = \varepsilon(t)$.
(1) Find the zero-state response $y_f(t)$ of the system.
(2) The systems consisting of $h_1(t)$ and $h_2(t)$ are shown in ▶ Figure Q3-4a and b for $h_1(t) = 0.5\,[h(t) + h(-t)]$, $h_2(t) = 0.5\,[h(t) - h(-t)]$. Find $y_{f1}(t)$ and $y_{f2}(t)$.

Solution. (1) From the calculus property of convolution, the zero-state response $y_f(t)$
is

$$y_f(t) = f'(t) * \int_{-\infty}^t h(t)dt = \left[\int_0^t e^{-\tau}d\tau\right]\varepsilon(t) = \left(1 - e^{-t}\right)\varepsilon(t) .$$

(2) Letting the impulse response of the system in ▶ Figure Q3-4a be $h_a(t)$,

$$h_a(t) = h_1(t) - h_2(t) = 0.5\,[h(t) + h(-t)] - 0.5\,[h(t) - h(-t)] = h(-t) = e^t\varepsilon(-t) .$$

Thus,

$$y_{f1}(t) = f(t) * h_a(t) = \varepsilon(t) * e^t\varepsilon(-t) = \int_{-\infty}^t e^\tau\varepsilon(-\tau)d\tau .$$

When $t < 0$, we have

$$y_{f1}(t) = \int_{-\infty}^t e^\tau d\tau = e^t .$$

When $t \geq 0$, we have

$$y_{f1}(t) = \int_{-\infty}^t e^\tau d\tau = 1 ,$$

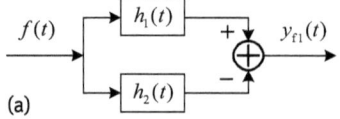

(a)

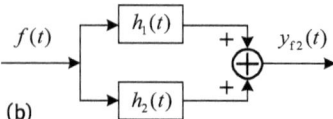

(b)

Fig. Q3-4

and thus,

$$y_{f1}(t) = e^t \varepsilon(-t) + \varepsilon(t) .$$

Letting the impulse response of the system in ▶ Figure Q3-4b be $h_b(t)$,

$$h_b(t) = h_1(t) + h_2(t) = 0.5 [h(t) + h(-t)] + 0.5 [h(t) - h(-t)] = h(t) ,$$

and thus,

$$y_{f2}(t) = y_f(t) = \left(1 - e^{-t}\right) \varepsilon(t) .$$

3.7 Learning tips

The time domain response is usually the final result of system analysis, and the reader should pay attention to the following points of knowledge:
(1) The introduction of operator can transform a differential equation into a pseudo-algebraic equation, and thus the process of solving differential equation can be simplified.
(2) The transfer operator can directly reflect the structure and characteristics of a system itself.
(3) The impulse response is the first element or soul in the time domain analysis methods.
(4) Neither the impulse response nor the transfer operator have anything to do with the response and the excitation of a system; they are relevant to the structure and parameters of the system itself.
(5) The zero-state response can be obtained through excitation and impulse response convolution. The essence is that an excitation $f(t)$ can be changed into a continuous sum of impulse signal $\delta(t)$, and a zero-state response $y_f(t)$ can be changed into a continuous sum of impulse response $h(t)$.

3.8 Problems

Problem 3-1. The differential equation of an LTI system is $y'(t) + 3y(t) = f(t)$. Knowing $y(0_+) = \frac{3}{2}$ and $f(t) = 3\varepsilon(t)$, find the natural response and forced response of this system.

Problem 3-2. The differential equation of an LTI system is $y''(t) + 4y'(t) + 4y(t) = 2f'(t) + 8f(t)$. Find the complete response and point out the natural response and forced response in it when $f(t) = e^{-t}$, $y(0_+) = 3$ and $y'(0_+) = 4$.

Problem 3-3. The differential equation of a system is $y''(t) + 3y'(t) + 2y(t) = f(t)$. Find the natural response and forced response of the system for each situation of which excitations and starting conditions are the following:
(1) $f(t) = \varepsilon(t)$, $y(0_-) = 1$, $y'(0_-) = 2$; and ⠀⠀⠀(2) $f(t) = e^{-2t}\varepsilon(t)$, $y(0_-) = 1$, $y'(0_-) = 2$.

Problem 3-4. The differential equation of an LTI system is $y''(t) + 3y'(t) + 2y(t) = f'(t) + 3f(t)$. If $f(t) = \varepsilon(t)$, $y(0_-) = 1$ and $y'(0_-) = 2$, find the complete response and point out the natural response and forced response, and the zero-input and zero-state responses in it.

Problem 3-5. The differential equation of a system is $y''(t) + 2y'(t) + 5y(t) = 2f'(t) + 4f(t)$, and $y(0_-) = 2$ and $y'(0_-) = -2$. Find the zero-input response $y_x(t)$ of this system.

Problem 3-6. The differential equation of a system is $y''(t) + 4y'(t) + 3y(t) = f(t)$. If the starting conditions are $y(0_-) = 1$ and $y'(0_-) = 2$, find the zero-input response $y_x(t)$ of this system.

Problem 3-7. The differential equation of a system is $y''(t) + 3y'(t) + 2y(t) = f'(t) + 5f(t)$. Knowing $f(t) = e^{-3t}\varepsilon(t)$, $y_f(0_+) = 1$ and $y'_f(0_+) = 2$, find the zero-state response $y_f(t)$.

Problem 3-8. In the circuit is shown in ► Figure P3-8, the switch is at "1" and the circuit is in a steady state for $t < 0$. At moment $t = 0$ the switch is from position "1" to "2".
(1) Find values of $u_C(0_+)$ and $i(0_+)$.
(2) Find the complete response of $u_C(t)$ and point out the free and forced responses, and the zero-input and zero-state responses in it.

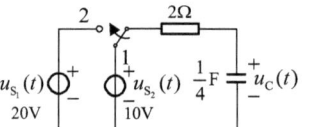

Fig. P3-8

Problem 3-9. For the circuit shown in ► Figure P3-9, it is known that $L = 2$ H, $C = \frac{1}{4}$ F, $R_1 = 1\,\Omega$ and $R_2 = 5\,\Omega$; the starting voltage of the capacitor $u_C(0_-) = 3$ V, the starting current of the inductor $i_L(0_-) = 1$ A; the excitation current source is $i_S(t) = \varepsilon(t)$. Find the zero-input response $i_{L_x}(t)$ and the zero-state response $i_{L_f}(t)$ of the inductor current $i_L(t)$.

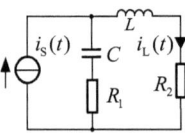

Fig. P3-9

Problem 3-10. For the circuit shown in ► Figure P3-10, find the transfer operators of $i(t)$ and $u(t)$ to the excitation $f(t)$.

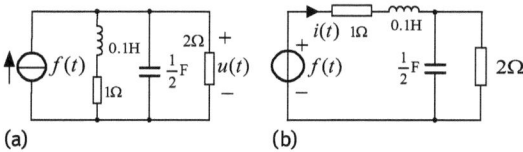

Fig. P3-10

(a)   (b)

Problem 3-11. The transfer operator $H(p)$ and the state or condition at moment 0_+ of a system are known. Find the zero-input response of the system.

(1) $H(p) = \frac{p+3}{p^2+3p+2}$, $y_x(0_+) = 1$, $y_x'(0_+) = 2$

(2) $H(p) = \frac{p+3}{p^2+2p+2}$, $y_x(0_+) = 1$, $y_x'(0_+) = 2$

(3) $H(p) = \frac{3p+1}{p(p+1)^2}$, $y_x(0_+) = y_x'(0_+) = 0$, $y_x''(0_+) = 1$

Problem 3-12. The zero-state response of an LTI system to an input $f(t) = 2e^{-3t}\varepsilon(t)$ is $y_f(t)$, and to $f'(t)$ it is $y_{fd}(t) = -3y_f(t) + e^{-2t}\varepsilon(t)$. Find the impulse response of the system $h(t)$.

Problem 3-13. The zero-state response of a system to an input $f(t)$ is $y_f(t) = \int_{t-2}^{\infty} e^{t-\tau}f \cdot (\tau-1)d\tau$ Find the impulse response of the system $h(t)$.

Problem 3-14. The zero-state response and the input of a system are plotted in ▶ Figure P3-14. Find the zero-state response $y_{f1}(t)$ of the system to another input $f_1(t) = \sin \pi t[\varepsilon(t) - \varepsilon(t-1)]$.

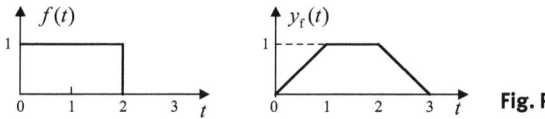

Fig. P3-14

Problem 3-15. The impulse response and excitation of a matched filter in a communication system are related by $h(t) = f(T-t)$, where T is the duration time of $f(t)$. If $f(t) = \varepsilon(t) - \varepsilon(t-T)$. Find the zero-state response of the matched filter $y_f(t)$.

Problem 3-16. The step response of an LTI system $g(t) = (2e^{-2t}-1)\varepsilon(t)$ is known. Find the zero-state response to each excitation shown in ▶ Figure P3-16 by the properties of convolution.

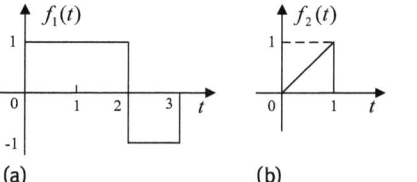

(a)   (b)   Fig. P3-16

Problem 3-17. The system shown in ▶ Figure P3-17 consists of several subsystems; the impulse response of each subsystem is $h_1(t) = \varepsilon(t)$ (integrator), $h_2(t) = \delta(t - 1)$ (unit delayer), $h_3(t) = -\delta(t)$ (inverter), $h_4(t) = 3\delta(t)$ (multiplier). Find the step response $g(t)$ and impulse response $h(t)$ of total system and draw their waveforms.

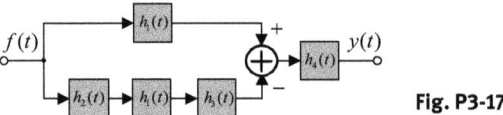

Fig. P3-17

Problem 3-18. The system shown in ▶ Figure P3-18 consists of several subsystems; the impulse responses of each subsystem are, respectively, $h_a(t) = \delta(t - 1)$ and $h_b(t) = \varepsilon(t) - \varepsilon(t - 3)$. Find the impulse response $h(t)$ of the total system.

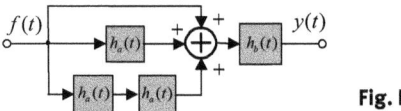

Fig. P3-18

Problem 3-19. In the circuit shown in ▶ Figure P3-19, $i_S(t)$ is the input and $u_L(t)$ is the output. Find the step response $g(t)$ and the impulse response $h(t)$.

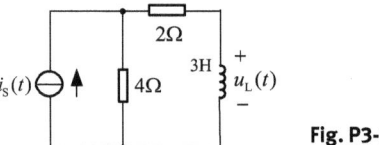

Fig. P3-19

Problem 3-20. In the circuit shown in ▶ Figure P3-20, $f(t)$ is the input and $u_C(t)$ is the output. Find the impulse response $h(t)$.

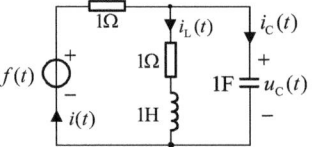

Fig. P3-20

4 Analysis of continuous-time systems excited by periodic signals in the real frequency domain

Questions: Besides characteristics of the time domain, a signal also has characteristics of the frequency domain. Then how can we analyze a system response to a periodic signal in the frequency domain?

Solution: Seek basic signals that can represent a variety of periodic signals. → Find the response of a system to the basic signals in the frequency domain. → Obtain the frequency responses to other periodic signals using the same approaches as for to the basic signals.

Results: Fourier series, system functions and harmonic response summation.

The time domain analysis method of an LTI system has some advantages, such as the clear physical conception, the intuitive results, etc., so it is the basic method of system analysis. However, it is also flawed. For example, it has too many concepts, needs to determine the boundary conditions (the state values of a system at moments 0_- and 0_+), involves more complicated calculations, etc. As a result, it is not commonly used in practice.

In the time domain analysis, because the signal $f(t)$ is a function of a time variable, we are interested in the change relations of the signal magnitude and speed and delay over time. Therefore, the analysis methods of signals and systems naturally are developed from the time variable. We must note the fact that the size (amplitude) and delay (phase) of a signal are also directly related to another variable frequency. In other words, the amplitude and phase of a signal are the functions of the frequency variable. So, we want to know whether can analyze signals and systems based on the frequency variable, and the answer is affirmative. The *real* frequency domain or, simply, the frequency domain analysis method will be introduced as a method that is a totally different analysis method from the time domain analysis method. Furthermore, it can provide another method for the analysis of signals and systems. A comparison of these two methods is shown in ▶ Figure 4.1.

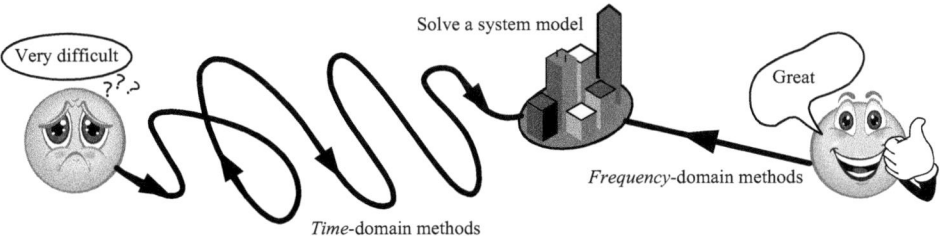

Fig. 4.1: Comparison between time and frequency domain methods.

https://doi.org/10.1515/9783110419535-004

As we know, all signals can be divided into two kinds, i.e. periodic and nonperiodic ones, so, this chapter will focus on the analysis of the responses of a system to periodic signals in the frequency domain. The analysis method for aperiodic signals will be discussed in Chapter 5.

4.1 Orthogonal functions

4.1.1 The orthogonal function set

As defined in the calculus course, if two functions $f_1(t)$ and $f_2(t)$ defined in an interval $[t_1, t_2]$ hold the relation

$$\int_{t_1}^{t_2} f_1(t)f_2(t)\mathrm{d}t = 0 \,, \tag{4.1-1}$$

we can say that $f_1(t)$ and $f_2(t)$ are orthogonal to each other on the interval $[t_1, t_2]$.

If two complex functions $f_1(t)$ and $f_2(t)$ defined in an interval $[t_1, t_2]$ are related by

$$\int_{t_1}^{t_2} f_1(t)f_2^*(t)\mathrm{d}t = \int_{t_1}^{t_2} f_1^*(t)f_2(t)\mathrm{d}t = 0 \,, \tag{4.1-2}$$

we can also say that $f_1(t)$ and $f_2(t)$ are orthogonal to each other in the interval $[t_1, t_2]$; $f^*(t)$ is the conjugate function of $f(t)$.

Suppose a function set consisting of n functions $f_1(t), f_2(t), \ldots, f_n(t)$ satisfies the following equation in an interval $[t_1, t_2]$:

$$\int_{t_1}^{t_2} f_i(t)f_j(t)\mathrm{d}t = \begin{cases} 0 & (i \neq j) \\ k_i & (i = j) \end{cases} \,. \tag{4.1-3}$$

We say that this function set is an orthogonal function set; k_i are constants.

If a complex function set $\{f_n(t)\}$ $(n = 1, 2, \ldots)$ in an interval $[t_1, t_2]$ holds the equation

$$\int_{t_1}^{t_2} f_i(t)f_j^*(t)\mathrm{d}t = \begin{cases} 0 & (i \neq j) \\ k_i & (i = j) \end{cases} \,, \tag{4.1-4}$$

we say that it is an orthogonal complex function set.

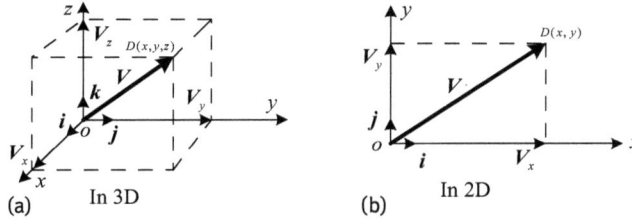

Fig. 4.2: Coordinate representation of a vector.

Except for $f_1(t), f_2(t), \ldots, f_n(t)$ in an orthogonal function set, if no function $y(t)$ $(0 < \int_{t_1}^{t_2} y^2(t)\mathrm{d}t < \infty)$ can satisfy the following equation:

$$\int_{t_1}^{t_2} f_i(t)y(t)\mathrm{d}t = 0 \quad (i = 1, 2, \ldots, n),$$

this function set is called a complete orthogonal set. This means that no other function can be orthogonal to all functions in this set.

The concept of orthogonal functions is similar to that of orthogonal (perpendicular) vectors. For example, three-dimensional coordinate axes are three vectors that are orthogonal to each other, this meaning that their projection lengths are all zero, and each vector cannot be represented by others (linear independent). However, in the three-dimensional space constructed with the three orthogonal vectors, any vector can be represented by their linear combination. For example, a vector V starting from the origin can be written as

$$V = V_x + V_y + V_z,$$

where V_x, V_y and V_z are, respectively, projections of V in three-dimensional coordinate axes, as shown in ▸ Figure 4.2a. Of course, V can also be represented by unit vectors i, j and k,

$$V = xi + yj + zk,$$

where x, y and z are the vertex coordinates of V, and the vertex is represented by D.

A vector in a two-dimensional coordinate system also has a similar concept, as shown in ▸ Figure 4.2b.

A function can also have an infinite dimensional existence space as a vector in conception. Therefore, the purpose of introducing the concept of orthogonal functions is to express a function linearly using an orthogonal function set.

Note that when a linear combination of all components in an orthogonal vector set is used to represent a vector, the vector set must be complete. Similarly, if a linear combination of all functions in an orthogonal function set is used to represent any function (or signal), the function set must also be complete.

4.1.2 Trigonometric function set

A function set $\{1,\ \cos\omega_0 t,\ \cos 2\omega_0 t,\ \ldots,\ \cos n\omega_0 t,\ \ldots,\ \sin\omega_0 t,\ \sin 2\omega_0 t,\ \ldots,\ \sin n\omega_0 t,\ \ldots\}$ defined over an interval $[t_0, t_0 + T]$ $(T = \frac{2\pi}{\omega_0})$ is orthogonal because it satisfies equations

$$\int_{t_0}^{t_0+T} \cos n\omega_0 t \cos m\omega_0 t\, dt = \begin{cases} 0 & (m \neq n) \\ \frac{T}{2} & (m = n) \end{cases}, \tag{4.1-5}$$

$$\int_{t_0}^{t_0+T} \sin n\omega_0 t \sin m\omega_0 t\, dt = \begin{cases} 0 & (m \neq n) \\ \frac{T}{2} & (m = n \neq 0) \end{cases}, \tag{4.1-6}$$

$$\int_{t_0}^{t_0+T} \sin n\omega_0 t \cos m\omega_0 t\, dt = 0 \quad \text{(for all } m \text{ and } n\text{)}. \tag{4.1-7}$$

The trigonometric function set is important and is used widely because of the following advantages:
- Trigonometric functions are a kind of basic function.
- Trigonometric functions relate simultaneously to the two physical quantities time and frequency.
- Trigonometric functions are easily produced, transmitted and processed.
- A trigonometric function through an LTI system is still one with the same frequency, and only the amplitude and phase may change.

4.1.3 Imaginary exponential function set

The imaginary exponential function set $\{e^{jn\omega_0 t}\}$ $(n = 0, \pm 1, \pm 2, \ldots)$ is also a complete orthogonal function set over the interval $[t_0, t_0 + T]$ $(T = \frac{2\pi}{\omega_0})$, because it satisfies the equation

$$\int_{t_0}^{t_0+T} e^{jm\omega t}.(e^{jn\omega t})^*\, dt = \int_{t_0}^{t_0+T} e^{j(m-n)\omega t}\, dt = \begin{cases} 0 & (m \neq n) \\ T & (m = n) \end{cases}. \tag{4.1-8}$$

The result from equation (4.1-8) is not surprising, because an imaginary exponential function can be represented by trigonometric functions from Euler's relation. The objective of discussing the concept of the orthogonal function is to introduce the Fourier series in the next section.

4.2 Fourier series

Fourier series (FS) is a periodic function analysis tool that was proposed by the French mathematician Fourier in 1807.

4.2.1 Trigonometric form of Fourier series

If an arbitrary periodic function $f(t)$ with a period T and an angular frequency $\omega_0 = \frac{2\pi}{T}$ can satisfy the Dirichlet conditions:
- it is continuous, or has a finite number of first-class discontinuities in one period;
- it has a finite number of maxima or minima in one period;
- it is bounded, namely, it is absolutely integrable $\int_{t_0}^{t_0+T} |f(t)| dt < \infty$,

so, $f(t)$ can be expanded in the trigonometric form of Fourier series:

$$f(t) = a_0 + a_1 \cos \omega_0 t + a_2 \cos 2\omega_0 t + \cdots + b_1 \sin \omega_0 t + b_2 \sin 2\omega_0 t + \cdots$$

or

$$f(t) = a_0 + \sum_{n=1}^{\infty} (a_n \cos n\omega_0 t + b_n \sin n\omega_0 t), \quad n = 1, 2, \ldots \quad (4.2\text{-}1)$$

where a_0, a_n and b_n are called Fourier coefficients.

Equation (4.2-1) can be considered as the definition of the Fourier series. It tells us that an arbitrary periodic signal can be represented by a linear combination of infinite orthogonal trigonometric functions. Thus, this is just the application example of signal decomposability.

Note that although the Dirichlet conditions only are the sufficient conditions for a periodic function to be expanded in Fourier series, the periodic signals in practice can all satisfy them. Hence, all subsequent periodic signals can be expanded in Fourier series without a special statement.

We can find the Fourier coefficients by using the orthogonality of the trigonometric function set:

Find a_0. Integrate to both sides of equation (4.2-1) term by term over an interval $[-\frac{T}{2}, \frac{T}{2}]$,

$$\int_{-\frac{T}{2}}^{\frac{T}{2}} f(t) dt = \int_{-\frac{T}{2}}^{\frac{T}{2}} a_0 dt + \sum_{n=1}^{\infty} \left[\int_{-\frac{T}{2}}^{\frac{T}{2}} a_n \cos n\omega_0 t dt + \int_{-\frac{T}{2}}^{\frac{T}{2}} b_n \sin n\omega_0 t dt \right].$$

For arbitrary n, we obtain

$$\int_{-\frac{T}{2}}^{\frac{T}{2}} \sin n\omega_0 t dt = 0.$$

If $n \neq 0$, we should have

$$\int_{-\frac{T}{2}}^{\frac{T}{2}} \cos n\omega_0 t \, dt = 0 .$$

Thus,

$$a_0 = \frac{1}{T} \int_{-\frac{T}{2}}^{\frac{T}{2}} f(t) dt . \qquad (4.2\text{-}2)$$

Find a_n. Multiply both sides of equation (4.2-1) by $\cos n\omega_0 t$, and integrate them term by term over the interval $\left[-\frac{T}{2}, \frac{T}{2}\right]$, that is,

$$\int_{-\frac{T}{2}}^{\frac{T}{2}} f(t) \cos n\omega_0 t \, dt = \int_{-\frac{T}{2}}^{\frac{T}{2}} a_0 \cos n\omega_0 t \, dt$$

$$+ \sum_{n=1}^{\infty} \left[\int_{-\frac{T}{2}}^{\frac{T}{2}} a_n \cos n\omega_0 t \cos n\omega_0 t \, dt + \int_{-\frac{T}{2}}^{\frac{T}{2}} b_n \sin n\omega_0 t \cos n\omega_0 t \, dt \right] .$$

According to the trigonometric function characteristics and orthogonality, we have

$$a_n = \frac{2}{T} \int_{-\frac{T}{2}}^{\frac{T}{2}} f(t) \cos n\omega_0 t \, dt . \qquad (4.2\text{-}3)$$

Find b_n. Multiply to both sides of equation (4.2-1) by $\sin n\omega_0 t$, and integrate them term by term over the interval $[-\frac{T}{2}, \frac{T}{2}]$, that is,

$$b_n = \frac{2}{T} \int_{-\frac{T}{2}}^{\frac{T}{2}} f(t) \sin n\omega_0 t \, dt . \qquad (4.2\text{-}4)$$

Because equation (4.2-1) contains both sin and cosine functions, its physical concept is not clear, and we need to combine the sin and cosine terms into a cosine form.

Supposing there is a right triangle in the fourth quadrant of a two-dimensional coordinate system, the opposite side of φ_n is $-b_n$ on the vertical axis, the adjacent side is a_n on the horizontal axis and the hypotenuse is c_n, thus

$$c_n = \sqrt{a_n^2 + b_n^2} \quad (n = 1, 2, \dots) , \qquad (4.2\text{-}5)$$

$$\cos \varphi_n = \frac{a_n}{\sqrt{a_n^2 + b_n^2}} = \frac{a_n}{c_n} \quad (n = 1, 2, \dots) , \qquad (4.2\text{-}6)$$

$$\sin \varphi_n = \frac{-b_n}{\sqrt{a_n^2 + b_n^2}} = \frac{-b_n}{c_n} \quad (n = 1, 2, \ldots), \tag{4.2-7}$$

$$\operatorname{tg}\varphi_n = \frac{\sin \varphi_n}{\cos \varphi_n} = \frac{-b_n}{a_n} \quad \rightarrow \quad \varphi_n = -\arctan \frac{b_n}{a_n}. \tag{4.2-8}$$

Using the sum angle formula of trigonometric functions, equation (4.2-1) can be transformed further as

$$f(t) = a_0 + \sum_{n=1}^{\infty} (a_n \cos n\omega_0 t + b_n \sin n\omega_0 t)$$

$$= a_0 + \sum_{n=1}^{\infty} \sqrt{a_n^2 + b_n^2} \left(\frac{a_n}{\sqrt{a_n^2 + b_n^2}} \cos n\omega_0 t - \frac{-b_n}{\sqrt{a_n^2 + b_n^2}} \sin n\omega_0 t \right)$$

$$= a_0 + \sum_{n=1}^{\infty} c_n (\cos \varphi_n \cos n\omega_0 t - \sin \varphi_n \sin n\omega_0 t)$$

$$= c_0 + \sum_{n=1}^{\infty} c_n \cos (n\omega_0 t + \varphi_n),$$

that is,

$$f(t) = c_0 + c_1 \cos (\omega_0 t + \varphi_1) + c_2 \cos (2\omega_0 t + \varphi_2) + \cdots$$

$$= c_0 + \sum_{n=1}^{\infty} c_n \cos (n\omega_0 t + \varphi_n), \tag{4.2-9}$$

where

$$c_0 = a_0 \tag{4.2-10}$$

is the DC component. If $c_0 > 0$, then $\varphi_0 = 0$; while if $c_0 < 0$, then $\varphi_0 = \pi$. c_n and φ_n are the magnitude and the phase of the nth cosine component.

Equation (4.2-9) shows that any periodic signal satisfying the Dirichlet conditions can be decomposed into the sum of a constant and infinite cosine components with different frequencies and phases. The first constant term c_0 is the average value of $f(t)$ in a period and represents the DC component of the periodic signal. The second term $c_1 \cos(\omega_0 t + \varphi_1)$ is called the fundamental wave or the first harmonic, whose angular frequency is the same as ω_0 of the original signal $f(t)$; the amplitude and the initial phase are, respectively, c_1 and φ_1. The third term $c_2 \cos(2\omega_0 t + \varphi_2)$ is called the second harmonic; its frequency is the double of ω_0, the amplitude and the initial phase are, respectively, c_2 and φ_2. The remaining terms can be written in the same manner. The term $c_n \cos(n\omega_0 t + \varphi_n)$ is the nth harmonic, where c_n and φ_n are the amplitude and the initial phase angle, respectively, of the nth harmonic.

Equation (4.2-9) can be considered as the standard trigonometric form of the Fourier series and has the following significance

(1) A periodic signal can be decomposed into the algebraic sum of a constant and infinite cosine signals with different frequencies and phases, or a periodic electronic signal can be decomposed into the algebraic sum of a DC component and infinite harmonics.

(2) The initial phase of each cosine signal or each harmonic is just φ_n.

These two points will be used in plotting the frequency spectrum of a periodic signal.

4.2.2 Relations between function symmetries and Fourier coefficients

When finding the Fourier series of a periodic signal, it is necessary to obtain a_0, a_n and b_n by integration three times, so the work is more complicated. From related research, we discover that the symmetry of signal itself can simplify the computing work for coefficients.

(1) If $f(t)$ is even as shown in ▶ Figure 4.2, $f(t) \cos n\omega_0 t$ is even and $f(t) \sin n\omega_0 t$ is odd, we have

$$a_0 = \frac{1}{T} \int_{-\frac{T}{2}}^{\frac{T}{2}} f(t)dt = \frac{2}{T} \int_0^{\frac{T}{2}} f(t)dt \,, \tag{4.2-11}$$

$$a_n = \frac{2}{T} \int_{-\frac{T}{2}}^{\frac{T}{2}} f(t) \cos n\omega_0 t dt = \frac{4}{T} \int_0^{\frac{T}{2}} f(t) \cos n\omega_0 t dt \,, \tag{4.2-12}$$

$$b_n = 0 \,. \tag{4.2-13}$$

We can see that the Fourier series of an even signal includes only the cosine components

$$f(t) = a_0 + \sum_{n=1}^{\infty} (a_n \cos n\omega_0 t) \,. \tag{4.2-14}$$

Whether or not the DC component a_0 is existent depends on whether or not $f(t)$ is symmetric around the horizontal axis. For example, the DC component of $f(t)$ in ▶ Figure 4.3a exists, but it does not in ▶ Figure 4.3b.

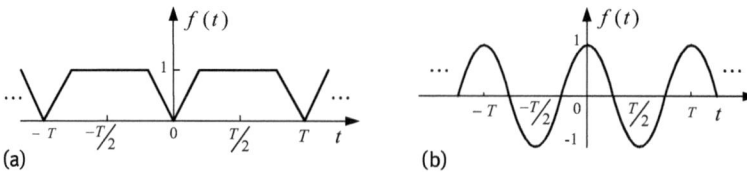

(a)   (b)

Fig. 4.3: Periodic even signals.

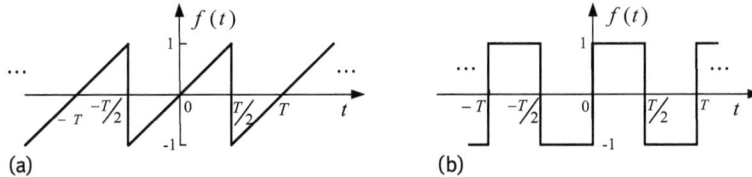

Fig. 4.4: Periodic odd signals.

(2) If $f(t)$ is an odd signal as shown in ▶ Figure 4.4, $f(t) \cos n\omega_0 t$ is odd and $f(t) \sin n\omega_0 t$ is even, we have

$$a_n = 0 \quad (n = 0, 1, 2, \ldots), \tag{4.2-15}$$

$$b_n = \frac{4}{T} \int_0^{\frac{T}{2}} f(t) \sin n\omega_0 t \, dt \quad (n = 1, 2, \ldots). \tag{4.2-16}$$

We can see that there are only the sine signals rather than any cosine signals and the DC components in the Fourier series of an odd signal, that is,

$$f(t) = \sum_{n=1}^{\infty} (b_n \sin n\omega_0 t). \tag{4.2-17}$$

(3) If $f(t)$ is an odd harmonic signal, it can be proved that its Fourier series expansion contains a fundamental wave and odd harmonics but not even harmonics and DC components. This is also the reason that it is named the odd harmonic signal.
If the waveform that is shifted by a half period along the time axis and is reversed around the axis is the same as the original one of a signal, we can call this signal the odd harmonic signal, as shown in ▶ Figure 4.4b. The odd harmonic signal satisfies the expression

$$f(t) = -f\left(t \pm \frac{T}{2}\right). \tag{4.2-18}$$

This property is also called *half period mirror symmetry*.

(4) If $f(t)$ is an even harmonic signal, it can be proved that its Fourier series expansion contains the DC component and even harmonics but not the fundamental wave and odd harmonics.
If the waveform that is shifted by a half period along time axis is the same as the original one of a signal, we can call this signal the even harmonic signal. The even harmonic signal satisfies the expression

$$f(t) = f\left(t \pm \frac{T}{2}\right). \tag{4.2-19}$$

This property is also called *the half period overlap*.

In summary, the even and the odd characteristics of a signal can determine what components (DC, sine and cosine components) appear in its Fourier series. The even and odd harmonic characteristics of a signal can affect whether or not the harmonic terms exist in its Fourier series, such as odd harmonics (including the fundamental wave) and even harmonics.

Note: A periodic signal may satisfy both even/odd symmetry and even/odd harmonic symmetry conditions at the same time. As a result, when we want to know whether the Fourier series of a signal includes the sine or the cosine components, we can first use even/odd symmetry to determine it, and then we can use even/odd harmonic characteristics to determine whether the Fourier series includes the odd or the even harmonics.

Example 4.2-1. Find the Fourier series of $f(t)$, which is a square wave in ▶ Figure 4.3b.

Solution. Because $f(t)$ satisfies the odd and the odd harmonic symmetries at the same time, $a_0 = 0$, $a_n = 0$ and there are no even harmonics, then

$$b_n = \frac{4}{T} \int_0^{\frac{T}{2}} f(t) \sin n\omega_0 t dt = \frac{4}{T} \int_0^{\frac{T}{2}} \sin n\omega_0 t dt = -\frac{4}{T} \frac{1}{n\omega_0} \cos n\omega_0 t \Big|_0^{\frac{T}{2}}$$

$$= -\frac{4}{T} \frac{1}{n\omega_0} \left(\cos n\omega_0 \frac{T}{2} - 1 \right) = \frac{2}{n\pi} (1 - \cos n\pi) = \begin{cases} 0 & (n = 2m) \\ \frac{4}{n\pi} & (n = 2m + 1) \end{cases},$$

so,

$$f(t) = \frac{4}{\pi} \left[\sin \omega_0 t + \frac{1}{3} \sin 3\omega_0 t + \frac{1}{5} \sin 5\omega_0 t + \cdots + \frac{1}{n} \sin n\omega_0 t + \cdots \right]$$

$$n = 1, 3, 5, \ldots$$

Example 4.2-2. Find the Fourier series of $f(t)$, which is a symmetric square wave in ▶ Figure 4.5a.

Solution. Because $f(t)$ satisfies even symmetry and odd harmonic symmetry at the same time, $b_n = 0$ and there are no even harmonics, then

$$a_0 = \frac{2}{T} \int_0^{\frac{T}{2}} f(t) dt = 0,$$

$$a_n = \frac{4}{T} \int_0^{\frac{T}{2}} f(t) \cos n\omega_0 t dt = \frac{4}{T} \int_0^{\frac{T}{4}} \cos n\omega_0 t dt - \frac{4}{T} \int_{\frac{T}{4}}^{\frac{T}{2}} \cos n\omega_0 t dt$$

$$= \frac{4}{T} \frac{1}{n\omega_0} \left(2 \sin n\omega_0 \frac{T}{4} - \sin n\omega_0 \frac{T}{2} \right)$$

$$= \frac{4}{n\pi} \sin \frac{n\pi}{2} = \frac{4}{n\pi} (-1)^{\frac{n-1}{2}}, \quad n = 1, 3, 5, \ldots$$

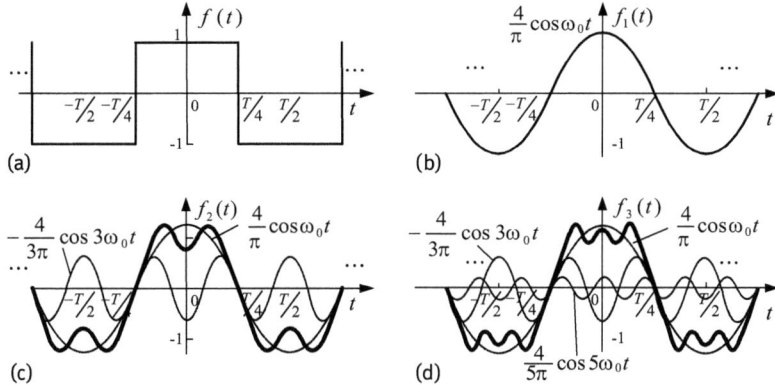

Fig. 4.5: Structures of a symmetric square wave.

We then have

$$f(t) = \sum_{n=1}^{\infty} a_n \cos n\omega_0 t$$

$$= \frac{4}{\pi}\left[\cos \omega_0 t - \frac{1}{3}\cos 3\omega_0 t + \frac{1}{5}\cos 5\omega_0 t - \frac{1}{7}\cos 7\omega_0 t + \cdots \right.$$

$$\left. \cdots + \frac{1}{n}(-1)^{\frac{n-1}{2}}\cos n\omega_0 t + \dots \right] \quad n = 1,3,5,\dots$$

The Fourier series with finite terms of a symmetric square wave $f(t)$ is plotted in ▶ Figure 4.5. ▶ Figure 4.5b is the waveform of which the Fourier series only contains the fundamental wave ($f_1(t) = \frac{4}{\pi}\cos \omega_0 t$). The waveform of which the Fourier series contains two terms, such as the fundamental wave and the third harmonic ($f_2(t) = \frac{4}{\pi}\cos \omega_0 t - \frac{4}{3\pi}\cos 3\omega_0 t$), is shown in ▶ Figure 4.5c. The waveform of which the Fourier series contains the fundamental wave and its third and fifth harmonics, ($f_3(t) = \frac{4}{\pi}\cos \omega_0 t - \frac{4}{3\pi}\cos 3\omega_0 t + \frac{4}{5\pi}\cos 5\omega_0 t$) is shown in ▶ Figure 4.5d. It can be seen that

(1) The more terms (number of harmonics) a Fourier series contains, the greater similarity between the series and the original $f(t)$. However, the variation of the peak values of the series waveform remains the same, which means that the variation will not change with the number of harmonic waves. This fact is called the Gibbs phenomenon.

(2) The amplitudes of high frequency components (harmonics) are smaller, and they mainly impact on the step edges of a pulse. The amplitudes of low frequency components are larger, and they mainly impact on the top form of a pulse, which is the main part of a square wave. In other words, the details of a waveform are mainly depicted by high frequency components, and the shape of a waveform is mainly determined by low frequency components in the waveform.

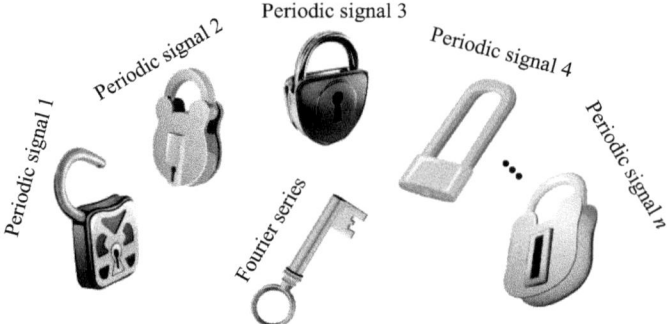

Fig. 4.6: Fourier series function diagram.

(3) When any frequency component in a Fourier series changes, the waveform of the signal will be more or less distorted.

(4) If $f_3(t)$ is given by ▸ Figure 4.5, it will be transformed into $f_2(t)$ if cos $5\omega_0 t$ is filtered by a lowpass filter. Moreover, $f_2(t)$ will be transformed into $f_1(t)$ if term cos $3\omega_0 t$ is also filtered. Obviously, a lowpass filter can filter ripples produced by high frequency components in the curve and smoothen the curve. This filtering concept is very important in communication principles, and readers should carefully learn this.

From the Fourier series, we know that arbitrary periodic signals in real engineering can be expressed as a unified form that is an algebraic sum of trigonometric functions (sinusoidal signals). Its significance is that we do not need to find the corresponding methods for thousands of periodic signals but only need to research one kind – the sinusoidal signal. As a universal method, the Fourier series can be figuratively compared to a master key that can open various periodic signal locks, as shown in ▸ Figure 4.6.

4.2.3 Exponential form of the Fourier series

If the imaginary exponential function set $\left\{e^{jn\omega_0 t}\right\}$ is chosen as the orthogonal function set, a periodic signal $f(t)$ can be written as the exponential form of the Fourier series.
Euler's relation is substituted into equation (4.2-1), which yields

$$
\begin{aligned}
f(t) &= a_0 + \sum_{n=1}^{\infty} \left(a_n \frac{e^{jn\omega_0 t} + e^{-jn\omega_0 t}}{2} + b_n \frac{e^{jn\omega_0 t} - e^{-jn\omega_0 t}}{2j} \right) \\
&= a_0 + \sum_{n=1}^{\infty} \left(\frac{a_n - jb_n}{2} e^{jn\omega_0 t} + \frac{a_n + jb_n}{2} e^{-jn\omega_0 t} \right).
\end{aligned}
\tag{4.2-20}
$$

Let

$$F_n = \frac{a_n - jb_n}{2} = |F_n| e^{j\varphi_n} , \tag{4.2-21}$$

where

$$|F_n| = \frac{1}{2}\sqrt{a_n^2 + b_n^2}, \quad \varphi_n = -\arctan\frac{b_n}{a_n} .$$

Obviously, the conjugate F_n^* of F_n is $F_n^* = \frac{a_n + jb_n}{2}$, and therefore, equation (4.2-20) becomes

$$f(t) = a_0 + \sum_{n=1}^{\infty} \left(F_n e^{jn\omega_0 t} + F_n^* e^{-jn\omega_0 t} \right) , \tag{4.2-22}$$

and

$$F_n = \frac{a_n - jb_n}{2} = \frac{1}{T}\int_{-\frac{T}{2}}^{\frac{T}{2}} f(t)\cos n\omega_0 t\,dt - j\frac{1}{T}\int_{-\frac{T}{2}}^{\frac{T}{2}} f(t)\sin n\omega_0 t\,dt$$

$$= \frac{1}{T}\int_{-\frac{T}{2}}^{\frac{T}{2}} f(t)(\cos n\omega_0 t - j\sin n\omega_0 t)dt .$$

Changing the algebraic form of the complex number into exponential form, we have

$$F_n = \frac{1}{T}\int_{-\frac{T}{2}}^{\frac{T}{2}} f(t)e^{-jn\omega_0 t}dt . \tag{4.2-23}$$

Similarly,

$$F_n^* = \frac{a_n + jb_n}{2} = \frac{1}{T}\int_{-\frac{T}{2}}^{\frac{T}{2}} f(t)e^{jn\omega_0 t}dt . \tag{4.2-24}$$

Comparing equations (4.2-23) and (4.2-24), we find that a difference between them is the minus sign in front of the integer n, and we have

$$F_n^* = F_{-n} . \tag{4.2-25}$$

Substituting equation (4.2-25) into equation (4.2-22), and changing F_{-n} into F_n by changing the summing direction, we obtain

$$f(t) = a_0 + \sum_{n=1}^{\infty}(F_n e^{jn\omega_0 t} + F_{-n}e^{-jn\omega_0 t}) = a_0 + \sum_{n=1}^{\infty} F_n e^{jn\omega_0 t} + \sum_{n=1}^{\infty} F_{-n}e^{-jn\omega_0 t}$$

$$= a_0 + \sum_{n=1}^{\infty} F_n e^{jn\omega_0 t} + \sum_{n=-1}^{-\infty} F_n e^{jn\omega_0 t} \tag{4.2-26}$$

$$= \sum_{n=-\infty}^{-1} F_n e^{jn\omega_0 t} + a_0 + \sum_{n=1}^{\infty} F_n e^{jn\omega_0 t} .$$

If a_0 is written as $F_0 e^{j\varphi_0} e^{j0\omega_0 t}$, and $\varphi_0 = 0$, $F_0 = a_0$, we have

$$f(t) = \sum_{n=-\infty}^{\infty} F_n e^{jn\omega_0 t} \quad n = 0, \pm 1, \pm 2, \ldots . \tag{4.2-27}$$

Equation (4.2-27) is the complex exponential form of the Fourier series.

Comparing equations (4.2-21), (4.2-6), (4.2-7) and (4.2-8), we obtain the coefficient relations between the trigonometric and the exponential forms of the Fourier series as follows:

$$F_0 = c_0 = a_0 , \tag{4.2-28}$$

$$|F_n| = \frac{1}{2} c_n \quad n \neq 0 , \tag{4.2-29}$$

$$\varphi_n = -\text{arctg}\frac{b_n}{a_n} . \tag{4.2-30}$$

Accordingly, the standard trigonometric and the complex exponential forms are related by

$$f(t) = \sum_{n=-\infty}^{+\infty} F_n e^{jn\omega_0 t} = \sum_{n=-\infty}^{+\infty} |F_n| e^{j(n\omega_0 t + \varphi_n)} = F_0 + \sum_{n=1}^{+\infty} 2|F_n| \cos(n\omega_0 t + \varphi_n) . \tag{4.2-31}$$

It can be seen that the DC components in both forms of Fourier series are the same, whereas the magnitude of each exponential harmonic coefficient is equal to a half of the value in the trigonometric form. Moreover, the initial phases of harmonic components of both forms equal each other. However, the ranges of n in two forms are different.

From equation (4.2-31) we know that although in the complex exponential form of Fourier series $f(t)$ can be constituted by a series of complex exponential signals like $F_n e^{jn\omega_0 t}$ which distribute in a range of frequency axis from $-\infty$ to $+\infty$, only two terms located at frequency points $-n\omega_0$ and $+n\omega_0$ among of these signals can form a harmonic component after they are added, and the separate term located at the point $-n\omega_0$ or $+n\omega_0$ is not a harmonic component, but is merely a kind of mathematical representation. This concept can be described by

$$\left(F_n e^{jn\omega_0 t} + F_{-n} e^{-jn\omega_0 t}\right) = \left(|F_n| e^{j\varphi_n} e^{jn\omega_0 t} + |F_n| e^{-j\varphi_n} e^{-jn\omega_0 t}\right)$$

$$= c_n \cos(n\omega_0 t + \varphi_n) \quad n \neq 0 .$$

Because F_n originates from $a_n \cos(n\omega_0 t + \varphi_n)$ and $b_n \sin(n\omega_0 t + \varphi_n)$, it means that F_n is related to two sinusoidal functions, so according to equation (4.2-31) the corresponding sinusoidal term of F_n can be written as

$$f_n(t) = 2|F_n| \cos(n\omega_0 t + \varphi_n) \quad n = 1, 2, 3, \ldots . \tag{4.2-32}$$

The above two different forms of Fourier series show that a periodic signal with arbitrary waveform can be regarded as a combination of numerous basic continuous signals (sinusoidal or complex exponential signals), namely, an algebraic sum consists of numerous harmonics with ω_0 as the basic frequency. Therefore, we come to the following conclusion:

Periodic signals with different shapes are different from each other only because they have different fundamental frequencies and different harmonic amplitudes and phases.

Example 4.2-3. Find the exponential form of Fourier series for $f(t)$, which is a square wave as shown in ▶ Figure 4.4b.

Solution.

$$F_n = \frac{1}{T} \int_{-\frac{T}{2}}^{\frac{T}{2}} f(t)e^{-jn\omega_0 t}\mathrm{d}t$$

$$= -\frac{1}{T} \int_{-\frac{T}{2}}^{0} e^{-jn\omega_0 t}\mathrm{d}t + \frac{1}{T} \int_{0}^{\frac{T}{2}} e^{-jn\omega_0 t}\mathrm{d}t = \frac{1}{T}\frac{1}{jn\omega_0} e^{-jn\omega_0 t}\Big|_{-\frac{T}{2}}^{0} - \frac{1}{T}\frac{1}{jn\omega_0} e^{-jn\omega_0 t}\Big|_{0}^{\frac{T}{2}}$$

$$= \frac{1}{T}\frac{1}{jn\omega_0}\left(1 - e^{jn\omega_0 \frac{T}{2}}\right) - \frac{1}{T}\frac{1}{jn\omega_0}\left(e^{-jn\omega_0 \frac{T}{2}} - 1\right) = \frac{1}{T}\frac{1}{jn\omega_0}\left(2 - e^{jn\omega_0 \frac{T}{2}} - e^{-jn\omega_0 \frac{T}{2}}\right)$$

$$= \frac{2}{jnT\omega_0}\left(1 - \cos n\omega_0 \frac{T}{2}\right) = \frac{1}{jn\pi}(1 - \cos n\pi) = \begin{cases} 0, & n \text{ is even} \\ \frac{2}{jn\pi}, & n \text{ is odd} \end{cases}$$

then

$$f(t) = \sum_{n=-\infty}^{\infty} F_n e^{jn\omega_0 t} = \frac{2}{j\pi}\left(\cdots - \frac{1}{3}e^{-j3\omega_0 t} - e^{-j\omega_0 t} + e^{j\omega_0 t} + \frac{1}{3}e^{j3\omega_0 t} + \cdots\right).$$

The expansion of $f(t)$ only contains the fundamental wave and odd harmonics, and it also verifies the characteristic of the odd harmonic function.

4.2.4 Properties of the Fourier series

To facilitate writing, we usually use the symbol $\overset{\mathscr{FS}}{\longleftrightarrow}$ to describe the relation between a function and its Fourier series.

1. Linearity

If

$$f_1(t) \overset{\mathscr{FS}}{\longleftrightarrow} F_{1n}, \quad f_2(t) \overset{\mathscr{FS}}{\longleftrightarrow} F_{2n},$$

then

$$a_1 f_1(t) + a_2 f_2(t) \overset{\mathscr{FS}}{\longleftrightarrow} a_1 F_{1n} + a_2 F_{2n}. \tag{4.2-33}$$

2. Time shifting

If

$$f(t) \overset{\mathcal{F}\mathcal{S}}{\longleftrightarrow} F_n \,,$$

then

$$f(t - t_0) \overset{\mathcal{F}\mathcal{S}}{\longleftrightarrow} F_n e^{-jn\omega_0 t_0} \,. \qquad (4.2\text{-}34)$$

Proof. Because

$$f(t) = \sum_{n=-\infty}^{\infty} F_n e^{jn\omega_0 t} \,,$$

then

$$f(t - t_0) = \sum_{n=-\infty}^{\infty} F_n e^{jn\omega_0(t - t_0)} = \sum_{n=-\infty}^{\infty} F_n e^{-jn\omega_0 t_0} e^{jn\omega_0 t} \,,$$

hence,

$$f(t - t_0) \overset{\mathcal{F}\mathcal{S}}{\longleftrightarrow} F_n e^{-jn\omega_0 t_0} \,.$$

$\square$

Example 4.2-4. The Fourier coefficient of the periodic signal $f_1(t)$ shown in ▶ Figure 4.7a is F_n. Represent the Fourier coefficients of the signals shown in ▶ Figure 4.7b–d with F_n.

Solution. Because

$$f_2(t) = f_1\left(t - \frac{T}{2}\right),$$

from time shifting we have

$$f_2(t) \overset{\mathcal{F}\mathcal{S}}{\longleftrightarrow} e^{-jn\frac{T}{2}\omega_0} F_n = (-1)^n F_n \,.$$

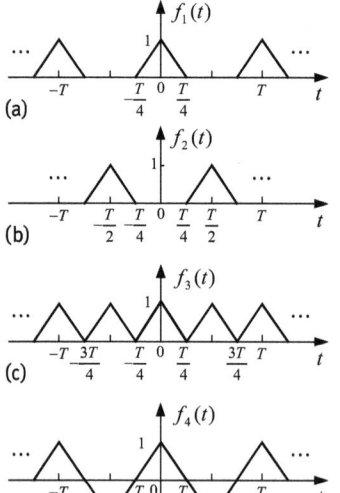

Fig. 4.7: E4.2-4.

Based on the given conditions, we have

$$f_3(t) = f_1(t) + f_2(t) \, ,$$

so

$$f_3(t) \overset{\mathscr{FS}}{\longleftrightarrow} F_n + (-1)^n F_n \, ,$$

and

$$f_4(t) = f_1(t) - f_2(t) \, ,$$

and then

$$f_4(t) \overset{\mathscr{FS}}{\longleftrightarrow} F_n - (-1)^n F_n \, .$$

3. Time reversal

If

$$f(t) \overset{\mathscr{FS}}{\longleftrightarrow} F_n \, ,$$

then

$$f(-t) \overset{\mathscr{FS}}{\longleftrightarrow} F_{-n} \, . \tag{4.2-35}$$

Proof. Because

$$f(t) = \sum_{n=-\infty}^{\infty} F_n e^{jn\omega_0 t} \, ,$$

then

$$f(-t) = \sum_{n=-\infty}^{\infty} F_n e^{-jn\omega_0 t} \overset{m=-n}{=} \sum_{m=-\infty}^{\infty} F_{-m} e^{jm\omega_0 t} \, ,$$

so

$$f(-t) \overset{\mathscr{FS}}{\longleftrightarrow} F_{-n} \, .$$

$\square$

4. Differential

If

$$f(t) \overset{\mathscr{FS}}{\longleftrightarrow} F_n \, ,$$

then

$$f'(t) \overset{\mathscr{FS}}{\longleftrightarrow} (jn\omega_0) F_n \, . \tag{4.2-36}$$

Proof. Because

$$f(t) = \sum_{n=-\infty}^{\infty} F_n e^{jn\omega_0 t} \, ,$$

then

$$f'(t) = \sum_{n=-\infty}^{\infty} (jn\omega_0) F_n e^{jn\omega_0 t} \, ,$$

so

$$f'(t) \overset{\mathscr{FS}}{\longleftrightarrow} (jn\omega_0) F_n \, .$$

$\square$

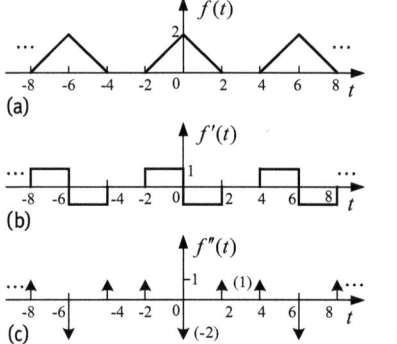

Fig. 4.8: E4.2-5.

Example 4.2-5. Find the Fourier series of the trigonometric wave shown in ► Figure 4.8a.

Solution. Taking the derivative of $f(t)$ two times, we obtain $f'(t)$ and $f''(t)$. Their waveforms of them are shown in ► Figure 4.8b and c. Assume that the Fourier coefficients of $f(t)$ and $f''(t)$ are, respectively, F_n and F_{2n}, we have

$$F_{2n} = \frac{1}{T} \int_{-\frac{T}{2}}^{\frac{T}{2}} f''(t)e^{-jn\omega_0 t}dt = \frac{1}{6} \int_{-3}^{3} [\delta(t+2) - 2\delta(t) + \delta(t-2)] e^{-jn\frac{\pi}{3}t}dt$$

$$= \frac{1}{6}\left(e^{jn\frac{2\pi}{3}} - 2 + e^{-jn\frac{2\pi}{3}}\right) = \frac{1}{3}\left(\cos\frac{2n\pi}{3} - 1\right) = -\frac{2}{3}\sin^2\frac{n\pi}{3}$$

According to the differential property, we have

$$F_{2n} = (jn\omega_0)^2 F_n ,$$

hence,

$$F_n = \frac{F_{2n}}{(jn\omega_0)^2} = \frac{1}{(jn\omega_0)^2}\left(-\frac{2}{3}\sin^2\frac{n\pi}{3}\right) = \frac{6}{(n\pi)^2}\sin^2\frac{n\pi}{3} .$$

The Fourier series of the periodic trigonometric wave is

$$f(t) = \sum_{n=-\infty}^{\infty} F_n e^{jn\frac{\pi}{3}t} = \frac{6}{\pi^2}\sum_{n=-\infty}^{\infty}\frac{\sin^2\frac{n\pi}{3}}{n^2}e^{jn\frac{\pi}{3}t} .$$

5. Energy conservation

The energy conservation properties of a periodic signal $f(t)$ in the time and frequency domains can be described by Parseval's relation, that is,

$$P = \overline{f^2(t)} = \frac{1}{T}\int_{t_0}^{t_0+T} f^2(t)dt = a_0^2 + \frac{1}{2}\sum_{n=1}^{\infty}(a_n^2 + b_n^2) = c_0^2 + \frac{1}{2}\sum_{n=1}^{+\infty}c_n^2 = \sum_{n=-\infty}^{+\infty}|F_n|^2 . \quad (4.2\text{-}37)$$

Equation (4.2-37) states that the whole average power in a periodic signal is equal to the sum of the average powers in all its harmonic components (including the DC component). Parseval's relation is often applied in the communication principles to find the SNR.

For a real signal, $F_n = F_{-n}^*$, therefore,

$$P = \sum_{n=-\infty}^{+\infty} |F_n|^2 = F_0^2 + 2 \sum_{n=1}^{+\infty} |F_n|^2 \, . \tag{4.2-38}$$

The change relation of P with $n\omega_0$ is usually called the power spectrum of a periodic signal.

4.3 Frequency spectrum

4.3.1 Concept of frequency spectrum

From equation (4.2-9) we can see that the relation between a periodic signal $f(t)$ and its variable t can be also expressed by an algebraic sum of a series of harmonics (cosine signals here), so a value of $f(t)$ at any moment t_0 is equal to the algebraic sum of all values of the harmonics at this time. By careful observation we also find that both the amplitudes and initial phases of all harmonics to represent $f(t)$ are different, furthermore, their frequencies are also different and change according to the rule by which they increase progressively with integer multiples of ω_0. Obviously, if the amplitude and the initial phase of each harmonic are defined as dependent variables, then they can be regarded as functions of the independent variable $n\omega_0$, which is the angular frequency of each harmonic. Note: The word "frequency" will subsequently refer to both frequency and angular frequency if no special instructions.

Thus, the Fourier series gives an important hint: The harmonic amplitudes and initial phases of any a periodic signal can be expressed as functions of frequency. Thus, we know that we can analyze a periodic signal based on the relations between the harmonic amplitudes or initial phases and frequency instead of using the traditional methods in the time domain.

Because components $F_n e^{jn\omega_0 t}$ in the complex exponential form of Fourier series appear at all harmonic frequency points on the frequency axis, and their complex amplitudes $F_n = |F_n| e^{j\varphi_n}$ (such as equation (4.2-21)) only relate to positions of all harmonics at the frequency axis (i.e. points $-n\omega_0$ and $+n\omega_0$) rather than time t, hence F_n is called the *frequency spectrum function* of a periodic signal or, simply, the spectrum. The relation of amplitudes $|F_n|$ of F_n with variable $n\omega$ is called the amplitude frequency characteristic or *amplitude spectrum*. The relation of the initial phases φ_n of F_n with $n\omega$ is called the phase frequency characteristic or *phase spectrum*. In this way, the spectrum function $F(n\omega_0)$ of any a periodic signal $f(t)$ can be defined as

$$F(n\omega_0) \overset{\text{def}}{=} F_n = \frac{1}{T} \int_{-T/2}^{T/2} f(t)e^{-jn\omega_0 t}dt \quad n = 0, \pm 1, \pm 2, \pm 3, \ldots . \tag{4.3-1}$$

When $n = 0$, $F(n\omega_0) = F(0) = \frac{1}{T}\int_{-T/2}^{T/2} f(t)dt$ represents the average value of a periodic signal in one cycle, which is the DC component a_0 or c_0. When $n \neq 0$, $F(n\omega_0)$ represents the complex amplitude of each harmonic. The graph of $F(n\omega_0)$ is very similar to the light spectrum, this is the reason for calling it the frequency spectrum.

Since $F_n = |F_n|e^{j\varphi_n}$ and $F_{-n} = |F_n|e^{-j\varphi_n}$, the amplitude spectrum $|F(n\omega_0)| = F_n$ is even and the phase spectrum φ_n is odd. Hence, they can be easily drawn.

Due to the introduction of $F(n\omega_0)$, equation (4.2-27) can be written as

$$f(t) = \sum_{n=-\infty}^{\infty} F(n\omega_0)e^{jn\omega_0 t} . \tag{4.3-2}$$

Thus, a periodic signal $f(t)$ and its spectrum function $F(n\omega_0)$ or F_n can be connected by the Fourier series. We can describe this relationship as

$$f(t) \overset{\mathscr{F}\mathscr{S}}{\longleftrightarrow} F(n\omega_0) \tag{4.3-3}$$

or

$$f(t) \overset{\mathscr{F}\mathscr{S}}{\longleftrightarrow} F_n . \tag{4.3-4}$$

Although the frequency spectrum is defined based on the complex exponential form of the Fourier series, the concept is also suitable for the trigonometric form. According to equation (4.2-9) the frequency spectrum can be also plotted. The specific plotting steps of two forms of frequency spectrum are as follows:

1. Trigonometric form (see Example 4.3-1)
Step 1: Transform the trigonometric form of the Fourier series into the standard form of equation (4.2-5), namely, $f(t) = c_0 + \sum_{n=1}^{\infty} c_n \cos(n\omega_0 t + \varphi_n)$.
Step 2: Draw a coordinate system of the frequency domain. The amplitude c_n and the frequency ω are, respectively, expressed as the vertical and horizontal axes. Then draw the scales with an interval ω_0 on the horizontal.
Step 3: Draw a vertical line segment with length c_0 at the origin and vertical line segments with length c_n at interval $n\omega_0$ in sequence, so that the amplitude spectrum has been completed.
Step 4: Draw a coordinate system of the frequency domain with the phase φ_n as the vertical axis again.
Step 5: At points $n\omega_0$ on the axis ω, draw vertical line segments with corresponding length φ_n, then the phase spectrum has been plotted. If $c_0 < 0$, $\varphi_0 = \pi$, otherwise, if $c_0 > 0$, $\varphi_0 = 0$.

Note the following:
(1) Because the cosine series is considered as the standard form of the Fourier series, that is, equation (4.2-9), if there are several sine components in this series, they should be transformed into cosine form, for example, $\sin(n\omega_0 t \pm \varphi_n) = \cos(n\omega_0 t \pm \varphi_n - \frac{\pi}{2})$, its initial phase is $\varphi_n = \pm\varphi_n - \frac{\pi}{2}$.
(2) If the harmonic component is $-\cos(n\omega_0 t \pm \alpha_n)$, it should be transformed into the form $\cos(n\omega_0 t \pm \alpha_n \mp \pi)$ in which the initial phase is $\varphi_n = \pm\alpha_n \mp \pi$, where the selection of a positive or negative sign in front of π should satisfy $-\pi < \varphi_n < \pi$.
(3) If $\alpha_n = 0$ in item (2), φ_n can be valued as $\pm\pi$, and $\varphi_n = +\pi$ in this book.
(4) If $b_n = 0$ in the expression (4.2-1), the relations of a_0, a_n over $n\omega_0$ can be plotted directly, namely, the spectrum.
(5) If $a_n = 0$ in the expression (4.2-1), the relations of a_0, b_n over $n\omega_0$ can be plotted after the sine functions are converted to the cosine functions, namely, the spectrum.

2. Complex exponential form
Step 1: From $F(n\omega_0) = F_n = \frac{1}{T}\int_{-T/2}^{T/2} f(t)e^{-jn\omega_0 t}\mathrm{d}t$ we find F_n and transform it into the form $F_n = |F_n|\,e^{j\varphi_n}$.
Step 2: Draw line segments $|F_n|$ at $-n\omega_0$ and $+n\omega_0$, respectively, on the frequency axis in the coordinate plane of the amplitude spectrum.
Step 3: Draw line segment φ_n at point $+n\omega_0$ on the frequency axis in the coordinate plane of the phase spectrum, and draw line segment $-\varphi_n$ at $-n\omega_0$.

To sum up, the conclusions are as follows:
(1) The frequency spectrum of a periodic signal can be shown in complex exponential and trigonometric two forms.
(2) The complex exponential spectrum is bilateral, and the trigonometric form is unilateral.
(3) Every amplitude value of harmonic wave in the bilateral amplitude spectrum is one half of corresponding value in the unilateral amplitude spectrum, but the DC component values in the bilateral and the unilateral spectrums are the same. The bilateral amplitude spectrum waveform is even symmetric.
(4) The waveform of unilateral phase spectrum and the waveform appearing on the right side of the vertical axis of a bilateral phase spectrum are the same; the waveform of a bilateral phase spectrum is odd symmetric.
(5) Usually the amplitude and phase spectrums cannot be plotted in one picture at the same time unless the phase spectrum only has the two values of π and 0.
(6) The physical meaning of a unilateral spectrum is that it can truly illustrate the changing relations of the harmonic amplitudes and phases with frequency, but the bilateral spectrum is just a mathematical expression without any practical meaning because it includes the negative frequency components.

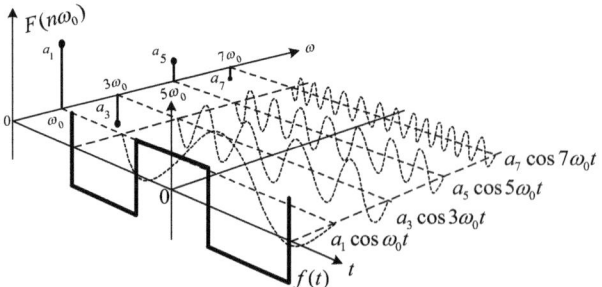

Fig. 4.9: Time domain and frequency domain waveforms of a symmetrical square wave.

(7) The complex exponential form is more commonly used than the trigonometric form, because its analysis and calculation are more convenient.

The comparison between two spectrums with different forms can be seen in Example 4.3-3. Obviously, if one kind has been pictured, the other can be obtained according to the above conclusions (3) and (4).

To facilitate understanding, the frequency spectrum can be qualitatively explained as follows:

The frequency spectrum of a periodic signal refers to the changing relations of the DC value, the amplitudes and phases of the fundamental wave, and all harmonic components of the signal with independent variable frequency.

In order to interpret more vividly the concept of the frequency spectrum, we redraw the waveforms in Example 4.2-2 in ▶ Figure 4.9.

4.3.2 Properties of the frequency spectrum

The plotting procedure and properties of the frequency spectrum of a periodic signal will be given by the following examples of three typical periodic signals.

Example 4.3-1. Plot the frequency spectrum of the periodic sawtooth wave $f(t)$ shown in ▶ Figure 4.10.

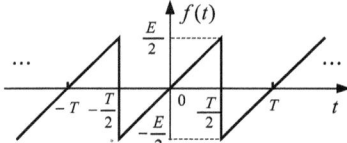

Fig. 4.10: Periodic sawtooth pulse signal.

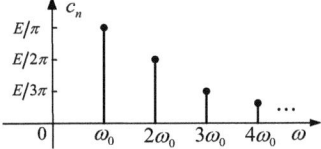

Fig. 4.11: Amplitude spectrum of a periodic sawtooth pulse signal.

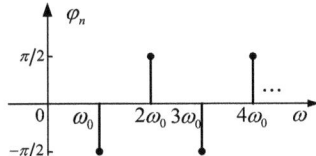

Fig. 4.12: Phase spectrum of a periodic sawtooth pulse signal.

Solution. $f(t)$ is odd, so $a_0 = 0$, $a_n = 0$, and we have

$$b_n = \frac{2}{T} \int_{-\frac{T}{2}}^{\frac{T}{2}} f(t) \sin n\omega_0 t \, dt = \frac{4E}{T^2} \int_0^{\frac{T}{2}} t \sin n\omega_0 t \, dt$$

$$= -\frac{4E}{n\omega_0 T^2} \left(t \cos n\omega_0 t \Big|_0^{\frac{T}{2}} - \int_0^{\frac{T}{2}} \cos n\omega_0 t \, dt \right)$$

$$= -\frac{4E}{n\omega_0 T^2} \frac{T}{2} \cos n\omega_0 \frac{T}{2} = \frac{E}{n\pi} (-1)^{n+1} .$$

The Fourier series of the periodic sawtooth wave $f(t)$ is

$$f(t) = \frac{E}{\pi} \left(\sin \omega_0 t - \frac{1}{2} \sin 2\omega_0 t + \frac{1}{3} \sin 3\omega_0 t - \frac{1}{4} \sin 4\omega_0 t + \cdots \right)$$

$$= \frac{E}{\pi} \sum_{n=1}^{\infty} \frac{1}{n} (-1)^{n+1} \sin n\omega_0 t = \frac{E}{\pi} \sum_{n=1}^{\infty} \frac{1}{n} (-1)^{n+1} \cos \left(n\omega_0 t - \frac{\pi}{2} \right)$$

The spectrum of the periodic sawtooth wave only contains sine components, and the amplitude of each harmonic is attenuated with speed $\frac{1}{n}$; its unilateral amplitude frequency and phase frequency spectrums are, respectively, pictured in ▶ Figure 4.11 and ▶ Figure 4.12.

Example 4.3-2. Find the front four nonzero terms in the amplitude spectrums of the half wave and full wave rectified signals with 50 Hz shown in ▶ Figure 4.13 and draw their amplitude spectrums.

Solution. Because the two waveforms are even, neither of them contain sine components.

For the halfwave rectified signal, we have

$$\omega_0 = 100\pi \, \text{rad/s}, \quad T = 0.02 \, \text{s} ,$$

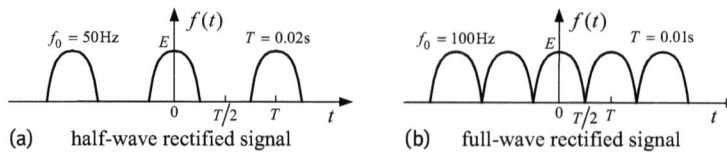

(a)   half-wave rectified signal   (b)   full-wave rectified signal

Fig. 4.13: E4.3-2 (1).

and

$$a_0 = \frac{1}{T} \int_{-\frac{T}{2}}^{\frac{T}{2}} f(t)\mathrm{d}t = 2 \times 50E \int_{0}^{\frac{0.02}{4}} \cos(100\pi t)\mathrm{d}t = \frac{E}{\pi},$$

$$a_n = \frac{2}{T} \int_{-\frac{T}{2}}^{\frac{T}{2}} f(t) \cos n\omega_0 t \mathrm{d}t = 4 \times 50E \int_{0}^{\frac{1}{200}} \cos(100\pi t) \cos(100 n\pi t)\mathrm{d}t .$$

Using Euler's relation, we have

$$a_n = \frac{2E \cos \frac{n\pi}{2}}{(1 - n^2)\pi},$$

when $n = 1$, with L'Hôpital's rule we obtain $a_1 = \frac{E}{2}$.
So,

$$f(t) = \frac{E}{\pi} + \frac{E}{2} \cos 100\pi t + \frac{2E}{3\pi} \cos 200\pi t - \frac{2E}{15\pi} \cos 400\pi t + \cdots$$

For the full wave rectified signal, we have

$$\omega_0 = 200\pi \, \mathrm{rad/s}, \quad T = 0.01 \, \mathrm{s},$$

and

$$a_0 = \frac{2E}{\pi}, \quad a_n = \frac{4E \cos n\pi}{(1 - 4n^2)\pi},$$

and finally,

$$f(t) = \frac{2E}{\pi} + \frac{4E}{3\pi} \cos 200\pi t - \frac{4E}{15\pi} \cos 400\pi t + \frac{4E}{35\pi} \cos 600\pi t + \cdots$$

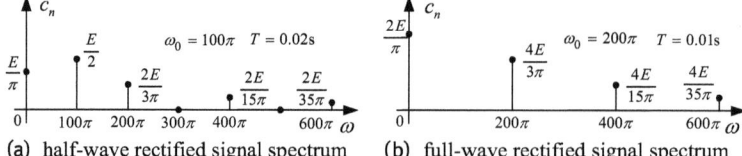

(a) half-wave rectified signal spectrum   (b) full-wave rectified signal spectrum

Fig. 4.14: E4.3-2 (2).

The amplitude spectrums of the half wave and full wave rectified waveforms are plotted in ▶ Figure 4.14.

From ▶ Figure 4.14, we have the following two points:

(1) The average value of the full wave rectification, namely, the DC component is twice that of the half wave. This is the reason why many electric devices can provide two switch positions of high (strong) and low (weak), such as an electric blower and an electric blanket.

(2) The frequency components of the full wave waveform are less than those of the half wave, which is good for the power supply filter.

Note: The coefficient a_n cannot be calculated by equation (4.1-5), because at this time the integrand function is not a complete cosine function, but only the cosine function with a half cycle.

Tips: This example can be also solved by the results of Example 5.3-8, that is, $c_n = \frac{2}{T}F(j\omega)|_{\omega=n\omega_0}$, $\omega_0 = \frac{2\pi}{T}$.

Example 4.3-3. Suppose that there is a periodic rectangular pulse signal $f(t)$ with pulse width τ, amplitude E and period T, as shown in ▶ Figure 4.15. Find the exponential and trigonometric forms of the Fourier series for this signal.

Solution. According to equation (4.2-25), we find

$$
F_n = \frac{1}{T}\int_{-\frac{\tau}{2}}^{\frac{\tau}{2}} f(t)e^{-jn\omega_0 t}dt = \frac{1}{T}E\int_{-\frac{\tau}{2}}^{\frac{\tau}{2}} e^{-jn\omega_0 t}dt = \frac{E}{T}\frac{1}{-jn\omega_0}e^{-jn\omega_0 t}\Big|_{-\frac{\tau}{2}}^{\frac{\tau}{2}}
$$

$$
= \frac{2E}{T}\frac{\sin\frac{n\omega_0\tau}{2}}{n\omega_0} = \frac{E\tau}{T}\frac{\sin\frac{n\omega_0\tau}{2}}{n\omega_0\frac{\tau}{2}} = \frac{E\tau}{T}sa\left(\frac{n\omega_0\tau}{2}\right).
$$

The F_n does not have an imaginary part but changes within positive and negative values. So, for $F_n > 0$, we have $\varphi_n = 0$, while for $F_n < 0$, $\varphi_n = \pi$. Because the phase spectrum is odd, $\varphi_n = -\pi$ on the left side of horizontal axis. In addition, $sa(\frac{n\omega_0\tau}{2})$ is zero when $\omega = \frac{2n\pi}{\tau}$; the phase spectrum is pictured in ▶ Figure 4.16d. The complex exponential form of the Fourier series of $f(t)$ is

$$
f(t) = \sum_{n=-\infty}^{\infty} \frac{E\tau}{T}Sa\left(\frac{n\omega_0\tau}{2}\right)e^{jn\omega_0 t} = \frac{E\tau}{T}\sum_{n=-\infty}^{\infty} Sa\left(\frac{n\omega_0\tau}{2}\right)e^{jn\omega_0 t}.
$$

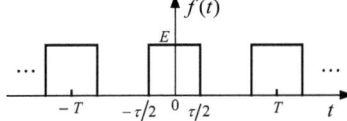

Fig. 4.15: Periodic rectangular pulse signal in E4.3-3.

If $f(t)$ is to be written in trigonometric form, based on odd and even properties, we have

$$a_0 = \frac{2}{T} \int_0^{\frac{T}{2}} f(t)dt = \frac{2}{T} \int_0^{\frac{T}{2}} Edt = \frac{E\tau}{T}, \quad b_n = 0.$$

$$a_n = \frac{4}{T} \int_0^{\frac{T}{2}} f(t) \cos n\omega_0 t\, dt = \frac{4E}{T} \int_0^{\frac{T}{2}} \cos n\omega_0 t\, dt = \frac{4E}{T} \frac{1}{n\omega_0} \sin n\omega_0 t\Big|_0^{\frac{T}{2}}$$

$$= \frac{4E}{T} \frac{1}{n\omega_0} \sin\left(\frac{n\omega_0\tau}{2}\right) = \frac{2E\tau}{T} Sa\left(\frac{n\omega_0\tau}{2}\right).$$

As a result,

$$f(t) = a_0 + \sum_{n=1}^{\infty} a_n \cos n\omega_0 t = \frac{E\tau}{T} + \frac{2E\tau}{T} \sum_{n=1}^{\infty} Sa\left(\frac{n\omega_0\tau}{2}\right) \cos n\omega_0 t.$$

The DC component is $c_0 = a_0 = \frac{E\tau}{T}$ and the nth harmonic amplitude is $c_n = a_n = \frac{2E\tau}{T} Sa(\frac{n\omega_0\tau}{2})$.

If the cycle value $T = 5\tau$, the trigonometric and complex exponential frequency spectrums of $f(t)$ respectively are shown in ▸ Figure 4.16a–d.

From ▸ Figure 4.16, in addition to the relative relations between the amplitude and phase of each frequency component of a periodic rectangular pulse signal, the following properties can also be found:
(1) The spectrum of a periodic rectangular wave is discrete. Spectral lines will only exist in some discrete frequency points, such as 0, ω_0, $2\omega_0$, etc. The interval in any two adjacent lines is always ω_0 ($\omega_0 = \frac{2\pi}{T}$), and therefore the greater period T, the smaller frequency interval ω_0 will be.
(2) The height or amplitude of each line is proportional to the pulse height E and width τ, and inversely proportional to period T.
(3) The height of each line will regularly change with the envelope $Sa(\frac{n\omega_0\tau}{2})$. For example, when $T = 5\tau$ and $E = 1$, we have $c_n = \frac{2}{5}\left|Sa(\frac{n\pi}{5})\right|$, so the fundamental wave amplitude is $c_1 = 0.37$ and the second harmonic amplitude is $c_2 = 0.30$; but when $n = 5m(m = 1, 2, 3, \ldots)$, the corresponding amplitudes of spectral lines all are zero.
(4) The spectrum of a periodic rectangular wave contains an infinite number of lines, which means that this signal can be decomposed into an infinite number of frequency components. The amplitude change trend of each component is to decrease with an increase of frequency.

In summary, the spectrum of a periodic signal has the following properties:
(1) Discreteness. Spectral lines exist only at points of integer times of the fundamental frequency, and height changes are nonperiodic.
(2) Harmonics. Spectral lines are uniformly spaced on the frequency axis with an interval of fundamental frequency, and there are no other frequency components

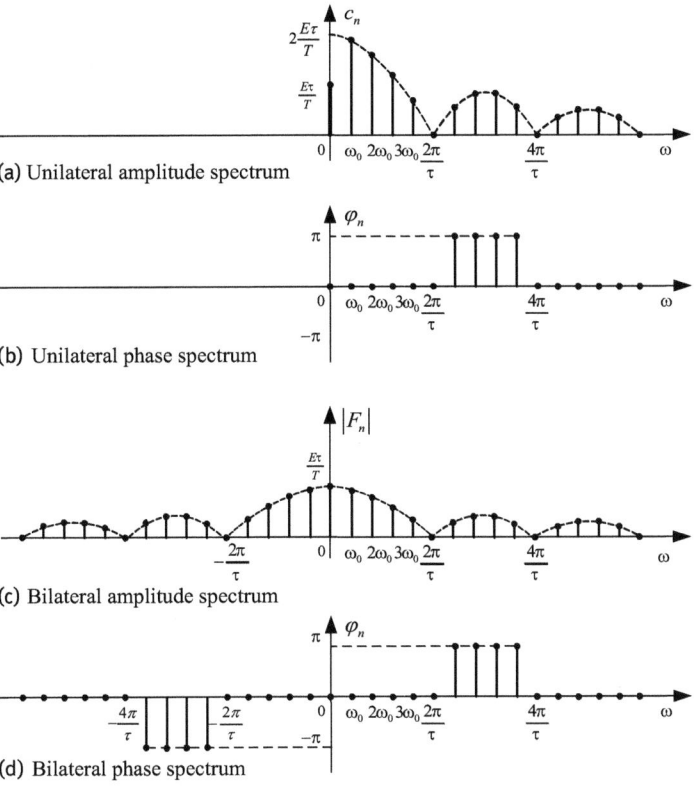

(a) Unilateral amplitude spectrum

(b) Unilateral phase spectrum

(c) Bilateral amplitude spectrum

(d) Bilateral phase spectrum

Fig. 4.16: The spectrums of a periodic rectangular pulse signal.

(spectral lines) except ones that are integral multiples of the fundamental frequency.

(3) Convergence. The general changing trend of each harmonic amplitude or line height is gradually damped with an increase in the harmonic number. The slower a signal waveform changes in the time domain, the faster those higher frequency components in its spectrum attenuate, and the fewer higher frequency components are in the spectrum. On the other hand, the faster a signal waveform changes in the time domain, the greater the number of high frequency components in the spectrum.

To facilitate understanding, the decomposition of a periodic signal with Fourier series can be compared with the decomposition phenomenon of a white light by a prism in a physics course. A beam of natural light can be decomposed into seven color lights with different wavelengths by a prism, such as red, orange, yellow, green, cyan, blue and violet (as shown in ▶ Figure 4.17), and a periodic signal can be decomposed into numerous harmonics with different frequencies by a "Fourier prism".

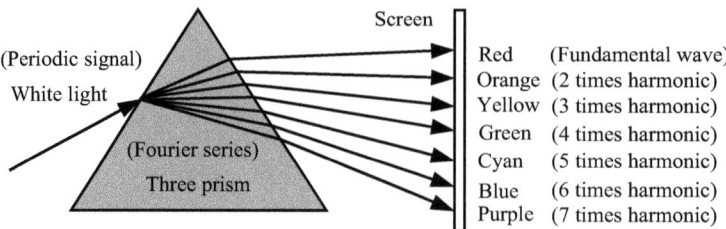

Fig. 4.17: Principle of decomposition with a triangular prism.

As mentioned above, the signal spectrum can show physical properties that are difficult to illustrate in the time domain. As a result, the Fourier series or the Fourier transform is a bridge between the time domain and frequency domain analyses. This is a significant analysis method and has an extremely important position in signal analysis.

4.4 Fourier series analysis

Although we have sought a master key that can be used to express all kinds of periodic signals, e.g. the Fourier series, this is the first step of system analysis. An ultimate goal is to find the response using Fourier series after a periodic signal passes through a linear system. Therefore, the second step of the system analysis is to find the response of a system to a basic signal – sinusoidal signal, followed by the response of the system to any periodic excitation.

4.4.1 System function

The steady state response of an LTI system to a sinusoidal signal such as excitation is still a sinusoidal signal with the same frequency. Therefore, the frequency variables in them can be ignored, and they can be represented by a phasor only related to the amplitude and phase of a signal. For example, a signal $f(t) = A\cos(\omega_0 t + \varphi)$ can be expressed as $\dot{F} = Ae^{j\varphi}$ or $\dot{F} = A\angle\varphi$.

Assuming that the excitation and the response of a system are, respectively, expressed as $\dot{F}$ and $\dot{Y}$, we can define as follows:

The ratio of the response and excitation phasors of a system is called the system transfer function or system function; it can be represented by a symbol $H(j\omega)$,

$$H(j\omega) = \frac{\dot{Y}}{\dot{F}} \,. \tag{4.4-1}$$

Usually, $H(j\omega)$ is a complex function, so it can be written as

$$H(j\omega) = |H(j\omega)|\, e^{j\varphi(\omega)} \,, \tag{4.4-2}$$

where $|H(j\omega)|$ is the magnitude of $H(j\omega)$ an even function about variable ω, $\varphi(\omega)$ is the phase of $H(j\omega)$ and an odd function about ω.

Because $H(j\omega)$ can reflect variations of amplitudes and phases of the system function for different frequencies, $H(j\omega)$ **is also called the frequency response characteristic of a system**; its magnitude and phase are called the amplitude frequency characteristic and the phase frequency characteristic, respectively. This concept is very useful in the study of communication systems.

Readers might ask: There is no term ω in the phasors $\dot{F}$ and $\dot{Y}$, so, why does ω exist in $H(j\omega)$? The reason is that $\dot{F}$ and $\dot{Y}$ obviously do not contain ω but this does not mean that they have nothing to do with it. Rather it is because both frequency values in excitation and response are the same, the excitation and the response can be simplified in the expressed form. In addition, $H(j\omega)$ is the concentrated reflection of the structure and performance of a system and should fit for sinusoidal excitations with different frequencies, that is, it is a function of ω, so, ω cannot be omitted in $H(j\omega)$, unlike $\dot{F}$ and $\dot{Y}$. Example 4.4-1 can help us to understand this problem.

From the above, we can see that the system function and the spectrum are similar in concept. In fact, we will find that the system function is just the spectrum of the impulse response $h(t)$ from the following chapters.

Although the system function $H(j\omega)$ is only a ratio of the output and input of a system, it can reflect the internal structure of a system, which is independent of the output and the input signals. Once a system is given, the system function $H(j\omega)$ is also determined and will not change with different excitations, and this concept is consistent with the $H(p)$ in Chapter 3. In fact, $H(j\omega)$ can be obtained by $H(p)$,

$$H(j\omega) = H(p)|_{p=j\omega} . \tag{4.4-3}$$

Equation (4.4-1) can be also written as

$$\dot{Y} = H(j\omega)\dot{F} . \tag{4.4-4}$$

Equation (4.4-4) tells us that the response of a system to a sinusoidal signal can be found by the product of the system function and the excitation phasor. This is the result that we wanted to reach in the second step mentioned above.

From the Fourier series, any a cosine or sine signal can be also known as a harmonic in form, and, therefore, we can also say that the response of a system to a harmonic can be obtained by means of the product of the system function and the harmonic phasor.

4.4.2 Analysis method

Periodic signals are defined over range $(-\infty, +\infty)$. Therefore, if a periodic signal is applied to a system, we can consider that the signal is accessed to the system when $t = -\infty$, and only the steady state response exists while the system is analyzed.

Based on step 1 of a periodic excitation being decomposed by Fourier series and step 2 of the response of system to a single frequency excitation (harmonic) being determined by equation (4.4-4), which was stated above, we can implement the final step of the analysis method, namely, to obtain the response of the system to any periodic signal. This can be realized by using the linearity property of the system to find the sum of all harmonic responses. The specific steps are as follows:

Step 1: Expand a given periodic signal as the standard form of the Fourier series. From the convergence of the spectrum, only limited terms in the Fourier series are usually necessary, such as c_0, $c_1 \cos(\omega_0 t + \varphi_1)$,, $c_n \cos(n\omega_0 t + \varphi_n)$. To facilitate the calculation, the best choice is to write each harmonic as a phasor form like $\dot{c}_n = c_n e^{j\varphi_n}$ or $\dot{c}_n = c_n \angle \varphi_n$.

Step 2: According to the knowledge about circuits or the concept of the transfer operator, find the system function $H(j\omega)$ and its limited terms of harmonics,

$$H(j\omega) = |H(j\omega)|\, e^{j\varphi(\omega)} = \left\{ |H(j0)|\, e^{j\varphi(0)},\, |H(j\omega_0)|\, e^{j\varphi(\omega_0)},\, |H(j2\omega_0)|\, e^{j\varphi(2\omega_0)},\, \dots \right.$$
$$\left. \dots,\, |H(jn\omega_0)|\, e^{j\varphi(n\omega_0)},\, \dots \right\}.$$

Let $\dot{F} = \dot{c}_n$ and substitute each harmonic component $\dot{c}_n$ into equation (4.4-4), then find responses $y_0(t), y_1(t), \dots, y_n(t)$ of the system to each $\dot{c}_n$, that is,

$$y_n(t) = c_n |H(jn\omega_0)| \cos(n\omega_0 t + \varphi_{Hn}), \tag{4.4-5}$$

where $\varphi_{Hn} = \varphi(n\omega_0) + \varphi_n$, namely, the phase of a harmonic response is equal to the sum of the phase of the system function and the initial phase of an excited harmonic.

Step 3: Add all harmonic responses to form the total response $y(t) = y_0(t) + y_1(t) + \dots + y_n(t)$, namely,

$$y(t) = \sum_{n=0}^{\infty} c_n |H(jn\omega_0)| \cos(n\omega_0 t + \varphi_{Hn}). \tag{4.4-6}$$

Example 4.4-1. A circuit and waveform of a voltage source $u_S(t)$ are shown in ▶ Figure 4.18a and b, and $T = 10\,\mu s$ is given. Find voltage $u_o(t)$ on resistor R.

Solution. (1) The Fourier series of the excitation source is

$$u_S(t) = \frac{80}{\pi^2} \left(\cos \omega_1 t + \frac{1}{9} \cos 3\omega_1 t + \frac{1}{25} \cos 5\omega_1 t + \dots \right),$$

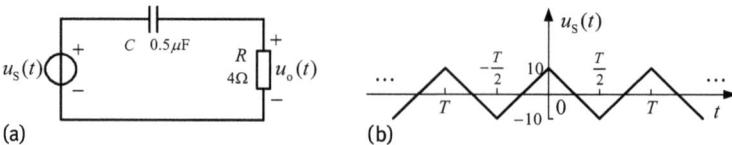

(a)   (b)

Fig. 4.18: E4.4-1.

where $\omega_1 = \frac{2\pi}{T} = 2\pi \times 10^5 \text{rad/s}$, so the harmonic phasors of the excitation are, respectively,

$$\dot{U}_{S1} = \frac{80}{\pi^2} \cos \omega_1 t = 8.11 \angle 0° \text{ V},$$

$$\dot{U}_{S2} = \frac{80}{\pi^2} \times \frac{1}{9} \cos 3\omega_1 t = 0.90 \angle 0° \text{ V},$$

$$\dot{U}_{S3} = \frac{80}{\pi^2} \times \frac{1}{25} \cos 5\omega_1 t = 0.324 \angle 0° \text{ V},$$

$$\vdots$$

(2) According to the split voltage formula, the system function can be written as

$$H(j\omega) = \frac{\dot{U}_O}{\dot{U}_S} = \frac{R}{R - j\frac{1}{\omega C}}.$$

(3) The phasor of the harmonic response on R generated by each excitation phasor can be obtained by equation (4.4-5)

$$\dot{U}_{O1} = H(j\omega_1)\dot{U}_{S1} = \frac{R}{R - j\frac{1}{\omega_1 C}}\dot{U}_{S1} = \frac{4}{4 - j3.18} \times 8.11 \angle 0° = 6.35 \angle 38.5° \text{ V},$$

$$\dot{U}_{O3} = H(j3\omega_1)\dot{U}_{S3} = \frac{R}{R - j\frac{1}{3\omega_1 C}}\dot{U}_{S3} = \frac{4}{4 - j1.06} \times 0.9 \angle 0° = 0.87 \angle 14.84° \text{ V},$$

$$\dot{U}_{O5} = H(j5\omega_1)\dot{U}_{S5} = \frac{R}{R - j\frac{1}{5\omega_1 C}}\dot{U}_{S5} = \frac{4}{4 - j0.636} \times 0.324 \angle 0° = 0.32 \angle 9.03° \text{ V},$$

$$\vdots$$

(4) Finally, from the superposition principle, the voltage response on R can be obtained by equation (4.4-6) as

$$u_O(t) = 6.35 \cos(\omega_1 t + 38.5°) + 0.87 \cos(3\omega_1 t + 14.84°) + 0.32 \cos(5\omega_1 t + 9.03°) + \cdots$$

Example 4.4-1 gives another more accurate and more vivid name: **harmonic response summation** for the analysis method of a system to a periodic signal in the frequency domain.

Comparing the content of this chapter with the sinusoidal steady state analysis way in the circuits analysis course, we can see that the phasor analysis method in sinusoidal circuits is the foundation of the Fourier series analysis method in this course. Just because the phasor analysis method has solved the response of system to a sine excitation, the Fourier series method can have scope for its power.

In this chapter, first, according to the signal decomposition, a periodic signal was represented by a basic signal – the sinusoidal signal. Then the response of system to the basic signal was found by means of the system function. Finally, the response of system to any periodic signal was obtained from the linearity of the system. This idea is shown in ▶ Figure 4.19 and is significant for the system analysis in subsequent chapters.

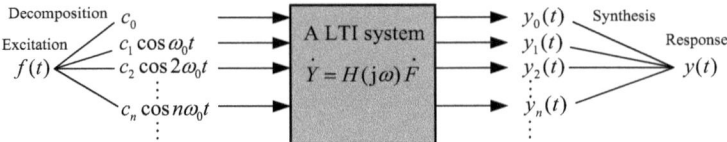

Fig. 4.19: Analysis idea of a system to a periodic signal.

4.5 Solved questions

Question 4-1. A period signal is $f(t) = 3\cos t + \sin\left(5t + \frac{\pi}{6}\right) - 2\cos\left(8t - \frac{2\pi}{3}\right)$. Plot its unilateral amplitude and phase spectrums.

Solution. From the known conditions, values of $f(t)$ exist at frequency points 1, 5, 8. The sine term should be changed into the cosine like $\sin\left(5t + \frac{\pi}{6}\right) = \cos\left(5t + \frac{\pi}{6} - \frac{\pi}{2}\right) = \cos\left(5t - \frac{\pi}{3}\right)$. Since the third term in $f(t)$ is minus, its phase should be changed as $\pi - \frac{2\pi}{3} = \frac{\pi}{3}$. The unilateral amplitude and phase spectrums are, respectively, pictured in ▶ Figure Q4-1a and b.

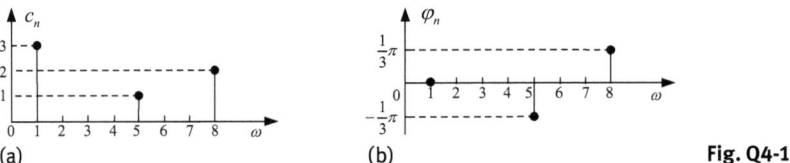

(a) (b) **Fig. Q4-1**

Question 4-2. A periodic signal is $f(t) = 2\sin\left(\frac{\pi}{2}t + \frac{\pi}{4}\right) - \cos\left(\frac{4\pi}{3}t - \frac{3\pi}{4}\right)$.
(1) Find its period T and angular frequency of the fundamental wave ω_0.
(2) Find all nonzero harmonics.
(3) Plot the unilateral amplitude and phase spectrums of this signal.

Solution. (1) Because $T_1 = 2\pi/(\pi/2) = 4$ s, $T_2 = 2\pi/(4\pi/3) = 1.5$ s, the period must be the least common multiple of 4 and 1.5, namely $T = m_1 T_1 = m_2 T_2$, $T_1/T_2 = m_1/m_2 = 8/3$, then $T = 3T_1 = 8T_2 = 12$ s and $\omega_0 = 2\pi/T = \pi/6$ (rad/s).
(2) From the expression of $f(t)$; the third and eighth harmonic components are nonzero.
(3) Transform the first term sine component into a cosine and change the second term from negative to positive. Then $f(t)$ will be transformed into the standard form

$$f(t) = 2\cos\left(\frac{\pi}{2}t - \frac{\pi}{4}\right) + \cos\left(\frac{4\pi}{3}t + \frac{\pi}{4}\right).$$

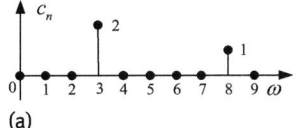
(a)

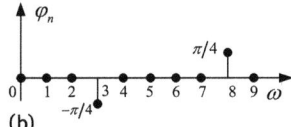
(b)   **Fig. Q4-2**

From the equation, the unilateral amplitude and the phase spectrums are plotted in ▶ Figure Q4-2.

Question 4-3. The period signal $f(t)$ is shown in ▶ Figure Q4-3. Find the DC component.

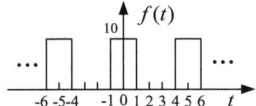
Fig. Q4-3

Solution. The DC component is equivalent to the coefficient a_0 in Fourier series, so we have

$$a_0 = \frac{1}{T_0} \int_{-T_0/2}^{T_0/2} f(t)\mathrm{d}t = \frac{1}{5} \int_{-1}^{1} 10\mathrm{d}t = 4 \, .$$

The DC component of the period signal $f(t)$ is 4.

Question 4-4. Judge which option is the spectrum component of the period signal shown in ▶ Figure Q4-4.
A. Cosine components in harmonics
B. Sine components in harmonics
C. Cosine components in odd harmonics
D. Sine components in odd harmonics

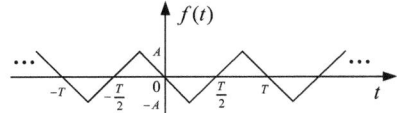
Fig. Q4-4

Solution. Because this signal meets expressions $f(t) = -f(-t)$ and $f(t) = -f(t \pm T/2)$, it is an odd harmonic function; the spectrum should only include sine components in odd harmonics, so the answer is D.

4.6 Learning tips

The periodic signal is a commonly used and important signal. Frequency domain analysis is not only the supplement for analysis in time domain, but also the means of display of the physical characteristics of a signal. Please pay attention to following points:

(1) Fourier series is a master key to analyze various periodic signals.
(2) The frequency spectrum is another description that is as important as the expression in time domain; it can reflect some invisible characteristics of signals in the time domain.
(3) The system function is an important function that can reflect the system structure and characteristics independently of excitation and response. It is also a link between excitation and response in the frequency domain.
(4) The response of a system to any periodic signal can be expressed as an algebraic sum of sinusoidal signals with different frequencies, namely, an algebraic sum of harmonics.

4.7 Problems

Problem 4-1. Point out the frequency components in the Fourier series of periodic signals shown in ▶ Figure P4-1 using the parity of the signal.

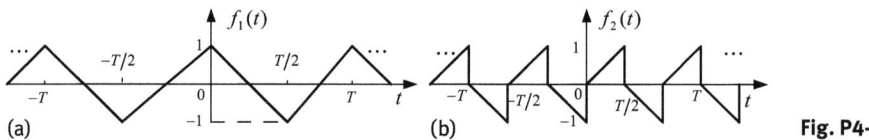

Fig. P4-1

Problem 4-2. Find the trigonometric and exponential forms of the Fourier series for signals shown in ▶ Figure P4-2 using the direct calculation method.

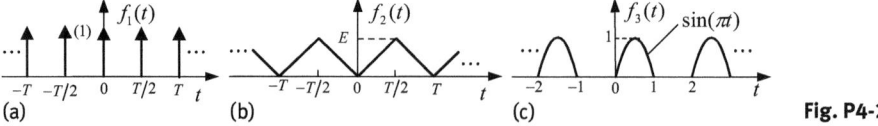

Fig. P4-2

Problem 4-3. Four signals with the same period are shown in ▶ Figure P4-3.
(1) Find the Fourier series of signal $f_1(t)$ using the direct calculation method.
(2) Find the Fourier series of $f_2(t), f_3(t)$ and $f_4(t)$ using the properties of Fourier series.

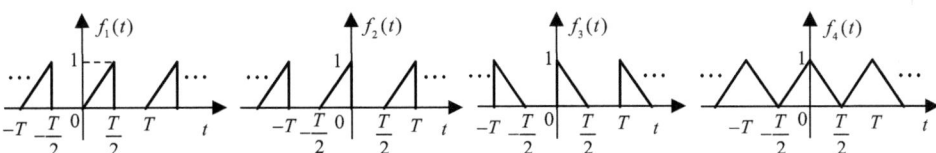

Fig. P4-3

Problem 4-4. Find the trigonometric forms of the Fourier series for the signals shown in ▶ Figure P4-4 using the differential property of the Fourier series.

Fig. P4-4

Problem 4-5. Plot the amplitude and phase spectrums of the following periodic signals.
(1) $f(t) = \frac{4}{\pi}\left(\cos \omega_0 t - \frac{1}{3}\cos 3\omega_0 t + \frac{1}{5}\cos 5\omega_0 t - \frac{1}{7}\cos 7\omega_0 t + \dots \right)$
(2) $f(t) = \frac{1}{2} - \frac{2}{\pi}\left(\sin 2\pi t + \frac{1}{2}\sin 4\pi t + \frac{1}{3}\sin 6\pi t + \frac{1}{4}\sin 8\pi t + \dots \right)$
(3) $f(t) = 1 - \sin \pi t + \cos \pi t + \frac{1}{\sqrt{2}}\cos\left(2\pi t + \frac{\pi}{6}\right)$

Problem 4-6. Let the Fourier series coefficients of $f(t), f_1(t), f_2(t)$ and $f_3(t)$ be, respectively, $F_n, F_{1n}, F_{2n}, F_{3n}$, and $f_1(t) = f^*(t), f_2(t) = f(t)\cos \omega_0 t, f_3(t) = f(t)\sin \omega_0 t$.
Prove
(1) $F_{1n} = F_{-n}^*$, (2) $F_{2n} = \frac{1}{2}(F_{n+1} + F_{n-1})$, (3) $F_{3n} = \frac{1}{2j}(F_{n-1} - F_{n+1})$.

Problem 4-7. If the Fourier coefficients of a complex function $f(t)$ and its conjugate function $f^*(t)$ are, respectively, A_n and B_n, prove $B_n = A_{-n}$.

Problem 4-8. Find the Fourier series of the half wave cosine signal in ▶ Figure P4-8 and plot its amplitude and phase spectrums.

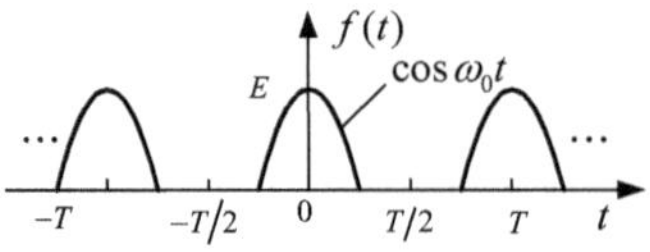

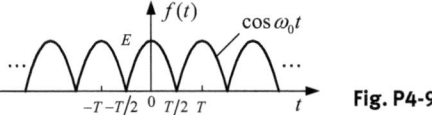

Fig. P4-8

Problem 4-9. Find the Fourier series of the full wave cosine signal in ▸ Figure P4-9 and plot its amplitude and phase spectrums.

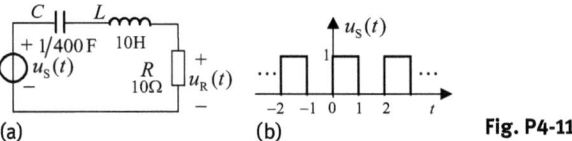

Fig. P4-9

Problem 4-10. A circuit is shown in ▸ Figure P4-10 and $u_S(t) = 6 + 10\cos(10^3 t) + 6\cos(2 \times 10^3 t)$ V is known. Find the capacitor voltage $u_C(t)$.

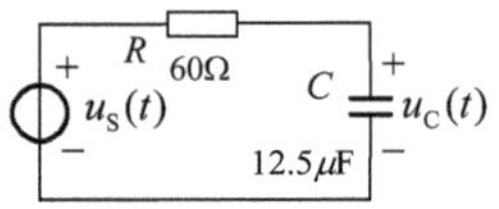

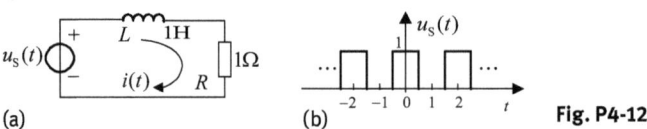

Fig. P4-10

Problem 4-11. In the circuit in ▸ Figure P4-11a, voltage source $u_S(t)$ is a periodic signal and its waveform is shown in ▸ Figure P4-11b. Find the voltage $u_2(t)$ on the resistor. (Do not consider harmonics above three orders.)

(a)   (b)

Fig. P4-11

Problem 4-12. If the periodic square wave voltage shown in ▸ Figure P4-12a is applied to an RL circuit, find the first four harmonics of $i(t)$.

(a)   (b)

Fig. P4-12

5 Analysis of continuous-time systems excited by nonperiodic signals in the real frequency domain

Questions: We have found a master key or a general tool to analyze periodic signals in frequency domain, which is the Fourier series. Similarly, can we also find a master key to analyze nonperiodic signals?

Solution: Seek the relation between a periodic signal and a nonperiodic signal → Find the analysis methods to a nonperiodic signal using the ways and results for a periodic signal.

Results: Fourier transform and system function. The Fourier transform of the zero-state response is equal to the product of Fourier transform of the excitation and the system function.

5.1 The concept of Fourier transform

The methods for solving for the response of a system to a periodic signal in the frequency domain were presented in Chapter 4. Their core is that an arbitrary periodic signal that satisfies the Dirichlet conditions can be developed as an algebraic sum of infinite sinusoidal signals (harmonics) with the Fourier series (usually, limited terms are considered), then subresponses corresponding to each harmonic are found, and finally they are superposed to form a complete response.

So, the questions are how to get the zero state responses to various nonperiodic signals? Can a general analysis method like the Fourier series for all nonperiodic signals be found? The answer is "yes".

Through the observations to periodic signal waveforms we can find that if a repeated cycle T of a periodic signal becomes infinite, its waveform is never repeated and will evolve into a nonperiodic signal. This finding provides a new approach for analysis to nonperiodic signals with the aid of the analysis methods for periodic signals.

After analysis of the spectrum of a periodic signal, we can see that the interval between two adjacent lines ω_0 will tend to be infinitesimal when the period T tends to infinity, the original discrete spectrum will become continuous and, at the same time, the amplitudes of frequency components (harmonics) will also approach infinitesimal but still keep a certain ratio. To describe this kind of frequency characteristic of an aperiodic signal, we will introduce the concept of spectrum density in this section.

From equation (4.3-1) if T tends to infinity, then $F(n\omega_0)$ or F_n will tend to zero; obviously, it is not the spectrum function that we know. However, if the T on the denominator on the right side of equation (4.3-1) is moved to the left, then $\frac{F(n\omega_0)}{\frac{1}{T}}$ is an

https://doi.org/10.1515/9783110419535-005

indefinite form of $\frac{0}{0}$, and its limit may exist when T tends to infinity,

$$\frac{F(n\omega_0)}{\frac{1}{T}} = \int_{-T/2}^{T/2} f(t)e^{-jn\omega_0 t}dt \, , \tag{5.1-1}$$

where $\frac{1}{T} = f_0 = \frac{\omega_0}{2\pi}$. Therefore, equation (5.1-1) becomes

$$\frac{2\pi F(n\omega_0)}{\omega_0} = \int_{-T/2}^{T/2} f(t)e^{-jn\omega_0 t}dt \, . \tag{5.1-2}$$

Equation (5.1-2) is considered as the size of the complex amplitude on the unit frequency, which has the obvious meaning of density. If T tends to infinite, then ω_0 tends to be infinitely small, the discrete variable $n\omega_0$ becomes a continuous variable ω, and $F(n\omega_0)$ will change from a discrete function to a continuous function. Therefore, we define that the limit of equation (5.1-2) is a spectral density function, which is denoted by $F(j\omega)$, that is,

$$F(j\omega) \stackrel{\text{def}}{=} \lim_{T\to\infty} \frac{2\pi F(n\omega_0)}{\omega_0} = \lim_{T\to\infty} \int_{-T/2}^{T/2} f(t)e^{-jn\omega_0 t}dt = \int_{-\infty}^{\infty} f(t)e^{-j\omega t}dt \, . \tag{5.1-3}$$

Then equation (4.3-2) can be rewritten as

$$f(t) = \lim_{T_0\to\infty} \frac{1}{2\pi} \sum_{n=-\infty}^{\infty} \frac{2\pi F(n\omega_0)}{\omega_0} e^{jn\omega_0 t} \cdot \omega_0 \, . \tag{5.1-4}$$

Under the conditions $n\omega_0 \to \omega$, $\sum_{n=-\infty}^{\infty} \to \int_{-\infty}^{\infty}$ and $\omega_0 \to d\omega$, equation (5.1-4) will become a typical integral expression

$$f(t) = \frac{1}{2\pi} \int_{-\infty}^{\infty} F(j\omega)e^{j\omega t}d\omega \, . \tag{5.1-5}$$

Equation (5.1-5) shows that $f(t)$ can be treated as a continuous sum of complex exponential signals $e^{j\omega t}$, whose frequencies are infinite density and amplitudes $\frac{d\omega}{2\pi}F(j\omega)$ are infinitesimal and terms are infinite. Thus, equations (5.1-3) and (5.1-5) are known as the Fourier transform pair,

$$F(j\omega) = \mathcal{F}[f(t)] = \int_{-\infty}^{\infty} f(t)e^{-j\omega t}dt \, , \tag{5.1-6}$$

$$f(t) = \mathcal{F}^{-1}[F(j\omega)] = \frac{1}{2\pi} \int_{-\infty}^{\infty} F(j\omega)e^{j\omega t}d\omega \, . \tag{5.1-7}$$

Equation (5.1-6) is called the Fourier transform and equation (5.1-7) is called the inverse Fourier transform. The $F(j\omega)$ is known as the Fourier transform of a signal $f(t)$, and $f(t)$

is called the original function of $F(j\omega)$. The symbol "$\mathscr{F}$" represents the Fourier transform operation, and "$\mathscr{F}^{-1}$" represents the operation of the inverse Fourier transform. The $f(t)$ and $F(j\omega)$ can be related by symbol "$\overset{\mathscr{F}}{\longleftrightarrow}$", that is,

$$f(t) \overset{\mathscr{F}}{\longleftrightarrow} F(j\omega) \,. \tag{5.1-8}$$

The Fourier transform is a type of linear transform and has the characteristic of one to one correspondence. equation (5.1-6) is also called the decomposition formula, and equation (5.1-7) is the synthetic formula. Obviously, the Fourier transform reflects the decomposition and synthesis features of signals once again.

Generally speaking, $F(j\omega)$ is a complex function and can be written as

$$F(j\omega) = R(\omega) + jI(\omega) = |F(j\omega)| \, e^{j\varphi(\omega)} \,, \tag{5.1-9}$$

where $|F(j\omega)|$ and $\varphi(\omega)$ are, respectively, the magnitude and phase of $F(j\omega)$, and $R(\omega)$ and $I(\omega)$ are, respectively, its real and imaginary parts.

Because $|F(j\omega)|$ describes the relative size of each frequency component density of an aperiodic signal $f(t)$, $\varphi(\omega)$ describes the phase relationships between various frequency component densities of $f(t)$. Thus, to facilitate research, $F(j\omega)$ is also known as the frequency spectrum of $f(t)$, curves $|F(j\omega)| \sim \omega$ and $\varphi(\omega) \sim \omega$ are called, respectively, the amplitude spectrum and the phase spectrum of a nonperiodic signal. In this way, $F(j\omega)$ and $F(n\omega_0)$ are unified in name. However, readers must remember that their meanings are also different.

It should be pointed that the above process of derivation for the Fourier transform focused on its physical concept. Strict mathematical derivation shows that the sufficient condition for the existence of the Fourier transform for a signal $f(t)$ is that $f(t)$ must be absolutely integrable,

$$\int_{-\infty}^{\infty} |f(t)| \, dt < \infty \,. \tag{5.1-10}$$

For some signals that are not absolutely integrable, such as DC signals, symbol signals, etc., the Fourier transforms can be found with the limit method.

The Fourier series states that a periodic signal $f(t)$ can be expressed as a discrete sum of imaginary exponential signals, which is

$$f(t) = \sum_{n=-\infty}^{\infty} F(n\omega_0)e^{jn\omega_0 t} \,.$$

The Fourier transform states that an aperiodic signal $f(t)$ can be expressed as a continuous sum of imaginary exponential signals, which is

$$f(t) = \frac{1}{2\pi} \int_{-\infty}^{\infty} F(j\omega)e^{j\omega t} d\omega \,.$$

The evolution process from a periodic signal to an aperiodic signal is a change process from a discrete sum to a continuous sum in form. Moreover, the Fourier transform is an extension of the Fourier series.

Like a period signal, a nonperiodic signal can be also decomposed into a sum of cosine components with different frequencies, which contains all frequency components from zero to infinity, that is, the frequency is a continuous variable. This conclusion can be deduced by

$$
f(t) = \frac{1}{2\pi} \int_{-\infty}^{\infty} F(j\omega)e^{j\omega t}d\omega = \frac{1}{2\pi} \int_{-\infty}^{\infty} |F(j\omega)| e^{j[\omega t + \varphi(\omega)]}d\omega
$$

$$
= \frac{1}{2\pi} \int_{-\infty}^{\infty} |F(j\omega)| \cos[\omega t + \varphi(\omega)]d\omega + \frac{j}{2\pi} \int_{-\infty}^{\infty} |F(j\omega)| \sin[\omega t + \varphi(\omega)]d\omega .
$$

(5.1-11)

We know

$$
F(j\omega) = \int_{-\infty}^{\infty} f(t)e^{-j\omega t}dt = \int_{-\infty}^{\infty} f(t)\cos\omega tdt - j \int_{-\infty}^{\infty} f(t)\sin\omega tdt ,
$$

and then, we have

$$
|F(j\omega)| = \left[\left(\int_{-\infty}^{\infty} f(t)\cos\omega tdt\right)^2 + \left(\int_{-\infty}^{\infty} f(t)\sin\omega tdt\right)^2\right]^{1/2} ,
$$

$$
\varphi(\omega) = -\arctan\frac{\int_{-\infty}^{\infty} f(t)\sin\omega tdt}{\int_{-\infty}^{\infty} f(t)\cos\omega tdt} .
$$

It can be seen that $|F(j\omega)|$ is an even function of ω, and $\varphi(\omega)$ is odd, so the first term and the second one on right side of equation (5.1-11), respectively, are

$$
\frac{1}{2\pi} \int_{-\infty}^{\infty} |F(j\omega)| \cos[\omega t + \varphi(\omega)]d\omega = \frac{1}{\pi} \int_{0}^{\infty} |F(j\omega)| \cos[\omega t + \varphi(\omega)]d\omega ,
$$

$$
\frac{j}{2\pi} \int_{-\infty}^{\infty} |F(j\omega)| \sin[\omega t + \varphi(\omega)]d\omega = 0 ,
$$

Therefore, equation (5.1-11) can be written as

$$
f(t) = \frac{1}{\pi} \int_{0}^{\infty} |F(j\omega)| \cos[\omega t + \varphi(\omega)]d\omega .
$$

(5.1-12)

A comparison of equations (5.1-12) and (4.2-5) yields

$$f(t) = c_0 + \sum_{n=1}^{\infty} c_n \cos(n\omega_0 t + \varphi_n) ;$$

it is not difficult to find similarities and differences between them.

It needs to be explained that the Fourier transform $F(j\omega)$ of a signal $f(t)$ defined in this book emphasizes the imaginary number as its independent variable, but it is also defined as $F(\omega)$ in some books (such as «Communication Principles»), where the idea is to define the independent variable on the real frequency axis, in order to understand easily the physical concepts of signals and noises in communication technology. From the aspect of application, there are no essential differences between the two definitions and they are usually interchangeable. One reason that $F(j\omega)$ is applied is conventional, another one is it can more clearly reflect the relationship between the Fourier transform and the Laplace transform. In the Laplace transform, when the real part $\sigma = 0$ in the independent variable $s = \sigma + j\omega$, the Laplace transform $F(s)$ of $f(t)$ will become its Fourier transform $F(j\omega)$, that is, **the Fourier transform is a special case of the Laplace transform for a signal.** In other words, the Fourier transform is just the Laplace transform when $\sigma = 0$.

From equation (5.1-3), the Fourier transform $F(j\omega)$ was derived from the Fourier series coefficient F_n by using the limit method. So, what is the specific relationship between them?

Assuming that $f_T(t)$ is a periodic signal, if $f(t)$ is the model of an arbitrary periodic waveform which is cut from the waveform of $f_T(t)$, then its Fourier transform should be

$$F(j\omega) = \mathcal{F}[f(t)] = \int_{-T/2}^{T/2} f(t)e^{-j\omega t}dt .$$

Comparing above the equation with the Fourier series coefficient formula of a periodic signal,

$$F_n = \frac{1}{T} \int_{-\frac{T}{2}}^{\frac{T}{2}} f(t)e^{-jn\omega_0 t}dt ,$$

we obtain

$$F_n = \frac{1}{T} F(j\omega)|_{\omega=n\omega_0} . \qquad (5.1\text{-}13)$$

Equation (5.1-13) shows that the Fourier coefficient F_n, namely, the spectrum $F(n\omega_0)$ of a periodic signal, can be found by the Fourier transform $F(j\omega)$ of any a periodic waveform in the periodic signal.

5.2 Fourier transforms of typical aperiodic signals

The Fourier transforms (spectrums) of some typical aperiodic signals are given below.

5.2.1 Gate signals

The rectangular pulse $f(t)$ with width τ and amplitude E shown in ▸ Figure 5.1 is called the gate signal when $E = 1$; it is denoted as $g_\tau(t)$. The Fourier transform pair is

$$f(t) = \begin{cases} E & (|t| \le \frac{\tau}{2}) \\ 0 & (|t| > \frac{\tau}{2}) \end{cases} \xleftrightarrow{\mathscr{F}} F(j\omega) = E\tau \cdot Sa\left(\frac{\omega\tau}{2}\right). \tag{5.2-1}$$

Proof. From equation (5.1-6), we have

$$F(j\omega) = \int_{-\infty}^{\infty} f(t)e^{-j\omega t}dt = \int_{-\frac{\tau}{2}}^{\frac{\tau}{2}} Ee^{-j\omega t}dt = -\frac{E}{j\omega}e^{-j\omega t}\bigg|_{-\frac{\tau}{2}}^{\frac{\tau}{2}} = \frac{2E}{\omega}\sin\frac{\omega\tau}{2} = E\tau \cdot Sa\left(\frac{\omega\tau}{2}\right).$$

That is,

$$|F(j\omega)| = E\tau \cdot \left|Sa\left(\frac{\omega\tau}{2}\right)\right|, \tag{5.2-2a}$$

$$\varphi(\omega) = \begin{cases} 0 & (F(j\omega) \ge 0) \\ \pi & (F(j\omega) < 0) \end{cases}. \tag{5.2-2b}$$

Therefore, we obtain

$$g_\tau(t) \xleftrightarrow{\mathscr{F}} \tau \cdot Sa\left(\frac{\omega\tau}{2}\right), \tag{5.2-3}$$

where $Sa(x) = \frac{\sin x}{x}$, or, $Sa(t) = \frac{\sin t}{t}$ is called the sampling function, which is an important function in the communication principles course. Obviously, the sampling function is even and has a feature of $\int_0^\infty Sa(t)dt = \frac{\pi}{2}$. $\square$

Because $F(j\omega)$ is a real function, it can be expressed by a curve $F(j\omega) \sim \omega$, as shown in ▸ Figure 5.2. When $F(j\omega)$ is positive, the phase is 0, and when it is negative, the phase is π.

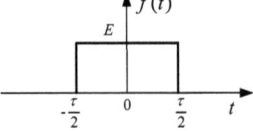

Fig. 5.1: Rectangular pulse signal.

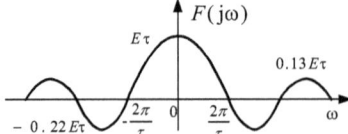

Fig. 5.2: Rectangular pulse signal spectrum.

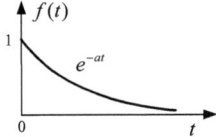

Fig. 5.3: Unilateral exponential signal.

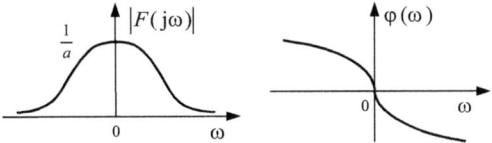

Fig. 5.4: Amplitude and phase spectrums of an unilateral exponential signal.

5.2.2 Unilateral exponential signals

The unilateral exponential signal is sketched in ▶ Figure 5.3 and the Fourier transform pair is

$$f(t) = \begin{cases} e^{-at} & (t \geq 0), (a > 0) \\ 0 & (t < 0) \end{cases} \xleftrightarrow{\mathcal{F}} F(j\omega) = \frac{1}{a + j\omega} . \tag{5.2-4}$$

Proof. The spectrum of a unilateral exponential signal is

$$F(j\omega) = \int_{-\infty}^{\infty} f(t)e^{-j\omega t}dt = \int_{0}^{\infty} e^{-at}e^{-j\omega t}dt = -\frac{1}{a + j\omega}e^{-(a+j\omega)t}\Big|_{0}^{\infty} = \frac{1}{a + j\omega} .$$

That is,

$$|F(j\omega)| = \frac{1}{\sqrt{a^2 + \omega^2}} , \tag{5.2-5a}$$

$$\varphi(\omega) = -\arctan\left(\frac{\omega}{a}\right) . \tag{5.2-5b}$$

They are illustrated in ▶ Figure 5.4. □

5.2.3 Bilateral exponential signals

The Fourier transform pair of a bilateral exponential signal as shown in ▶ Figure 5.5 is

$$f(t) = \begin{cases} e^{-at} & (t \geq 0), (a > 0) \\ e^{at} & (t < 0) \end{cases} \xleftrightarrow{\mathcal{F}} F(j\omega) = \frac{2a}{a^2 + \omega^2} \tag{5.2-6}$$

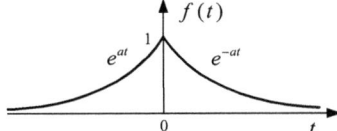

Fig. 5.5: Bilateral exponential signal.

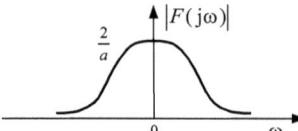

Fig. 5.6: Amplitude spectrum of a bilateral exponential signal.

Proof. The spectrum of the bilateral exponential signal is

$$F(j\omega) = \int_{-\infty}^{\infty} f(t)e^{-j\omega t}dt = \int_{-\infty}^{0} e^{at}e^{-j\omega t}dt + \int_{0}^{\infty} e^{-at}e^{-j\omega t}dt$$

$$= \frac{1}{a - j\omega} e^{(a-j\omega)t}\Big|_{-\infty}^{0} - \frac{1}{a + j\omega} e^{(a+j\omega)t}\Big|_{0}^{\infty} = \frac{2a}{a^2 + \omega^2} .$$

That is,

$$|F(j\omega)| = \frac{2a}{a^2 + \omega^2} , \qquad (5.2\text{-}7a)$$

$$\varphi(\omega) = 0 . \qquad (5.2\text{-}7b)$$

Equation (5.2-7a) is sketched in ▶ Figure 5.6. ☐

5.2.4 Unit DC signals

A DC signal with size 1 is called the unit DC signal and is shown in ▶ Figure 5.7; its expression is

$$f(t) = 1 \quad (-\infty < t < \infty) .$$

Because it is not absolutely integrable, its frequency spectrum cannot be directly found by equation (5.1-6). However, it can be considered as a limit form of a bilateral exponential signal when $a \to 0$, so, the frequency spectrum can be also considered as a limit form of the spectrum of the bilateral exponential signal as $a \to 0$. We have

$$F(j\omega) = \lim_{a \to 0} \frac{2a}{a^2 + \omega^2} = \begin{cases} \infty & (\omega = 0) \\ 0 & (\omega \neq 0) \end{cases} .$$

Obviously, the spectrum function of a unit DC signal is an impulse function. Now, the weight of the impulse function needs to be determined. As we know, it should be an

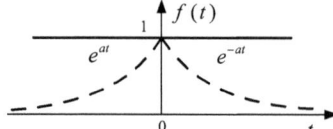

Fig. 5.7: A unit DC signal.

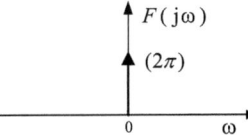

Fig. 5.8: Spectrum of a unit DC signal.

integral

$$\int_{-\infty}^{\infty} \frac{2a}{a^2 + \omega^2} d\omega = \int_{-\infty}^{\infty} \frac{2}{1 + \left(\frac{\omega}{a}\right)^2} d\left(\frac{\omega}{a}\right) = 2 \arctan\frac{\omega}{a}\Big|_{-\infty}^{\infty} = 2\pi,$$

and, therefore, the spectrum of a unit DC signal is

$$F(j\omega) = 2\pi\delta(\omega).$$

That is,

$$f(t) = 1 \xleftrightarrow{\mathcal{F}} F(j\omega) = 2\pi\delta(\omega). \tag{5.2-8}$$

The spectrum is plotted in ▶ Figure 5.8.

5.2.5 Unit impulse signals

The spectrum function of a unit impulse signal, as shown in ▶ Figure 5.9, is

$$F(j\omega) = \int_{-\infty}^{\infty} \delta(t)e^{-j\omega t}dt = \int_{-\infty}^{\infty} \delta(t)dt = 1,$$

So, we have

$$\delta(t) \xleftrightarrow{\mathcal{F}} F(j\omega) = 1 \tag{5.2-9}$$

The spectrum is displayed in ▶ Figure 5.10. Obviously, the spectrum occupies the frequency range from $-\infty$ to $+\infty$, and amplitude values of all frequency components of the spectrum are the same. It explains why the impulse signal contains rich frequency components or contributions of these components to the signal are the same. Therefore the spectrum is also called the uniform spectrum or the *white spectrum*.

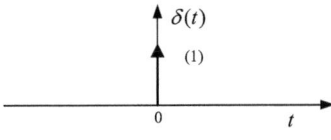

Fig. 5.9: A unit impulse signal.

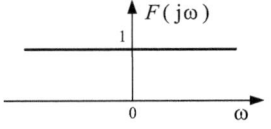

Fig. 5.10: Spectrum of a unit impulse signal.

5.2.6 Signum signals

Because the signum or sign signal as shown in ▶ Figure 5.11 is not also absolutely integrable, its frequency spectrum cannot be directly found by equation (5.1-6) either, but it can be obtained by the limit form of the spectrum $F_1(j\omega)$ of $f_1(t)$, which is shown in ▶ Figure 5.12 as $a \to 0$.

Since

$$f_1(t) = \begin{cases} e^{-at} & (t > 0) \\ -e^{at} & (t < 0) \end{cases},$$

its frequency spectrum function is

$$F_1(j\omega) = \mathcal{F}[f_1(t)] = \int_{-\infty}^{\infty} f_1(t)e^{-j\omega t}dt = \int_{-\infty}^{0} \left(-e^{at}\right)e^{-j\omega t}dt + \int_{0}^{\infty} e^{-at}e^{-j\omega t}dt$$

$$= -\frac{1}{a - j\omega} + \frac{1}{a + j\omega} = -\frac{2\omega}{a^2 + \omega^2}j.$$

Therefore, the spectrum function of the signum signal is

$$F(j\omega) = \mathcal{F}[\text{sgn}(t)] = \lim_{a \to 0} F_1(j\omega) = -\lim_{a \to 0} \frac{2\omega j}{a^2 + \omega^2} = \frac{2}{j\omega}.$$

So, we have

$$f(t) = \text{sgn}(t) \xrightarrow{\mathcal{F}} F(j\omega) = \frac{2}{j\omega}. \tag{5.2-10}$$

Because

$$F(j\omega) = |F(j\omega)| e^{j\varphi(\omega)},$$

we have

$$|F(j\omega)| = \frac{2}{|\omega|}, \tag{5.2-11a}$$

$$\varphi(\omega) = \begin{cases} -\frac{\pi}{2} & (\omega > 0) \\ \frac{\pi}{2} & (\omega < 0) \end{cases}. \tag{5.2-11b}$$

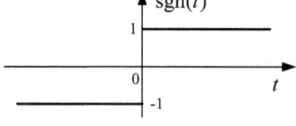

Fig. 5.11: A sign signal.

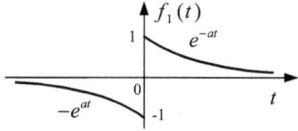

Fig. 5.12: $f_1(t)$.

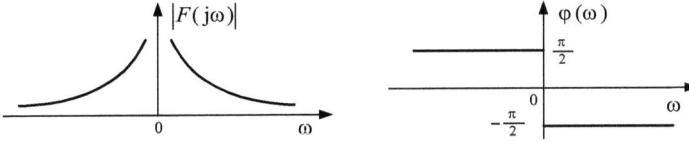

Fig. 5.13: Amplitude and phase spectrums of a sign signal.

The amplitude spectrum and the phase spectrum are illustrated in ▸ Figure 5.13.

Note: Strictly speaking, the Fourier transform of the signum should be $F(j\omega) = \frac{2}{j\omega} + k\delta(\omega)$, where the impulse term represents a DC component in the signum. Because the average value of the signal is zero in the time domain, that is, $k = 0$, $F(j\omega) = \frac{2}{j\omega}$.

5.2.7 Unit step signals

The spectrum function of a unit step signal is

$$F(j\omega) = \frac{1}{j\omega} + \pi\delta(\omega) .$$

Thus, we have

$$\varepsilon(t) \overset{\mathcal{F}}{\longleftrightarrow} \frac{1}{j\omega} + \pi\delta(\omega) . \tag{5.2-12}$$

The proof of equation (5.2-12) and the spectrum waveform can be seen in Example 5.3-1.

Note: The DC, the signum, the step signals and the periodic signals are all power signals. Their Fourier transforms all contain impulse signal terms, and this is a general characteristic of power signals and can be used to test if a signal is a power signal in the frequency domain.

Tab. 5.1: The Fourier transforms of typical signals.

No.	Name	Expression	Spectrum
1	Gate signal	$g_\tau(t)$	$\tau Sa(\frac{\omega\tau}{2})$
2	Unilateral exponential	$e^{-at}\varepsilon(t)$ $(a > 0)$	$\frac{1}{a+j\omega}$
3	Bilateral exponential	$e^{-a|t|}\varepsilon(t)$ $(a > 0)$	$\frac{2a}{a^2+\omega^2}$
4	Unit DC	1	$2\pi\delta(\omega)$
5	Unit impulse	$\delta(t)$	1
6	Sign signal	$\text{sgn}(t)$	$\frac{2}{j\omega}$
7	Unit step signal	$\varepsilon(t)$	$\frac{1}{j\omega} + \pi\delta(\omega)$
8	Triangle pulse signal	$\Delta_{2\tau}(t) = \left(1 - \frac{|t|}{\tau}\right)\left[\varepsilon\left(t + \frac{\tau}{2}\right) - \varepsilon\left(t - \frac{\tau}{2}\right)\right]$	$\tau Sa^2\left(\frac{\omega\tau}{2}\right)$
9	Sine signal	$\sin(\omega_0 t)$	$j\pi[\delta(\omega + \omega_0) - \delta(\omega - \omega_0)]$
10	Cosine signal	$\cos(\omega_0 t)$	$\pi[\delta(\omega + \omega_0) + \delta(\omega - \omega_0)]$

Finally, the Fourier transforms of typical signals are summarized in Table 5.1 for reference.

5.3 Properties of the Fourier transform

It is helpful for calculation of the Fourier transform of a signal to learn properties of it.

5.3.1 Linearity

If

$$\mathcal{F}[f_1(t)] = F_1(j\omega), \quad \mathcal{F}[f_2(t)] = F_2(j\omega) ,$$

so,

$$\mathcal{F}[a_1 f_1(t) + a_2 f_2(t)] = a_1 F_1(j\omega) + a_2 F_2(j\omega) . \tag{5.3-1}$$

Example 5.3-1. Find the frequency spectrum of the unit step signal shown in ► Figure 5.14a.

Solution. Because

$$\varepsilon(t) = \frac{1}{2} \operatorname{sgn}(t) + f_1(t) ,$$

according to the linearity property, we have

$$F(j\omega) = \mathcal{F}\left[\frac{1}{2} \operatorname{sgn}(t) + f_1(t)\right] = \frac{1}{2} \cdot \frac{2}{j\omega} + \frac{1}{2} \cdot 2\pi\delta(\omega) = \frac{1}{j\omega} + \pi\delta(\omega) .$$

That is,

$$|F(j\omega)| = \frac{1}{|\omega|} + \pi\delta(\omega) ,$$

$$\varphi(\omega) = \begin{cases} -\frac{\pi}{2} & (\omega > 0) \\ \frac{\pi}{2} & (\omega < 0) \end{cases} .$$

The amplitude spectrum and the phase spectrum are plotted in ► Figure 5.15a and b.

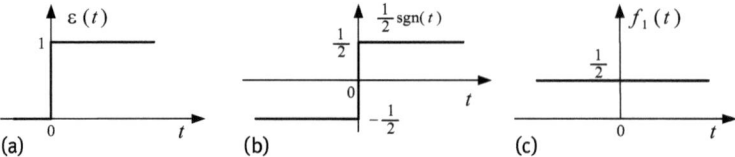

(a)　(b)　(c)

Fig. 5.14: E5.3-1.

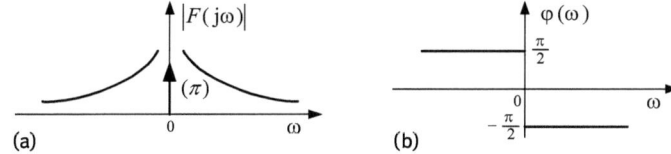

(a)

(b)

Fig. 5.15: Amplitude and phase spectrum of a unit step signal.

5.3.2 Time shifting

If

$$\mathcal{F}[f(t)] = F(j\omega) ,$$

then

$$\mathcal{F}[f(t - t_0)] = e^{-j\omega t_0} F(j\omega) , \tag{5.3-2}$$
$$\mathcal{F}[f(t + t_0)] = e^{j\omega t_0} F(j\omega) . \tag{5.3-3}$$

Proof.

$$\mathcal{F}[f(t - t_0)] = \int_{-\infty}^{\infty} f(t - t_0) e^{-j\omega t} dt \overset{t - t_0 = x}{=\!=\!=} \int_{-\infty}^{\infty} f(x) e^{-j\omega(x + t_0)} d(x + t_0)$$

$$= e^{-j\omega t_0} \int_{-\infty}^{\infty} f(x) e^{-j\omega x} dx = e^{-j\omega t_0} F(j\omega) .$$

Similarly, this is proved by

$$\mathcal{F}[f(t + t_0)] = e^{j\omega t_0} F(j\omega) .$$

Time shifting shows that if a signal in the time domain is delayed time t_1, then its amplitude spectrum stays the same, but the phase spectrum is shifted, $-\omega t_1$. The details are given below.   □

If

$$\mathcal{F}[f(t)] = F_1(j\omega), \quad \mathcal{F}[f(t - t_1)] = F_2(j\omega) ,$$

then

$$F_2(j\omega) = e^{-j\omega t_1} F_1(j\omega) = e^{-j\omega t_1} |F_1(j\omega)| e^{j\varphi_1(\omega)} = |F_1(j\omega)| e^{j[\varphi_1(\omega) - \omega t_1]} , \tag{5.3-4}$$

and

$$F_2(j\omega) = |F_2(j\omega)| e^{j\varphi_2(\omega)} . \tag{5.3-5}$$

Comparing equation (5.3-4) and equation (5.3-5), we have

$$|F_2(j\omega)| = |F_1(j\omega)| , \tag{5.3-6}$$
$$\varphi_2(\omega) = \varphi_1(\omega) - \omega t_1 . \tag{5.3-7}$$

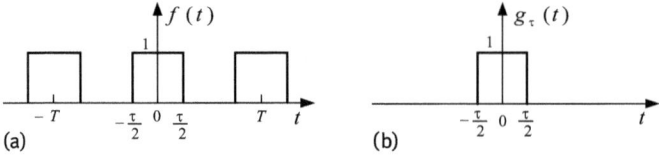

Fig. 5.16: E5.3-3.

Example 5.3-2. Find the Fourier transform of a signal $\delta(t - t_0)$.

Solution. We know that

$$\mathcal{F}[\delta(t)] = 1 \,,$$

so, we have

$$\mathcal{F}[\delta(t - t_0)] = e^{-j\omega t_0} \cdot 1 = e^{-j\omega t_0} \,.$$

Example 5.3-3. Find the spectrum of a signal $f(t)$ with three rectangular pulses shown in ▸ Figure 5.16a.

Solution. It is known that the spectrum of a gate function $g_\tau(t)$ shown in ▸ Figure 5.16b

$$G(j\omega) = \tau Sa\left(\frac{\omega\tau}{2}\right) \,,$$

and, then

$$f(t) = g_\tau(t) + g_\tau(t + T) + g_\tau(t - T) \,.$$

According to the time shifting property, we have

$$F(j\omega) = G(j\omega)\left(1 + e^{j\omega T} + e^{-j\omega T}\right) = (1 + 2\cos\omega T)\tau Sa\left(\frac{\omega\tau}{2}\right) \,.$$

5.3.3 Frequency shifting

If

$$\mathcal{F}[f(t)] = F(j\omega) \,,$$

then

$$\mathcal{F}[f(t)e^{j\omega_0 t}] = F[j(\omega - \omega_0)] \,, \tag{5.3-8}$$

$$\mathcal{F}[f(t)e^{-j\omega_0 t}] = F[j(\omega + \omega_0)] \,. \tag{5.3-9}$$

Proof.

$$\mathcal{F}[f(t)e^{j\omega_0 t}] = \int_{-\infty}^{\infty} f(t)e^{j\omega_0 t}e^{-j\omega t}dt = \int_{-\infty}^{\infty} f(t)e^{-j(\omega - \omega_0)t}dt = F[j(\omega - \omega_0)] \,.$$

Similarly, it can be proved by

$$\mathcal{F}[f(t)e^{-j\omega_0 t}] = F[j(\omega + \omega_0)] \,. \qquad \square$$

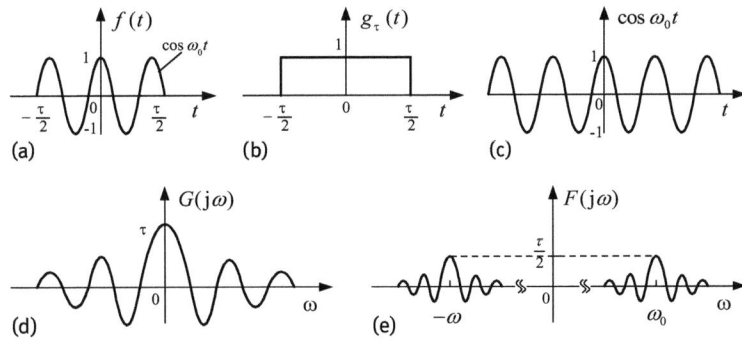

Fig. 5.17: E5.3-4.

Example 5.3-4. Find the spectrum of the high frequency pulse $f(t)$ shown in ► Figure 5.17a.

Solution. The high frequency pulse signal can be regarded as the product of the gate signal shown in ► Figure 5.17b and the cosine signal shown in ► Figure 5.17c, that is,

$$f(t) = g_\tau(t) \cdot \cos \omega_0 t = \frac{1}{2}[g_\tau(t)e^{j\omega_0 t} + g_\tau(t)e^{-j\omega_0 t}] \,.$$

Because $\mathscr{F}[g_\tau(t)] = G(j\omega) = \tau Sa\left(\frac{\omega\tau}{2}\right)$ as shown in ► Figure 5.17d is known, from linearity and frequency shifting, the spectrum of a high frequency pulse signal is

$$F(j\omega) = \frac{\tau}{2}\left\{Sa\left[\frac{\tau(\omega - \omega_0)}{2}\right] + Sa\left[\frac{\tau(\omega + \omega_0)}{2}\right]\right\} \,.$$

Its waveform is shown in ► Figure 5.17e.

5.3.4 Time scaling

If

$$\mathscr{F}[f(t)] = F(j\omega) \,,$$

then

$$\mathscr{F}[f(at)] = \frac{1}{|a|}F\left(j\frac{\omega}{a}\right) \,. \tag{5.3-10}$$

Proof. When $a > 0$, we have

$$\mathscr{F}[f(at)] = \int_{-\infty}^{\infty} f(at)e^{-j\omega t}dt \overset{at=x}{=} \frac{1}{a}\int_{-\infty}^{\infty} f(x)e^{-j\frac{\omega}{a}x}dx = \frac{1}{a}F\left(j\frac{\omega}{a}\right)$$

When $a < 0$, we have

$$\mathscr{F}[f(at)] = \int_{-\infty}^{\infty} f(at)e^{-j\omega t}dt \overset{at=x}{=} \frac{1}{a}\int_{\infty}^{-\infty} f(x)e^{-j\frac{\omega}{a}x}dx = -\frac{1}{a}F\left(j\frac{\omega}{a}\right) \,.$$

Integrating to the above two cases, the expression of the scaling transform is obtained by

$$\mathcal{F}[f(at)] = \frac{1}{|a|} F\left(j\frac{\omega}{a}\right) .$$

□

Equation (5.3-10) states that the width of a signal compressed to $\frac{1}{a}$ times of this original signal ($a > 1$) in the time domain is equivalent to its spectrum width being extended a times in the frequency domain, while its amplitude is reduced to the original $\frac{1}{a}$ times. Conversely, the width of a signal is extended to the original one a times ($0 < a < 1$) in the time domain, which is equivalent to its spectrum width being compressed $\frac{1}{a}$ times, while its amplitude is increased to the original one a times in the frequency domain. In a nutshell, **the compression effect of a signal in the time domain is equivalent to its expansion effect in the frequency domain; the expansion effect in the time domain is equivalent to that of compression in the frequency domain.** The property is explained in ▶ Figure 5.18 with a rectangular pulse signal.

In particular, when $a = -1$, we have

$$\mathcal{F}[f(-t)] = F(-j\omega) . \tag{5.3-11}$$

Equation (5.3-11) is also called the *time inversion theorem*.

A living example of scaling property is that when a sound recorder set works with a speed of twice the normal speed, the signal waveform in the time domain is compressed, the high frequency components are enhanced, so, squeal noise results. Otherwise, the sound becomes lower and deeper with a working speed of half of the normal

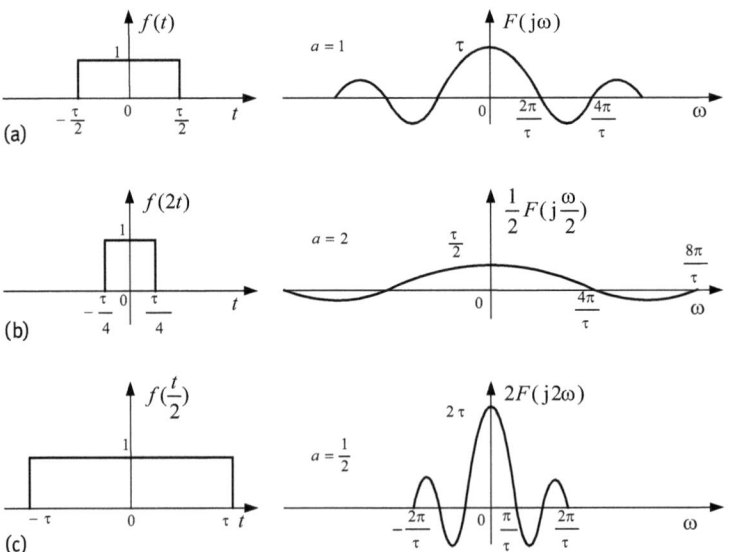

Fig. 5.18: Scaling transform for a rectangle pulse signal.

speed. This feature tells us that if a signal is compressed in the time domain to improve the speed of information transmission, a communication system must provide
a wider passband for it in the frequency domain.

5.3.5 Symmetry

If
$$\mathcal{F}[f(t)] = F(j\omega) ,$$
then
$$\mathcal{F}[F(jt)] = 2\pi f(-\omega) . \qquad (5.3\text{-}12)$$

Proof. Because

$$f(t) = \frac{1}{2\pi} \int_{-\infty}^{\infty} F(j\omega)e^{j\omega t}d\omega ,$$

$$f(-t) = \frac{1}{2\pi} \int_{-\infty}^{\infty} F(j\omega)e^{-j\omega t}d\omega ,$$

and exchanging the variables ω and t, we have

$$\int_{-\infty}^{\infty} F(jt)e^{-j\omega t}dt = 2\pi f(-\omega) ,$$

so,

$$\mathcal{F}[F(jt)] = 2\pi f(-\omega) .$$

We can see that $\mathcal{F}[F(jt)] = 2\pi f(\omega)$ from equation (5.3-12) when $f(t)$ is even. ☐

Example 5.3-5. Find the spectrum of a sampling signal $Sa(t) = \frac{\sin t}{t}$.

Solution. The spectrum of a gate signal $g_\tau(t)$ with amplitude 1 and width τ is

$$g_\tau(t) \xrightarrow{\mathcal{F}} \tau Sa\left(\frac{\omega\tau}{2}\right) .$$

Letting $\tau = 2$, we have

$$g_2(t) \xrightarrow{\mathcal{F}} 2Sa(\omega) .$$

That is,

$$\frac{1}{2}g_2(t) \xrightarrow{\mathcal{F}} Sa(\omega) .$$

According to the symmetry property and paying attention to the fact that $g_2(t)$ is even,
we have

$$Sa(t) \xrightarrow{\mathcal{F}} 2\pi\frac{1}{2}g_2(\omega) = \pi g_2(\omega) .$$

Therefore, the spectrum of a sampling signal is

$$F(j\omega) = \pi g_2(\omega) .$$

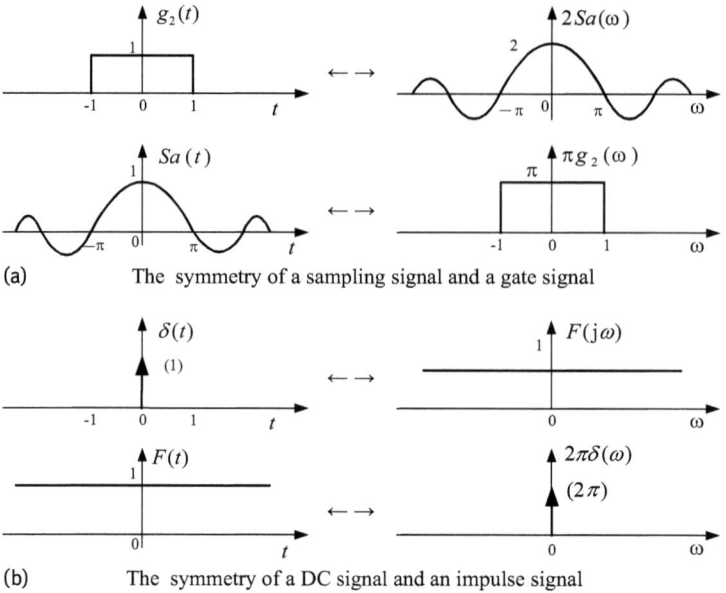

(a)   The symmetry of a sampling signal and a gate signal

(b)   The symmetry of a DC signal and an impulse signal

Fig. 5.19: Fourier transform symmetry.

We can see that the spectrum of a gate signal is a sampling signal, and the spectrum of a sampling signal is a gate signal [as shown in ▶ Figure 5.19a]. This is an important conclusion and states that for a time limited signal (like a gate signal in the time domain), the frequency range of its spectrum is infinite in the frequency domain (like a sampling signal in the frequency domain); for a frequency limited signal (like a gate signal in *frequency* domain), the time range of its original signal is infinite in the time domain (like a sampling signal in the time domain). It is very helpful to understand the ISI concept in the communication principles course.

Another analogical case is that the spectrum of a unit impulse signal is a DC signal, and the spectrum of a DC signal is an impulse signal, as shown in ▶ Figure 5.19b. That is,

$$\delta(t) \xleftrightarrow{\mathcal{F}} 1 \, ,$$

$$1 \xleftrightarrow{\mathcal{F}} 2\pi\delta(\omega) \, .$$

Features of signals in two domains based on the symmetry property are listed in Table 5.2.

Tab. 5.2: Features of signals in two domains based on the symmetry property.

NO.	Time domain		Frequency domain
1	Periodic	$\overset{\mathcal{F}}{\longrightarrow}$	Discrete
	Discrete	$\overset{\mathcal{F}}{\longleftarrow}$	Periodic
2	Continuous	$\overset{\mathcal{F}}{\longrightarrow}$	Nonperiodic
	Nonperiodic	$\overset{\mathcal{F}}{\longleftarrow}$	Continuous
3	Infinite lasting time	$\overset{\mathcal{F}}{\longrightarrow}$	Limited frequency band
	Limited lasting time	$\overset{\mathcal{F}}{\longleftarrow}$	Infinite frequency band

5.3.6 Properties of convolution

1. Convolution in the time domain

If

$$\mathcal{F}[f_1(t)] = F_1(j\omega) ,$$
$$\mathcal{F}[f_2(t)] = F_2(j\omega) ,$$

then

$$\mathcal{F}[f_1(t) * f_2(t)] = F_1(j\omega) F_2(j\omega) . \tag{5.3-13}$$

Proof.

$$\mathcal{F}[f_1(t) * f_2(t)] = \int_{-\infty}^{\infty} [f_1(t) * f_2(t)]e^{-j\omega t}dt = \int_{-\infty}^{\infty}\left[\int_{-\infty}^{\infty} f_1(\tau)f_2(t-\tau)d\tau\right]e^{-j\omega t}dt$$

$$= \int_{-\infty}^{\infty} f_1(\tau)\left[\int_{-\infty}^{\infty} f_2(t-\tau)e^{-j\omega t}dt\right]d\tau = \int_{-\infty}^{\infty} f_1(\tau)e^{-j\omega\tau}F_2(j\omega)d\tau$$

$$= F_2(j\omega)\int_{-\infty}^{\infty} f_1(\tau)e^{-j\omega\tau}d\tau = F_1(j\omega)F_2(j\omega) \qquad \square$$

Example 5.3-6. Find the Fourier transform of a signal $f(t) = \frac{\sin(t-t_0)}{2t-2t_0}$.

Solution. Because

$$f(t) = \frac{1}{2}Sa(t) * \delta(t - t_0) ,$$

and

$$Sa(t) \overset{\mathcal{F}}{\longleftrightarrow} \pi g_2(\omega) ,$$
$$\delta(t - t_0) \overset{\mathcal{F}}{\longleftrightarrow} e^{-j\omega t_0} .$$

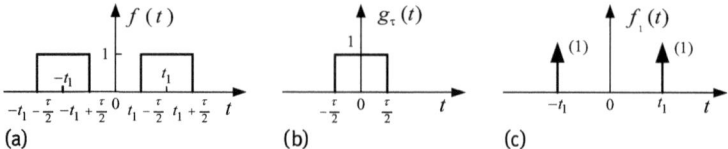

Fig. 5.20: E5.3-7.

Thus, according to the time domain convolution property the Fourier transform is

$$F(j\omega) = \frac{1}{2}\mathcal{F}[Sa(t)] \cdot \mathcal{F}[\delta(t - t_0)] = \frac{1}{2}\pi g_2(\omega) \cdot e^{-j\omega t_0} .$$

Example 5.3-7. Find the spectrum of the signal shown in ▶ Figure 5.20a.

Solution. The signal shown in ▶ Figure 5.20a can be regarded as the convolution of a gate signal shown in ▶ Figure 5.20b and the signal $f_1(t)$ shown in ▶ Figure 5.20c, namely,

$$f(t) = g_\tau(t) * f_1(t) = g_\tau(t) * [\delta(t + t_1) + \delta(t - t_1)] .$$

Therefore, according to the time domain convolution property, we have

$$F(j\omega) = \mathcal{F}[g_\tau(t)] \cdot \mathcal{F}[f_1(t)] = \tau Sa\left(\frac{\omega\tau}{2}\right) \cdot \left(e^{j\omega t_1} + e^{-j\omega t_1}\right) = 2\tau Sa\left(\frac{\omega\tau}{2}\right)\cos\omega t_1 .$$

2. Convolution in the frequency domain
If

$$\mathcal{F}[f_1(t)] = F_1(j\omega) ,$$
$$\mathcal{F}[f_2(t)] = F_2(j\omega) ,$$

then

$$\mathcal{F}[f_1(t) \cdot f_2(t)] = \frac{1}{2\pi}F_1(j\omega) * F_2(j\omega) . \qquad (5.3\text{-}14)$$

Proof.

$$\mathcal{F}^{-1}[F_1(j\omega) * F_2(j\omega)] = \frac{1}{2\pi}\int_{-\infty}^{\infty}\left[\int_{-\infty}^{\infty}F_1(j\eta)F_2[j(\omega - \eta)]d\eta\right]e^{j\omega t}d\omega$$

$$= \frac{1}{2\pi}\int_{-\infty}^{\infty}F_1(j\eta)\left\{\int_{-\infty}^{\infty}F_2[j(\omega - \eta)e^{j\omega t}]d\omega\right\}d\eta$$

$$= \int_{-\infty}^{\infty}F_1(j\eta)f_2(t)e^{j\eta t}d\eta = f_2(t)\int_{-\infty}^{\infty}F_1(j\eta)e^{j\eta t}d\eta$$

$$= 2\pi f_1(t)f_2(t)$$

□

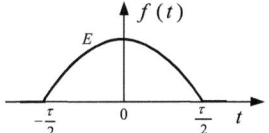

Fig. 5.21: E5.3-8.

Example 5.3-8. Find the spectrum of the cosine pulse signal shown in ▶ Figure 5.21.

$$f(t) = \begin{cases} E\cos\left(\frac{\pi t}{\tau}\right) & (|t| \le \frac{\tau}{2}) \\ 0 & (|t| > \frac{\tau}{2}) \end{cases}.$$

Solution. The signal $f(t)$ can be regarded as the product of a rectangular pulse signal $Eg_\tau(t)$ and a cosine signal $\cos(\frac{\pi t}{\tau})$, namely,

$$f(t) = Eg_\tau(t)\cos\left(\frac{\pi t}{\tau}\right).$$

Moreover,

$$\mathcal{F}\left[\cos\left(\frac{\pi t}{\tau}\right)\right] = \frac{1}{2}\mathcal{F}\left[e^{-j\frac{\pi}{\tau}t} + e^{j\frac{\pi}{\tau}t}\right] = \frac{1}{2}\left[2\pi\delta\left(\omega + \frac{\pi}{\tau}\right) + 2\pi\delta\left(\omega - \frac{\pi}{\tau}\right)\right]$$
$$= \pi\delta\left(\omega + \frac{\pi}{\tau}\right) + \pi\delta\left(\omega - \frac{\pi}{\tau}\right).$$

Thus, according to the convolution property in the frequency domain, we have

$$\mathcal{F}[f(t)] = \mathcal{F}\left[g_\tau(t)\cdot\cos\left(\frac{\pi t}{\tau}\right)\right] = \frac{1}{2\pi}\mathcal{F}[g_\tau(t)] * \mathcal{F}\left[\cos\left(\frac{\pi t}{\tau}\right)\right]$$
$$= \frac{1}{2\pi}\left[E\tau Sa\left(\frac{\omega\tau}{2}\right)\right] * \left[\pi\delta\left(\omega + \frac{\pi}{\tau}\right) + \pi\delta\left(\omega - \frac{\pi}{\tau}\right)\right]$$
$$= \frac{1}{2}E\tau\left\{Sa\left[\frac{\tau}{2}\left(\omega + \frac{\pi}{\tau}\right)\right] + Sa\left[\frac{\tau}{2}\left(\omega - \frac{\pi}{\tau}\right)\right]\right\} = \frac{2E\tau}{\pi}\frac{\cos\left(\frac{\omega\tau}{2}\right)}{1 - \left(\frac{\omega\tau}{\pi}\right)^2}$$

5.3.7 Differentiation in the time domain

If

$$\mathcal{F}[f(t)] = F(j\omega),$$

then

$$\mathcal{F}\left[\frac{df(t)}{dt}\right] = j\omega F(j\omega). \tag{5.3-15}$$

Proof. Because

$$f(t) = \frac{1}{2\pi}\int_{-\infty}^{\infty} F(j\omega)e^{j\omega t}d\omega$$

and

$$\frac{df(t)}{dt} = \frac{1}{2\pi}\frac{d}{dt}\int_{-\infty}^{\infty}F(j\omega)e^{j\omega t}d\omega = \frac{1}{2\pi}\int_{-\infty}^{\infty}[j\omega F(j\omega)]e^{j\omega t}d\omega ,$$

so,

$$\mathcal{F}\left[\frac{df(t)}{dt}\right] = j\omega F(j\omega) .$$

This result can be also generalized to the following situation with an nth-order derivative:

$$\mathcal{F}\left[\frac{d^n f(t)}{dt^n}\right] = (j\omega)^n F(j\omega) . \tag{5.3-16}$$

The differential property shows that the effect of differentiating a signal in the *time* domain is equivalent to strengthening the sizes of high frequency components of the signal in the *frequency* domain. In fact, the effect is to sharpen the signal waveform in the *time* domain. □

Example 5.3-9. Find the spectrum of the triangle pulse $f(t)$ shown in ▶ Figure 5.22.

$$f(t) = \begin{cases} E\left(1 - \frac{2|t|}{\tau}\right) & (|t| < \frac{\tau}{2}) \\ 0 & (|t| > \frac{\tau}{2}) \end{cases} .$$

Solution. The first-order derivative function of the triangle pulse $f(t)$ is

$$f'(t) = \frac{2E}{\tau}\left[\varepsilon\left(t + \frac{\tau}{2}\right) - \varepsilon(t)\right] - \frac{2E}{\tau}\left[\varepsilon(t) - \varepsilon\left(t - \frac{\tau}{2}\right)\right] ,$$

and the second-order derivative function is

$$f''(t) = \frac{2E}{\tau}\left[\delta\left(t + \frac{\tau}{2}\right) + \delta\left(t - \frac{\tau}{2}\right) - 2\delta(t)\right] .$$

Thus, we have

$$\mathcal{F}\left[f''(t)\right] = \frac{2E}{\tau}\left(e^{j\omega\frac{\tau}{2}} + e^{-j\omega\frac{\tau}{2}} - 2\right) = \frac{4E}{\tau}\left[\cos\left(\frac{\omega\tau}{2}\right) - 1\right] = -\frac{8E}{\tau}\sin^2\frac{\omega\tau}{4} .$$

According to the differential property in the time domain, we have

$$\mathcal{F}\left[f''(t)\right] = (j\omega)^2 F(j\omega), \quad F(j\omega) = \mathcal{F}[f(t)] ,$$

so,

$$F(j\omega) = \frac{1}{(j\omega)^2}\mathcal{F}\left[f''(t)\right] = -\frac{1}{(j\omega)^2}\cdot\frac{8E}{\tau}\sin^2\frac{\omega\tau}{4} = \frac{E\tau}{2}Sa^2\left(\frac{\omega\tau}{4}\right) .$$

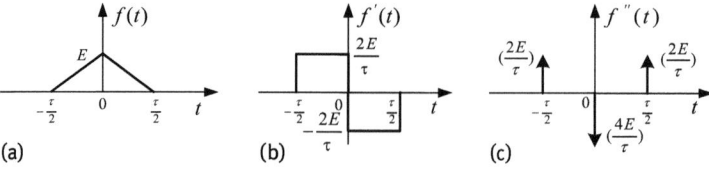

(a)   (b)   (c)

Fig. 5.22: E5.3-9.

5.3.8 Integration in the time domain

If

$$\mathcal{F}[f(t)] = F(j\omega) ,$$

then

$$\mathcal{F}\left[\int_{-\infty}^{t} f(\tau)d\tau\right] = \frac{1}{j\omega}F(j\omega) + F(0)\pi\delta(\omega) . \qquad (5.3\text{-}17)$$

Where, $F(0) = F(j\omega)|_{\omega=0} = \int_{-\infty}^{\infty} f(t)dt.$

Proof. We know that

$$f(t) * \varepsilon(t) = \int_{-\infty}^{t} f(\tau)d\tau ,$$

so,

$$\mathcal{F}\left[\int_{-\infty}^{t} f(\tau)d\tau\right] = \mathcal{F}\left[f(t) * \varepsilon(t)\right] = F(j\omega) \cdot \mathcal{F}[\varepsilon(t)] = F(j\omega)\left[\frac{1}{j\omega} + \pi\delta(\omega)\right] .$$

Moreover,

$$F(j\omega)\delta(\omega) = F(0)\delta(\omega) ,$$

so,

$$\mathcal{F}\left[\int_{-\infty}^{t} f(\tau)d\tau\right] = \frac{1}{j\omega}F(j\omega) + F(0)\pi\delta(\omega) .$$

The integral feature shows that a signal integrated in the *time* domain is equivalent to enhancing its low frequency components in the *frequency* domain and reducing its high frequency components. In fact, effect of the integral feature is to smooth the waveform of a signal in the time domain. This is the theoretical basis of Δ-Σ modulation technology in the communication principles course. □

Example 5.3-10. Find the spectrum of the signal $f(t)$ shown in ▶ Figure 5.23a.

$$f(t) = \begin{cases} 0 & (t < 0) \\ t & (0 \le t \le 1) \\ 1 & (t > 1) \end{cases}$$

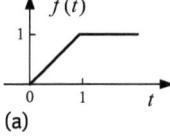
(a)

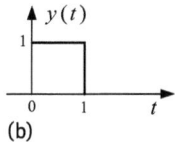
(b)

Fig. 5.23: E5.2-10.

Solution. $f(t)$ can be regarded as an integration of a rectangular pulse $y(t)$, as shown in ▶ Figure 5.23b, namely

$$f(t) = \int_{-\infty}^{t} y(\tau)d\tau ,$$

and

$$Y(j\omega) = \mathcal{F}[y(t)] = Sa\left(\frac{\omega}{2}\right)e^{-j\frac{\omega}{2}} .$$

It is known that

$$Y(0) = 1 ,$$

so, from the integration property, we have

$$F(j\omega) = \frac{1}{j\omega}Y(j\omega) + \pi Y(0)\delta(\omega) = \frac{1}{j\omega}Sa\left(\frac{\omega}{2}\right)e^{-j\frac{\omega}{2}} + \pi\delta(\omega) .$$

The properties of differentiation and integration in the time domain can be used to change differentiation or integration equation in the time domain into an algebraic equation in the frequency domain, which is very useful in the system analysis.

5.3.9 Modulation

If a signal $f(t)$ is multiplied by the cosine $\cos \omega_0 t$, what will be the result? Let $s(t) = f(t)\cos \omega_0 t$ and its spectrum be $S(j\omega)$. Then, based on Euler's relation and the frequency shifting property we have

$$S(j\omega) = \mathcal{F}[f(t)\cos \omega_0 t] = \mathcal{F}\left[\frac{1}{2}f(t)e^{j\omega_0 t} + \frac{1}{2}f(t)e^{-j\omega_0 t}\right]$$
$$= \frac{1}{2}F[j(\omega - \omega_0)] + \frac{1}{2}F[j(\omega + \omega_0)] \tag{5.3-18}$$

This shows that the result of $f(t)$ being multiplied by $\cos \omega_0 t$ in the time domain is equivalent to the spectrum $F(j\omega)$ of $f(t)$ being shifted to points $\pm\omega_0$ on the frequency axis in the frequency domain, and its shape remains the same and the amplitude value is only half of the original one. Thus, an oscillation signal whose amplitude changes with the signal $f(t)$ in the time domain results. This is called the modulation property (theorem) and is further illustrated by Example 5.3-4.

In communication technology, this feature is often used to accomplish a process called modulation, in which a low frequency signal $f(t)$ (called a modulating signal) is multiplied by a high frequency cosine signal (called a carrier). Thus, the waveform of $f(t)$ is modulated (placed) to the amplitude of the carrier $\cos \omega_c t$, and a high frequency signal $s(t)$ (called a modulated signal) forms, whose amplitude values contain waveform information of $f(t)$; in other words, the spectrum of the low frequency signal $f(t)$ is shifted to two higher frequency places $\pm\omega_c$ on the frequency axis. The modulation model is depicted in ▶ Figure 5.24.

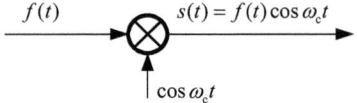

$$f(t) \quad \otimes \quad s(t) = f(t) \cos \omega_c t$$

$$\cos \omega_c t$$

Fig. 5.24: Modulation model.

The main purpose of modulation is to transform a low frequency signal into a high frequency signal, which can facilitate radioing with a size limited antenna or frequency division multiplexing. Therefore, the modulation theorem is an important theoretical foundation of radio communication and frequency division multiplexing technologies.

5.3.10 Conservation of energy

Energy calculation ways of a nonperiodic signal in the time and frequency domains are also related by Parseval's relation

$$E = \int_{-\infty}^{+\infty} |f(t)|^2 \, dt = \frac{1}{2\pi} \int_{-\infty}^{+\infty} |F(\omega)|^2 \, d\omega = \int_{-\infty}^{+\infty} |F(f)|^2 \, df . \tag{5.3-19}$$

The term $|F(f)|^2$ represents the energy on the unit band, which can reflect the relative size of the signal energy on different frequencies, that is, the energy distribution. Thus, it is regarded as the energy spectral density or, simply, ESD, and it is denoted as $E(f)$ or $E(\omega)$,

$$E(f) = |F(f)|^2 \quad \text{or} \quad E(\omega) = |F(\omega)|^2 \tag{5.3-20}$$

Obviously, the ESD is associated only with the amplitude spectrum and has nothing to do with the phase spectrum, and it is a nonnegative real even function. Thus, equation (5.3-19) can be written as

$$E = \frac{1}{\pi} \int_0^{+\infty} E(\omega) d\omega = 2 \int_0^{+\infty} E(f) df . \tag{5.3-21}$$

We can prove that for a gate signal with time width τ, the energy within the range from the first zero point $\left(\omega = \frac{2\pi}{\tau} \right)$ in its spectrum to the origin occupies around 90.3% of the total energy of this signal. This means that the main energy is concentrated on the range from the zero frequency point to the first spectral zero point. This point also fits for the periodic rectangle pulse and the triangle pulse, etc. As a result, in communication systems, only the low frequency components within $\omega = 0 \sim \frac{2\pi}{\tau}$ of a signal are transferred in general, which is usually called the effective frequency band width of the signal, denoted as B_ω and $B_\omega = \frac{2\pi}{\tau}$; its unit is rad/s, or $B_f = \frac{1}{\tau}$, with unit Hz. Obviously, the effective frequency bandwidth B_ω is inversely proportional to the pulse duration τ for a rectangle pulse, which means the shorter the signal duration in the

Tab. 5.3: Properties of Fourier transforms.

No.	Name		$f(t)$ (time domain)	$F(j\omega)$ (frequency domain)
1	Definition		$f(t) = \frac{1}{2\pi}\int_{-\infty}^{\infty} F(j\omega)e^{j\omega t}d\omega$	$F(j\omega) = \int_{-\infty}^{\infty} f(t)e^{-j\omega t}dt$
2	Linearity		$a_1 f_1(t) + a_2 f_2(t)$	$a_1 F_1(j\omega) + a_2 F_2(j\omega)$
3	Time shifting		$f(t \pm t_0)$	$e^{\pm j\omega t_0} F(j\omega)$
4	Frequency shifting		$f(t)e^{\pm j\omega_0 t}$	$F[j(\omega \mp \omega_0]$
5	Scaling transform		$f(at)(a \neq 0)$	$\frac{1}{\|a\|}F(j\frac{\omega}{a})$
6	Symmetry		$F(jt)$	$2\pi f(-\omega)$
7	Convolution theorem	Time domain	$f_1(t) * f_2(t)$	$F_1(j\omega)F_2(j\omega)$
		Frequency domain	$f_1(t) \cdot f_2(t)$	$\frac{1}{2\pi}F_1(j\omega) * F_2(j\omega)$
8	Differential in the time domain		$f^{(n)}(t)$	$(j\omega)^n F(j\omega)$
9	Integral in the time domain		$\int_{-\infty}^{t} f(\tau)d\tau$	$\frac{1}{j\omega}F(j\omega) + \pi F(0)\delta(\omega)$
10	Differential in the frequency domain		$(-jt)^n f(t)$	$F^{(n)}(j\omega)$
11	Integral in the frequency domain		$j\frac{f(t)}{t} + \pi f(0)\delta(t)$	$\int_{-\infty}^{\infty} F(j\Omega)d\Omega$
12	Parseval's relation		$E = \int_{-\infty}^{+\infty} \|f(t)\|^2\,dt = \frac{1}{2\pi}\int_{-\infty}^{+\infty} \|F(\omega)\|^2\,d\omega = \int_{-\infty}^{+\infty} \|F(f)\|^2\,df$	

time domain, the wider the frequency band becomes in the frequency domain. This conclusion fits well for power signals, so it is of great importance in the communication principles course.

Finally, we summarize the properties of the Fourier transform in Table 5.3.

5.4 Fourier transforms of periodic signals

We know that the Fourier transform was put forward to analyze nonperiodic signals. So, does the Fourier transform of a periodic signal exist?

As mentioned above, the sufficient condition for the existence of the Fourier transform of a signal is that it is absolutely integrable. However, general periodic signals cannot meet this condition, so their Fourier transforms cannot be found directly from the definition formula but can be obtained indirectly by using a singularity signal—the impulse signal.

If the spectrum of a periodic signal $f(t)$ with period T is $F_n = \frac{1}{T}\int_{-\frac{T}{2}}^{\frac{T}{2}} f(t)e^{-jn\omega_0 t}dt$, so its Fourier series is

$$f(t) = \sum_{n=-\infty}^{\infty} F_n e^{jn\omega_0 t}.$$

Fourier transformation on both sides of the equation leads to

$$\mathcal{F}\left[f(t)\right] = \mathcal{F}\left[\sum_{n=-\infty}^{\infty} F_n e^{jn\omega_0 t}\right] = \sum_{n=-\infty}^{\infty} F_n \mathcal{F}\left[e^{jn\omega_0 t}\right].$$

According to frequency shifting, we have

$$\mathcal{F}\left[e^{jn\omega_0 t}\right] = 2\pi\delta(\omega - n\omega_0).$$

So, the Fourier transform of $f(t)$ is

$$\mathcal{F}\left[f(t)\right] = 2\pi \sum_{n=-\infty}^{\infty} F_n \delta(\omega - n\omega_0). \tag{5.4-1}$$

Equation (5.4-1) shows that the Fourier transform or spectrum density of a periodic signal is composed by infinite shifted impulse signals that are located at harmonic frequency points $n\omega_0$ and have the weights $2\pi F_n$.

Example 5.4-1. Find the Fourier transform of a periodic impulse signal (also called the train of impulses or the unit comb) $\delta_T(t) = \sum_{n=-\infty}^{+\infty} \delta(t - nT)$ (see ▶ Figure 5.25a).

Solution. F_n of $\delta_T(t)$ is

$$F_n = \frac{1}{T} \int_{-\frac{T}{2}}^{\frac{T}{2}} \delta_T(t)e^{-jn\omega_0 t}dt = \frac{1}{T} \int_{-\frac{T}{2}}^{\frac{T}{2}} \delta(t)e^{-jn\omega_0 t}dt = \frac{1}{T}.$$

Equation (5.3-1) yields

$$\mathcal{F}\left[\delta_T(t)\right] = 2\pi \sum_{n=-\infty}^{\infty} F_n \delta(\omega - n\omega_0) = \frac{2\pi}{T} \sum_{n=-\infty}^{\infty} \delta(\omega - n\omega_0) = \omega_0 \sum_{n=-\infty}^{\infty} \delta(\omega - n\omega_0).$$

If $\delta_{\omega_0}(j\omega) = \sum_{n=-\infty}^{\infty} \delta(\omega - n\omega_0)$, then we have

$$\delta_T(t) \overset{\mathcal{F}}{\longleftrightarrow} \omega_0\delta_{\omega_0}(j\omega). \tag{5.4-2}$$

This states that the Fourier transform of a train of impulses with cycle T, intensity 1, in the time domain is still a train of impulses with cycle $\omega_0 = \frac{2\pi}{T}$ intensity $\omega_0 = \frac{2\pi}{T}$ in the frequency domain (see ▶ Figure 5.25b).This result can be used to prove the "sampling theorem".

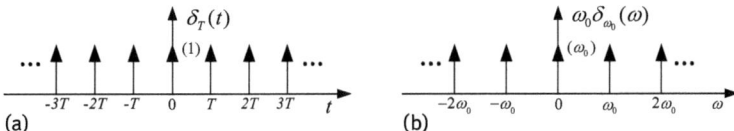

(a)   (b)

Fig. 5.25: E5.4-1.

It can be proved that if a signal is absolutely integrable, there is no any impulse component in its frequency spectrum; if there is impulse component in the frequency spectrum, this signal must be periodicity or have the DC component in the time domain.

5.5 Solutions for the inverse Fourier transform

If the Fourier transform $F(j\omega)$ of a signal $f(t)$ is known, in many cases the original signal $f(t)$ can be obtained by $F(j\omega)$, which means that the inverse Fourier transform operation needs to be discussed.

It was first thought that the inverse Fourier transform is calculated by using equation (5.1-3), $f(t) = \frac{1}{2\pi} \int_{-\infty}^{\infty} F(j\omega)e^{j\omega t}d\omega$, but this integral is actually too complex to calculate in general, therefore, this method is not very commonly used. Usually, the Fourier transform properties and the Fourier transforms of some typical signals can be used to find the inverse Fourier transforms.

Example 5.5-1. Find the original functions $f(t)$ of the following spectrums.
(1) $F(j\omega) = \omega^2$ (2) $F(j\omega) = \delta(\omega - 2)$ (3) $F(j\omega) = 2\cos\omega$

Solution. (1) According to the Fourier transform of $\delta(t)$ and the differential property, we have

$$\omega^2 = -(j\omega)^2 \times 1 \xrightarrow{\ \mathcal{F}\ } -\delta'(t) \,,$$

so,

$$f(t) = -\delta'(t) \,.$$

(2) According to the Fourier transform of the DC signal and the frequency shifting property, we have

$$1 \xrightarrow{\ \mathcal{F}\ } 2\pi\delta(\omega) \,,$$

$$\frac{1}{2\pi}e^{j2t} \xleftrightarrow{\ \mathcal{F}\ } \delta(\omega - 2) \,,$$

so,

$$f(t) = \frac{1}{2\pi}e^{j2t} \,.$$

(3) Because $\cos\omega_0 t = \frac{1}{2}(e^{j\omega_0 t} + e^{-j\omega_0 t})$, according to the Fourier transform of the DC signal and the time shifting property, we have

$$\cos 1t = \frac{1}{2}(e^{jt} + e^{-jt}) \xrightarrow{\ \mathcal{F}\ } \pi[\delta(\omega - 1) + \delta(\omega + 1)] \,,$$

and according to the symmetry property, we have

$$2\pi(\frac{1}{\pi}\cos\omega) \xrightarrow{\ \mathcal{F}\ } \delta(t - 1) + \delta(t + 1) \,,$$

so,

$$f(t) = \delta(t - 1) + \delta(t + 1) \,.$$

5.6 System analysis methods for aperiodic signals

As mentioned above, a nonperiodic signal can be converted into a continuous sum of imaginary exponential signals or sinusoidal signals by using the Fourier transform, and the spectrum density function of a signal is also introduced from it. So, what is the significance of this knowledge to system analysis? Or what are the advantages of solving a linear system model with the Fourier transform?

Usually, a nonperiodic signal exists only in a certain time interval. In order to illustrate expediently the method of solving the system response to a nonperiodic signal such as excitation, we assume that the starting state of the system is zero. Thus, we only discuss the problem of finding the zero-state response in this chapter.

In Chapter 4, the phasors' ratio of a response and an excitation was defined as the system function based on the characteristic of which a system's response to a sinusoidal signal is still a sinusoidal signal with the same frequency. Thus, a bridge between the excitation and the response is built by this function. Then, under the action of a nonperiodic signal, can the system's excitation and the response also be linked with the system function or not? Three aspects will be discussed.

5.6.1 Analysis method from system models

Suppose that the excitation and the zero-state response of a nth-order LTI system are, respectively, $f(t)$ and $y_f(t)$, the mathematical model of this system is

$$a_n \frac{\mathrm{d}^n y_f(t)}{\mathrm{d}t^n} + a_{n-1} \frac{\mathrm{d}^{n-1} y_f(t)}{\mathrm{d}t^{n-1}} + \cdots + a_1 \frac{\mathrm{d}y_f(t)}{\mathrm{d}t} + a_0 y_f(t)$$
$$= b_m \frac{\mathrm{d}^m f(t)}{\mathrm{d}t^m} + \cdots + b_1 \frac{\mathrm{d}f(t)}{\mathrm{d}t} + b_0 f(t). \quad (5.6\text{-}1)$$

Taking the Fourier transform on both sides of equation (5.6-1) and making $Y_f(j\omega) = \mathscr{F}[y_f(t)]$, $F(j\omega) = \mathscr{F}[f(t)]$, from the linearity and differential properties of the Fourier transform, we obtain

$$\left[a_n (j\omega)^n + a_{n-1} (j\omega)^{n-1} + \cdots + a_1 (j\omega) + a_0 \right] Y_f(j\omega)$$
$$= \left[b_m (j\omega)^m + b_{m-1} (j\omega)^{m-1} + \cdots + b_1 (j\omega) + b_0 \right] F(j\omega). \quad (5.6\text{-}2)$$

Thus, the Fourier transform of the zero-state response is

$$Y_f(j\omega) = \frac{b_m (j\omega)^m + b_{m-1} (j\omega)^{m-1} + \cdots + b_1 (j\omega) + b_0}{a_n (j\omega)^n + a_{n-1} (j\omega)^{n-1} + \cdots + a_1 (j\omega) + a_0} F(j\omega). \quad (5.6\text{-}3)$$

Like in Chapter 4, we define the ratio of the Fourier transforms of the zero-state response and the excitation as the system function, which is still expressed as

$$H(j\omega) \stackrel{\text{def}}{=} \frac{Y_f(j\omega)}{F(j\omega)} . \qquad (5.6\text{-}4)$$

Thus, the Fourier transform of the zero-state response of a system can be written as

$$Y_f(j\omega) = H(j\omega)F(j\omega) , \qquad (5.6\text{-}5)$$

where

$$H(j\omega) = \frac{b_m(j\omega)^m + b_{m-1}(j\omega)^{m-1} + \cdots + b_1(j\omega) + b_0}{a_n(j\omega)^n + a_{n-1}(j\omega)^{n-1} + \cdots + a_1(j\omega) + a_0} . \qquad (5.6\text{-}6)$$

Equation (5.6-6) states that the system function only depends on the structure and the element parameters of a system but does not relate to the excitation or the response signals. Obviously, we obtained a similar result with the case of applying a periodic signal to a system, which is not surprise to us, because a nonperiodic signal can be expressed as a continuous sum of sinusoidal signals by means of the Fourier transform, the excitation and response of a system are still sinusoidal signals.

Equation (4.4-1) is different from equation (5.6-4) on the surface, because the system function is defined by means of the Fourier transform (spectrum) when a system is acted upon by a nonperiodic signal, but it is defined by using the phasor under the action of a periodic signal. In fact, their essences are the same, because a phasor is actually another kind of simplified manifestation of the spectrum of a sinusoidal signal with a single frequency.

Equation (5.6-5) is just the contribution of the Fourier transform to the system analysis. That is, the Fourier transform of the zero-state response of an LTI system to an aperiodic excitation equals the product of the Fourier transform of the input signal and the system function.

5.6.2 Analysis with the system function

We know that if the excitation is $\delta(t)$, the corresponding zero-state response is the impulse response $h(t)$. We have

$$F(j\omega) = \mathcal{F}[\delta(t)] = 1 ,$$
$$Y_f(j\omega) = \mathcal{F}[h(t)] .$$

According to equation (5.6-5), we have

$$Y_f(j\omega) = F(j\omega)H(j\omega) = H(j\omega) ,$$

and we obtain

$$\mathcal{F}[h(t)] = H(j\omega) . \qquad (5.6\text{-}7)$$

Equation (5.6-7) reflects an important relationship between the system impulse response and the system function, that is, they are a Fourier transform pair. It can be expressed as

$$h(t) \xleftrightarrow{\ \mathcal{F}\ } H(j\omega)\,,$$

$$H(j\omega) = \mathcal{F}[h(t)] = \int_{-\infty}^{\infty} h(t)e^{-j\omega t}dt\,, \qquad (5.6\text{-}8)$$

$$h(t) = \mathcal{F}^{-1}[H(j\omega)] = \frac{1}{2\pi}\int_{-\infty}^{\infty} H(j\omega)e^{j\omega t}d\omega\,. \qquad (5.6\text{-}9)$$

The significance of the conclusion is as follows:

If the structure of a system is not intuitive or invisible, the system function cannot be obtained directly from constraint conditions of components and the system (circuit), but it can be found indirectly by the impulse response of the system.

From time domain analysis we know that the zero-state response $y_f(t)$ of an LTI system to an excitation $f(t)$ is the convolution of the impulse response $h(t)$ and $f(t)$,

$$y_f(t) = f(t) * h(t)\,. \qquad (5.6\text{-}10)$$

Letting $Y_f(j\omega) = \mathcal{F}[y_f(t)]$, $F(j\omega) = \mathcal{F}[f(t)]$, taking the Fourier transform on both sides of equation (5.6-10) at same time and using the convolution theorem, we have

$$\mathcal{F}[y_f(t)] = \mathcal{F}[f(t) * h(t)] = \mathcal{F}[f(t)] \cdot \mathcal{F}[h(t)]$$

or

$$Y_f(j\omega) = F(j\omega)H(j\omega)\,. \qquad (5.6\text{-}11)$$

This conclusion proves the correctness of equation (5.6-5) from another point of view.

5.6.3 Analysis with signal decomposition

The zero-state response of a system to a nonperiodic signal is deduced based on the decomposition and synthesis of a signal as follows.

From equation (5.1-7), an aperiodic signal $f(t)$ can be expressed as a linear combination of infinite imaginary exponential signals as $e^{j\omega t}$. Therefore, $e^{j\omega t}$ is a kind of basic signal, and first of all we must find the zero-state response $y_{f1}(t)$ of the system to signal $e^{j\omega t}$.

Because an excitation $f_1(t) = e^{j\omega t}$, from equation (5.6-10) we have

$$y_{f1}(t) = h(t) * e^{j\omega t} = \int_{-\infty}^{\infty} h(\tau)e^{j\omega(t-\tau)}d\tau = e^{j\omega t}\int_{-\infty}^{\infty} h(\tau)e^{-j\omega \tau}d\tau\,.$$

The term $\int_{-\infty}^{\infty} h(\tau)e^{-j\omega \tau}d\tau = \int_{-\infty}^{\infty} h(t)e^{-j\omega t}dt$ is the Fourier transform $H(j\omega)$ of $h(t)$, so

$$y_{f1}(t) = H(j\omega)e^{j\omega t}\,. \qquad (5.6\text{-}12)$$

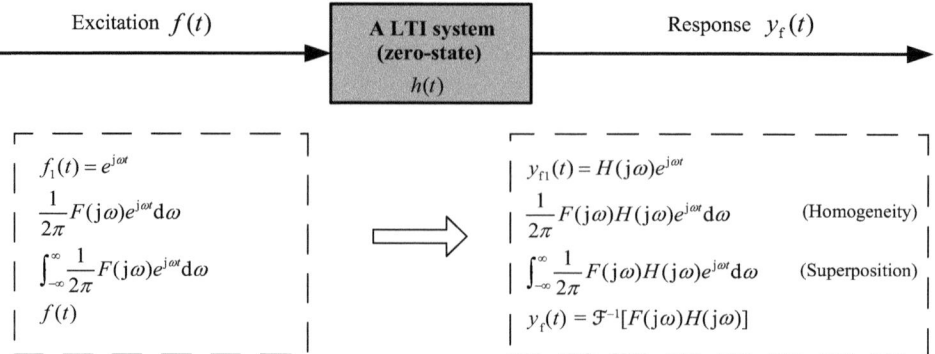

Fig. 5.26: Derivation process of the zero-state response of a system to an aperiodic signal.

Equation (5.6-12) shows that the zero-state response of a system to a basic signal – imaginary exponential $e^{j\omega t}$ is the product of the signal itself and a constant coefficient that has nothing to do with the time t, and this coefficient is just the Fourier transform of the impulse response $h(t)$ of the system – system function $H(j\omega)$. Accordingly, we have that the Fourier transform of the zero-state response of a system to a nonperiodic signal is equal to the product of the Fourier transform of this signal and the system function, i.e., $Y_f(j\omega) = H(j\omega)F(j\omega)$; the proof process is plotted in ▶ Figure 5.26. In fact, with a slight change, this process can also be used for system analysis in the s and z domains.

To sum up, the same conclusion is obtained from the three methods:

The zero-state response of an LTI system to an arbitrary aperiodic signal can be obtained by the inverse Fourier transform of the product of the system function and the Fourier transform of the excitation signal. Note that the system function is the Fourier transform of the impulse response.

Steps to find the zero-state response of a system to an aperiodic signal by using the Fourier transform are as follows:

Step 1: Find the Fourier transform $F(j\omega)$ of the excitation $f(t)$.

Step 2: According to the definition (or differential equation), circuit knowledge, transfer operator or impulse response, find the system function $H(j\omega)$.

Step 3: Find the product of $F(j\omega)$ and $H(j\omega)$, and obtain the Fourier transform $Y_f(j\omega)$ of the zero-state response.

Step 4: Find the inverse Fourier transform of $Y_f(j\omega)$ and obtain the zero-state response $y_f(t)$ of the system.

The system analysis methods for aperiodic signals are detailed by the following examples.

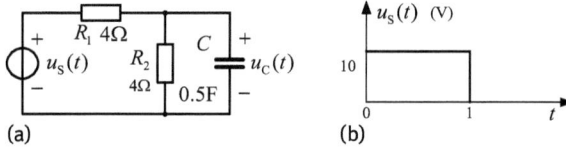

Fig. 5.27: E5.6-1.

Example 5.6-1. A circuit is shown in ▶ Figure 5.27a and a voltage source $u_S(t)$ that is a rectangle pulse is shown in ▶ Figure 5.27b. Find the zero-state response $u_C(t)$ of this circuit.

Solution. The excitation $u_S(t)$ can be regarded as the sum of two step signals as follows:

$$u_S(t) = 10\varepsilon(t) - 10\varepsilon(t-1) .$$

If

$$u_{S1}(t) = 10\varepsilon(t) ,$$

$$u_{S2}(t) = -10\varepsilon(t-1) ,$$

then

$$u_S(t) = u_{S1}(t) + u_{S2}(t) .$$

From the superposition principle, the responses $u_{C1}(t)$ and $u_{C2}(t)$ can be calculated corresponding to $u_{S1}(t)$ and $u_{S2}(t)$, respectively. The total response of the system is $u_C(t) = u_{C1}(t) + u_{C2}(t)$.

If the Fourier transform of $u_{S1}(t)$ is $U_{S1}(j\omega)$, then

$$U_{S1}(j\omega) = \mathcal{F}[u_{S1}(t)] = 10\left[\pi\delta(\omega) + \frac{1}{j\omega}\right] .$$

From the circuit diagram, using the constraint conditions of components and circuit, the system function is

$$H(j\omega) = \frac{\mathcal{F}[u_C(t)]}{\mathcal{F}[u_S(t)]} = \frac{U_C(j\omega)}{U_S(j\omega)} = \frac{\frac{R_2}{1+j\omega R_2 C}}{R_1 + \frac{R_2}{1+j\omega R_2 C}} = \frac{R_2}{R_1 + R_2} \cdot \frac{1}{1 + j\omega \frac{R_1 R_2}{R_1+R_2}C} .$$

Substituting the element parameters into the above expression, we have

$$H(j\omega) = \frac{1}{2} \cdot \frac{1}{1 + j\omega} .$$

According to equation (5.6-5) we have

$$U_{C1}(j\omega) = U_{S1}(j\omega) \cdot H(j\omega) = \frac{1}{2} \cdot \frac{1}{1 + j\omega} \times 10\left[\pi\delta(\omega) + \frac{1}{j\omega}\right]$$

$$= \frac{5\pi}{1 + j\omega}\delta(\omega) + \frac{5}{j\omega(1 + j\omega)} = 5\pi\delta(\omega) + \frac{5}{j\omega} - \frac{5}{1 + j\omega}$$

Because

$$\mathcal{F}^{-1}[5\pi\delta(\omega)] = \frac{5}{2},$$

$$\mathcal{F}^{-1}\left[\frac{5}{j\omega}\right] = \frac{5}{2}\,\mathrm{sgn}(t),$$

$$\mathcal{F}^{-1}\left[\frac{5}{1+j\omega}\right] = 5e^{-t}\varepsilon(t),$$

we have

$$u_{C1}(t) = \frac{5}{2} + \frac{5}{2}\,\mathrm{sgn}(t) - 5e^{-t}\varepsilon(t) = 5\varepsilon(t) - 5e^{-t}\varepsilon(t) = 5\left(1 - e^{-t}\right)\varepsilon(t).$$

Because

$$u_{S2}(t) = -u_{S1}(t-1),$$

and, according to the time shifting, we have

$$u_{C2}(t) = -u_{C1}(t-1) = -5\left[1 - e^{-(t-1)}\right]\varepsilon(t-1).$$

So, the complete zero-state response is

$$u_C(t) = u_{C1}(t) + u_{C2}(t) = 5\left(1 - e^{-t}\right)\varepsilon(t) - 5\left[1 - e^{-(t-1)}\right]\varepsilon(t-1).$$

Example 5.6-2. In the system shown in ▶ Figure 5.28, the excitation $f(t)$ is known, and the impulse response is $h(t) = \frac{1}{\pi t}$. Find the zero-state response $y_f(t)$ of the system.

Solution. Since the system structure is invisible, the system function must be obtained by the impulse response but cannot be obtained from the constraint conditions. Considering $F(j\omega)$ as the spectrum of $f(t)$, the system function is

$$H(j\omega) = \mathcal{F}[h(t)] = \mathcal{F}\left[\frac{1}{\pi}\cdot\frac{1}{t}\right] = \frac{1}{\pi}[-j\pi\,\mathrm{sgn}(\omega)] = -j\,\mathrm{sgn}(\omega),$$

and then

$$\begin{aligned}
Y_f(j\omega) = F(j\omega)H(j\omega)H(j\omega) &= F(j\omega)\cdot[-j\,\mathrm{sgn}(\omega)]\cdot[-j\,\mathrm{sgn}(\omega)] \\
&= F(j\omega)[-\mathrm{sgn}(\omega)\,\mathrm{sgn}(\omega)] \\
&= -F(j\omega).
\end{aligned}$$

So, the zero-state response is

$$y_f(t) = -f(t).$$

This system is an inverter.

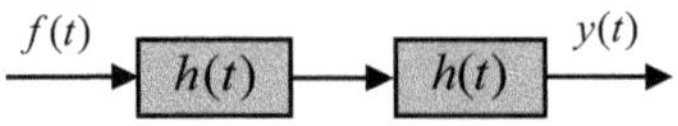

Fig. 5.28: E5.6-2.

From the above examples, it can be seen that in the system analysis in the frequency domain, changing the convolution in the time domain into the algebraic multiplication in the frequency domain can not only greatly simplify the process to find the zero-state response, but also give the lively physical concept of a signal – the spectrum. Especially the frequency response characteristics of a system, which can be illustrated by the waveform of the system function, play an important role in communication technology. In contrast to the time domain method, the main disadvantage of the frequency domain method is the need to apply two transforms, namely, the Fourier transform and the inverse Fourier transform.

5.7 System analysis methods for periodic signals

From Section 5.4 we know that the Fourier transform of a periodic signal also exists after the impulse function has been introduced. Then, can the zero-state response of a system to a periodic signal be obtained by the Fourier transform method? The answer is "yes".

Assume that the excitation is

$$f(t) = \sin \omega_0 t \,,$$

then

$$\mathscr{F}[f(t)] = F(j\omega) = j\pi \left[\delta \left(\omega + \omega_0\right) - \delta \left(\omega - \omega_0\right)\right] \,.$$

If the system function is of the form

$$H(j\omega) = |H(j\omega)| \, e^{j\varphi(\omega)} \,,$$

in places $\pm\omega_0$, we have

$$H\left(j\omega_0\right) = |H\left(j\omega_0\right)| \, e^{j\varphi(\omega_0)} \quad \text{and} \quad H\left(-j\omega_0\right) = |H\left(j\omega_0\right)| \, e^{-j\varphi(\omega_0)} \,.$$

Therefore, the spectrum of the zero-state response is

$$\begin{aligned} Y_f(j\omega) &= F(j\omega)H(j\omega) \\ &= j\pi H(j\omega) \cdot \left[\delta \left(\omega + \omega_0\right) - \delta \left(\omega - \omega_0\right)\right] \\ &= j\pi \left[H\left(-j\omega_0\right) \delta \left(\omega + \omega_0\right) - H\left(j\omega_0\right) \delta \left(\omega - \omega_0\right)\right] \\ &= j\pi \, |H\left(j\omega_0\right)| \left[e^{-j\varphi(\omega_0)} \delta \left(\omega + \omega_0\right) - e^{j\varphi(\omega_0)} \delta \left(\omega - \omega_0\right)\right] \,. \end{aligned}$$

Then the zero-state response of the system is

$$y_f(t) = \mathscr{F}^{-1}\left[Y_f(j\omega)\right] = |H\left(j\omega_0\right)| \sin \left[\omega_0 t + \varphi \left(\omega_0\right)\right] \,. \tag{5.7-1}$$

If the excitation is $f(t) = A \sin \left(\omega_0 t + \varphi\right)$, the zero-state response $y_f(t)$ can be directly written as

$$y_f(t) = A \, |H\left(j\omega_0\right)| \sin \left[\omega_0 t + \varphi \left(\omega_0\right) + \varphi\right] \,. \tag{5.7-2}$$

It can be explained by equation (5.7-2) that the zero-state response of a system to a sine signal is still the sine wave with the same frequency of the excitation, but its amplitude and phase are determined by the excitation and the system function together.

Similarly, the zero-state response of a system to $f(t) = A \cos(\omega_0 t + \varphi)$ is

$$y_f(t) = A \, |H(j\omega_0)| \cos\left[\omega_0 t + \varphi(\omega_0) + \varphi\right] . \tag{5.7-3}$$

It is not unexpected that equations (5.7-2) and (5.7-3) are of the same form, this is just the concrete embodiment of the time invariant and linearity properties of system, because $f(t) = A \sin(\omega_0 t + \varphi)$ results from the delay of $f(t) = A \cos(\omega_0 t + \varphi)$ by $1/4$ period.

Conclusion: For a common periodic signal, it must be expanded as the trigonometric form of the Fourier series first, and then the subzero-state responses corresponding to all its harmonics are calculated by equations (5.7-2) or (5.7-3), respectively, after which they are superimposed into the total zero-state response.

Note: If there is no subsequent explanation, the response of a system to a periodic signal refers to the zero-state response of the system.

Example 5.7-1. Find the response $y(t)$ of a system to an excitation $f(t) = 2 + \cos t + 5 \cos(3t + 20.6°)$. The system function $H(j\omega) = \frac{1}{1+j\omega}$ is known.

Solution. Since $H(j0) = 1$, $H(j1) = \frac{1}{1+j} = \frac{1}{\sqrt{2}} e^{-j45°}$, $H(j3) = \frac{1}{1+j3} = \frac{1}{\sqrt{10}} e^{-j71.6°}$, equation (5.7-3) is utilized three times, and it can be obtained by the superposition

$$y(t) = 2 + \frac{1}{\sqrt{2}} \cos(t - 45°) + \frac{5}{\sqrt{10}} \cos(3t - 51°) .$$

After comparing equations (5.7-3) and (4.4-5), $y_n(t) = c_n |H(jn\omega_0)| \cos(n\omega_0 t + \varphi_{Hn})$, it can be found that when we analyze a system response to a periodic signal, whether with the Fourier series or the Fourier transform, the solution steps are basically the same, where both need the decomposition operation for an excitation and the addition operation for sub responses.

5.8 The Hilbert transform

Is there a constraint relationship between the real and imaginary parts of the system function for a causal system? The Hilbert transform can answer this question.

We know that the impulse response $h(t)$ of a causal system satisfies the equation

$$h(t) = 0, t < 0 .$$

That is,

$$h(t) = h(t)\varepsilon(t) . \tag{5.8-1}$$

Assume that the system function $H(j\omega)$, which is also the Fourier transform of $h(t)$, can be decomposed into the real part $R(\omega)$ and the imaginary part $I(\omega)$, that is,

$$H(j\omega) = \mathcal{F}[h(t)] = R(\omega) + jI(\omega) .$$

Using the convolution theorem of the Fourier transform in the frequency domain in equation (5.8-1), yields

$$\mathcal{F}[h(t)] = \frac{1}{2\pi}\{\mathcal{F}[h(t)] * \mathcal{F}[\varepsilon(t)]\} .$$

So, we have

$$
\begin{aligned}
R(\omega) + jI(\omega) &= \frac{1}{2\pi}\left\{[R(\omega) + jI(\omega)] * \left[\pi\delta(\omega) + \frac{1}{j\omega}\right]\right\} \\
&= \frac{1}{2\pi}\left\{R(\omega) * \pi\delta(\omega) + I(\omega) * \frac{1}{\omega}\right\} + \frac{j}{2\pi}\left\{I(\omega) * \pi\delta(\omega) - R(\omega) * \frac{1}{\omega}\right\} \\
&= \left\{\frac{R(\omega)}{2} + \frac{1}{2\pi}\int_{-\infty}^{\infty}\frac{I(\lambda)}{\omega - \lambda}d\lambda\right\} + j\left\{\frac{I(\omega)}{2} - \frac{1}{2\pi}\int_{-\infty}^{\infty}\frac{R(\lambda)}{\omega - \lambda}d\lambda\right\}
\end{aligned}
$$

Comparing the two sides of the above equation, we have

$$R(\omega) = \frac{1}{\pi}\int_{-\infty}^{\infty}\frac{I(\lambda)}{\omega - \lambda}d\lambda , \tag{5.8-2}$$

$$I(\omega) = -\frac{1}{\pi}\int_{-\infty}^{\infty}\frac{R(\lambda)}{\omega - \lambda}d\lambda . \tag{5.8-3}$$

Equations (5.8-2) and (5.8-3) are known as the Hilbert transform pair. They give an interdependence relationship between the real part $R(\omega)$ and the imaginary part $I(\omega)$ of a system function $H(j\omega)$ with the causality property; that is, the real part $R(\omega)$ is only determined by the imaginary part $I(\omega)$, and the imaginary part $I(\omega)$ is also only determined by the real part $R(\omega)$.

To generalize the general situation, we define the convolution of a real signal $f(t)$ and a signal $\frac{1}{\pi t}$ as the Hilbert transform of $f(t)$, which is denoted as

$$\tilde{f}(t) \overset{\text{def}}{=} f(t) * \frac{1}{\pi t} = \frac{1}{\pi}\int_{-\infty}^{\infty}\frac{f(\tau)}{t - \tau}d\tau . \tag{5.8-4}$$

Its inverse transform is

$$f(t) \overset{\text{def}}{=} \tilde{f}(t) * \frac{-1}{\pi t} = \frac{-1}{\pi}\int_{-\infty}^{\infty}\frac{\tilde{f}(\tau)}{t - \tau}d\tau . \tag{5.8-5}$$

Considering $\mathcal{H}[\cdot]$ as the Hilbert transform character, we have

$$\tilde{f}(t) = \mathcal{H}[f(t)] ,$$
$$f(t) = \mathcal{H}^{-1}[\tilde{f}(t)] .$$

We can see an important feature of the Hilbert transform, which is that the transform pair exists in the same variable domain. Comparing it with the Fourier transform, the

Hilbert transform reflects the relationship between one function and another one in the same variable domain, whereas the Fourier transform reveals the relationship between two expressing forms of a function in the time domain and the frequency domain.

The Hilbert transform also has the following characteristics:

(1)
$$\mathcal{H}[\cos(\omega_0 t + \varphi)] = \sin(\omega_0 t + \varphi).$$ (5.8-6)

(2)
$$\mathcal{H}[\sin(\omega_0 t + \varphi)] = -\cos(\omega_0 t + \varphi).$$ (5.8-7)

(3) If the frequency band of $f(t)$ is limited by $|\omega| \le \omega_0$, then we have

$$\mathcal{H}[f(t) \cos \omega_0 t] = f(t) \sin \omega_0 t.$$ (5.8-8)
$$\mathcal{H}[f(t) \sin \omega_0 t] = -f(t) \cos \omega_0 t.$$ (5.8-9)

(4) If $F(j\omega)$ is the Fourier transform of $f(t)$, then the Fourier transform $\tilde{F}(j\omega)$ of $\tilde{f}(t)$ should be

$$\tilde{F}(j\omega) = \mathcal{F}[\tilde{f}(t)] = -jF(j\omega)\,\mathrm{sgn}\,\omega.$$ (5.8-10)

Equation (5.8-10) has an obvious physical meaning, that is, $\tilde{f}(t)$ can be obtained by a signal $f(t)$ passing a filter with the transfer function $-j\,\mathrm{sgn}\,\omega$. This filter is called the Hilbert filter or the 90° phase shifter and is an all pass system for all frequency components without magnitude damping.

(5) The original signal $f(t)$ can be restored because the Hilbert transform $\tilde{f}(t)$ of a real signal $f(t)$ is inverse transformed again. This is equivalent to the effect of $f(t)$ passing through a lag system with 90° first and then passing an advanced system by 90° again. The effect of twice Hilbert transforming to $f(t)$ is equivalent to the process of $f(t)$ passing through two 90° lag systems, and $f(t)$ is changed into the inverse phase signal $-f(t)$.

(6) The Hilbert transform can only change the phase spectrum rather than the amplitude spectrum of $f(t)$, which means that $f(t)$ and $\tilde{f}(t)$ have the same amplitude spectrum, energy spectrum or power spectrum and the same energy or power.

(7) $f(t)$ and $\tilde{f}(t)$ are orthogonal to each other, that is, $\int_{-\infty}^{\infty} f(t)\tilde{f}(t)dt = 0$.

(8) If $f(t)$ is even (odd), then $\tilde{f}(t)$ is odd (even).

Theoretically, the Hilbert transform is a kind of mathematical tool to obtain the corresponding orthogonal signal of a signal. It is also used to express a single side band signal or to transfer a bandpass signal into a low pass signal in the communication principles course.

5.9 Advantages and disadvantages of Fourier transform analysis

Usually, the system analysis methods based on the Fourier series and the Fourier transform are collectively referred to as the Fourier analysis method, which has obvious advantages as outlined below.

(1) The explicit physical meaning. The method is based on the frequency domain, so, it can directly give the frequency properties of signals and systems.

(2) The wide range of applications. Technologies such as the resonance circuit, filter, modulator and demodulator, sampling theorem, frequency conversion circuit, frequency division multiplexing way, equalization circuit, spectrum analysis way, etc., are all derived from it.

However, just as a coin has two sides, from the previous chapters we can see that it has two obvious shortcomings in the system analysis process for either nonperiodic or periodic signals:

(1) It cannot apply to all signals. Some signals are not absolutely integrable according to Dirichlet conditions, such as an exponential signal $e^{\alpha t}\varepsilon(t)$ ($\alpha > 0$), etc., and since their Fourier transforms do not exist, the Fourier transform method cannot be used for them. There are also some signals, such as the step signal, the DC signal, the sgn signal and so on, whose Fourier transforms cannot be directly obtained by the definition because they are also not absolutely integrable.

(2) It is only suitable for finding the zero-state response of system. Because the Fourier transform does not involve the boundary conditions of a signal or a system, the zero-input response of system cannot be given by this method.

So, it is natural to pose a question: Can we find a better method to overcome the shortcomings of the Fourier analysis method for system analysis? This leads us into the next chapter.

5.10 Solved questions

Question 5-1. Knowing the Fourier transform $F_1(j\omega)$ of $f_1(t)$, find the Fourier transform $F_2(j\omega)$ of $\int_{\infty}^{t} f_1[2(\tau - 1)]\,d\tau$.

Solution. According to the characteristics of Fourier transform, we have

$$f_1(t) \to F_1(j\omega), \quad f_1(2t) \to \frac{1}{2}F_1\left(\frac{j\omega}{2}\right), \quad f_1[2(t-1)] \to \frac{1}{2}F_1\left(\frac{j\omega}{2}\right)e^{-j\omega t},$$

$$\int_{\infty}^{t} f_1[2(\tau - 1)]\,d\tau \to \frac{\pi}{2}F_1(0)\delta(j\omega) + \frac{F_1(\omega/2)e^{-j\omega t}}{2j\omega}.$$

Question 5-2. Knowing $x(t) = \begin{cases} e^{-t}, & 0 \le t \le 1 \\ 0, & \text{other} \end{cases}$, find its Fourier transform using the characteristics of the Fourier transform.

Solution. According to the differential property, we have

$$x(t) = e^{-t}\left[\varepsilon(t) - \varepsilon(t-1)\right] ,$$

$$x'(t) = -e^{-t}\varepsilon(t) + e^{-t}\varepsilon(t-1) + \delta(t) - e^{-1}\delta(t-1) ,$$

$$j\omega X(j\omega) = \frac{-1}{1+j\omega} + \frac{e^{-1}e^{j\omega}}{1+j\omega} + 1 - e^{-1}e^{j\omega} .$$

So, the Fourier transform of $x(t)$ is

$$X(j\omega) = \frac{-1 + e^{-(1-j\omega)}}{j\omega(1+j\omega)} + \frac{1 - e^{-(1+j\omega)}}{j\omega} = \frac{1 - e^{-(1+j\omega)}}{1+j\omega} .$$

Question 5-3. Proof $\int_{-\infty}^{\infty} Sa^2(t)dt = \pi$ with the properties of the Fourier transform.

Solution. As we know, the Fourier transform pair is

$$Sa(t) \xleftrightarrow{\;\mathcal{F}\;} \pi G_2(\omega) .$$

Because

$$\int_{-\infty}^{\infty} f^2(t)dt = \frac{1}{2\pi} \int_{-\infty}^{\infty} |F(\omega)|^2 \, d\omega ,$$

and so,

$$\int_{-\infty}^{\infty} Sa^2(t)dt = \frac{1}{2\pi} \int_{-\infty}^{\infty} |\pi G_2(\omega)|^2 \, d\omega = \frac{1}{2\pi} \int_{-1}^{1} \pi^2 d\omega = \pi .$$

Question 5-4. A signal $f(t) = \varepsilon(t+1) - \varepsilon(t-3)$ is given, its Fourier transform is $F(j\omega)$. Find the integral $\int_{-\infty}^{\infty} 2F(j\omega)Sa(\omega)e^{j2\omega} d\omega$.

Solution. Letting $Sa(\omega) = \mathcal{F}[g_2(t)]$, according to convolution characteristics, yields

$$F(j\omega)Sa(\omega) = \mathcal{F}[f(t) * g_2(t)] .$$

A gate signal can be obtained by

$$g_2(t) = \mathcal{F}^{-1}[Sa(\omega)] = \frac{1}{2}[\varepsilon(t+1) - \varepsilon(t-1)] ,$$

and so,

$$\frac{1}{2\pi} \int_{-\infty}^{\infty} F(j\omega)Sa(\omega)e^{j\omega t} d\omega = f(t) * g_2(t) .$$

Obviously, the shape of $f(t) * g_2(t)$ is trapezoidal, and $[f(t) * g_2(t)]|_{t=2} = 1$ by using convolution, which is illustrated in ▶ Figure Q5-4. So,

$$\int_{-\infty}^{\infty} 2F(j\omega)Sa(\omega)e^{j2\omega} d\omega = 4\pi [f(t) * g_2(t)]|_{t=2} = 4\pi .$$

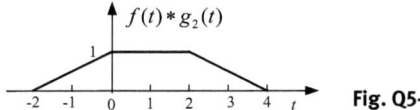

Fig. Q5-4

Question 5-5. Calculate the convolution $\frac{\sin(2\pi t)}{2\pi t} * \frac{\sin(8\pi t)}{8\pi t}$.

Solution. We have the Fourier transform pair

$$g_\tau(t) \overset{\mathcal{F}}{\longleftrightarrow} \tau Sa\left(\frac{\tau \omega}{2}\right).$$

From the symmetry of the Fourier transform we have

$$\tau Sa\left(\frac{\tau t}{2}\right) \overset{\mathcal{F}}{\longleftrightarrow} 2\pi g_\tau(\omega).$$

Thus, we obtain

$$\frac{\sin(2\pi t)}{2\pi t} \overset{\mathcal{F}}{\longleftrightarrow} \frac{1}{2}g_{4\pi}(\omega) \quad \text{and} \quad \frac{\sin(8\pi t)}{8\pi t} \overset{\mathcal{F}}{\longleftrightarrow} \frac{1}{8}g_{16\pi}(\omega).$$

According to the property of which convolution in the time domain means multiplication in the frequency domain, the expression of convolution in the frequency domain can be calculated by

$$\frac{\sin(2\pi t)}{2\pi t} * \frac{\sin(8\pi t)}{8\pi t} \overset{\mathcal{F}}{\longleftrightarrow} \frac{1}{16}g_{4\pi}(\omega)g_{16\pi}(\omega) = \frac{1}{16}g_{4\pi}(\omega).$$

Taking the inverse Fourier transform, the convolution can be found by

$$\frac{\sin(2\pi t)}{2\pi t} * \frac{\sin(8\pi t)}{8\pi t} = \frac{\sin(2\pi t)}{16}.$$

Question 5-6. The frequency response of a stable LTI system is $H(\omega) = \frac{1-e^{-(j\omega+1)}}{j\omega+1}$. Calculate its unit step response $g(t)$.

Solution. First, we obtain the impulse response $h(t)$ of the system by calculating the inverse Fourier transform of $H(\omega)$. Because

$$H(\omega) = \frac{1-e^{-(j\omega+1)}}{j\omega+1} = \frac{1}{j\omega+1} - \frac{-e^{-1}}{j\omega+1}e^{-j\omega},$$

calculating the inverse Fourier transform of $H(\omega)$ yields

$$h(t) = e^{-t}\varepsilon(t) - e^{-t}e^{-(t-1)}\varepsilon(t-1).$$

The unit step response $g(t)$ can be obtained by integrating to $h(t)$, s

$$g(t) = \int_\infty^t h(\tau)d\tau = h(t) * \varepsilon(t) = e^{-t}\varepsilon(t) * \varepsilon(t) - e^{-1}e^{-(t-1)}\varepsilon(t-1) * \varepsilon(t)$$

$$= \left(1 - e^{-t}\right)\varepsilon(t) - e^{-1}\left[1 - e^{-(t-1)}\right]\varepsilon(t-1).$$

Question 5-7. The excitation is $f(t) = e^{-\alpha t}\varepsilon(t)$ and the unit impulse response of the system is $h(t) = e^{-\beta t}\varepsilon(t)$. Calculate the zero-state response $y_f(t)$ of the system.

Solution. According to the definition of the zero-state response,

$$y_f(t) = f(t) * h(t) = e^{-\alpha t}\varepsilon(t) * e^{-\beta t}\varepsilon(t) .$$

Calculating and analyzing the expression of $y_f(t)$ above, we have
(1) When $\alpha = \beta$, the zero-state response is $y_f(t) = te^{-\alpha t}\varepsilon(t) = te^{-\beta t}\varepsilon(t)$.
(2) When $\alpha \neq \beta$, the zero-state response is $y_f(t) = \frac{1}{\beta-\alpha}\left(e^{-\alpha t} - e^{-\beta t}\right)\varepsilon(t)$.

Question 5-8. In the system shown in ▶ Figure Q5-8, $H_1(\omega) = e^{-j2\omega}$ and $h_2(t) = 1 + \cos\frac{\pi t}{2}$. Find the zero-state response $y_f(t)$, when the excitation is $f(t) = \varepsilon(t)$.

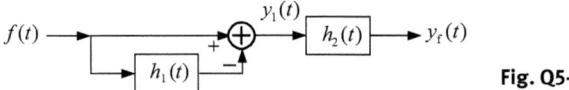

Fig. Q5-8

Solution. If $y_1(t) \overset{\mathcal{F}}{\longleftrightarrow} Y_1(\omega)$, the spectrum of $y_1(t)$ in the form of $\varepsilon(t)$ is

$$Y_1(\omega) = F(\omega) - F(\omega)H_1(\omega) = \left(\pi\delta(\omega) + \frac{1}{j\omega}\right)\left(1 - e^{-j2\omega}\right) ,$$

so,

$$y_1(t) = \varepsilon(t) - \varepsilon(t - 2) .$$

The zero-state response $y_f(t)$ is

$$y_f(t) = y_1(t) * h_2(t) = y_1'(t) * \int_0^t h_2(t)dt = [\delta(t) - \delta(t-2)] * \left[t + \frac{2}{\pi}\sin\frac{\pi}{2}t\right]$$

$$= t + \frac{2}{\pi}\sin\frac{\pi}{2}t - (t-2) - \frac{2}{\pi}\sin\frac{\pi}{2}(t-2) = 2 + \frac{2}{\pi}\sin\frac{\pi}{2}t - \frac{2}{\pi}\sin\left(\frac{\pi}{2}t - \pi\right)$$

$$= 2 + \frac{2}{\pi}\sin\frac{\pi}{2}t - \left[-\frac{2}{\pi}\sin\frac{\pi}{2}t\right] = 2 + \frac{4}{\pi}\sin\frac{\pi}{2}t .$$

5.11 Learning tips

The aperiodic signal is as important as the periodic signal, and readers should pay attention to the following points.
(1) A periodic signal can become aperiodic when its cycle T tends to infinity, and this is the foundation to introduce the Fourier transform.

(2) The gate and the sampling signals are an important Fourier transform pair. The important concept reflected by it is that the frequency range of a time limited signal (e.g., a gate signal in the time domain) is unlimited in the frequency domain (e.g., a sampling signal in the frequency domain), and the duration of a frequency limited signal (e.g., a gate signal in the frequency domain) is unlimited in the time domain (e.g., a sampling signal in the time domain). Simply, time is limited → frequency is unlimited, and frequency is limited → time is unlimited.

(3) The impulse response and the system function compose a Fourier transform pair.

(4) The Fourier transform of the zero-state response to a nonperiodic signal is equal to the product of the Fourier transform of the excitation and the system function.

5.12 Problems

Problem 5-1. Find the frequency spectrum of each signal in ▶ Figure P5-1.

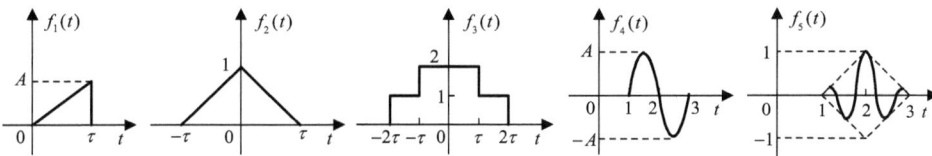

Fig. P5-1

Problem 5-2. Find the frequency spectrums and plot them for the following signals by the symmetry properties of the Fourier transform.

(1) $f_1(t) = \left[\frac{\sin(2\pi t)}{2\pi t}\right]^2$ (2) $f_2(t) = \frac{\sin 50(t-3)}{100(t-3)}$ (3) $f_3(t) = \frac{2}{4+t^2}$

Problem 5-3. Using the calculus properties of the Fourier transform, find the frequency spectrum of the following graphic signals.

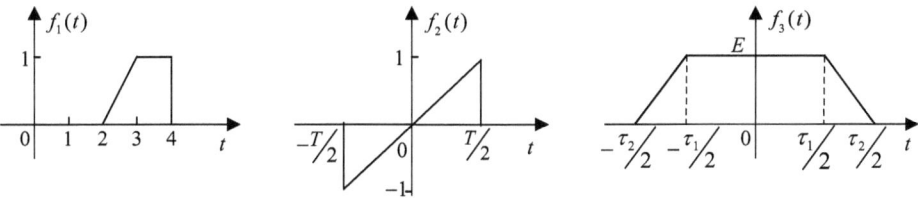

Fig. P5-3

Problem 5-4. The Fourier transform of $f(t)$ is $F(j\omega)$. Find the frequency spectrum of each signal.

(1) $f(2t - 5)$ (4) $(t - 4)f(-2t)$ (7) $[1 - \cos(4t)]f(t - 3)$

(2) $f(3 - 5t)$ (5) $t\frac{df(t)}{dt}$

(3) $tf(2t)$ (6) $f\left(\frac{t}{2} + 3\right)\cos(4t)$

Problem 5-5. Assume that the Fourier transform of $f(t)$ is $F(j\omega)$. Prove

(1) $F(0) = \int_{-\infty}^{\infty} f(t)dt$

(2) $f(0) = \frac{1}{2\pi}\int_{-\infty}^{\infty} F(j\omega)d\omega$

(3) $\int_{-\infty}^{\infty} |f(t)|^2 dt = \frac{1}{2\pi}\int_{-\infty}^{\infty} |F(j\omega)|^2 d\omega$

Calculate the following equations:

(1) $\int_{-\infty}^{\infty} \frac{1}{a^2+\omega^2} d\omega$

(2) $\int_{0}^{\infty} \frac{\sin^4 a\omega}{\omega^4} d\omega$

(3) $\int_{-\infty}^{\infty} Sa(\omega_0 t) dt$

Problem 5-6. Find the original functions $f(t)$ of following frequency spectrums.

(1) $\frac{1}{\omega^2}$;   (3) $e^{a\omega}\varepsilon(-\omega)$;

(2) $\delta(\omega + 100) - \delta(\omega - 100)$;   (4) $\frac{5e^{-j\omega}}{(j\omega-2)(j\omega+3)}$

Problem 5-7. Knowing the amplitude spectrum $|F(j\omega)|$ and the phase spectrum $\varphi(\omega)$ as shown in ▶ Figure P5-7, find the corresponding original function $f(t)$.

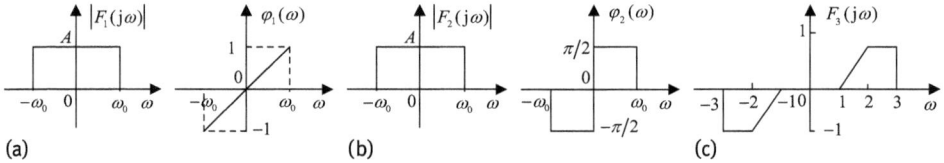

(a)   (b)   (c)

Fig. P5-7

Problem 5-8. Knowing the system function $H(j\omega) = \frac{-\omega^2+j4\omega+5}{-\omega^2+j3\omega+2}$ and the excitation $f(t) = e^{-3t}\varepsilon(t)$, find the zero-state response $y_f(t)$.

Problem 5-9. Knowing the system function $H(j\omega) = \frac{j\omega}{-\omega^2+j5\omega+6}$, the starting states of the system $y(0_-) = 2$, $y'(0_-) = 1$, excitation $f(t) = e^{-t}\varepsilon(t)$, find the full response $y(t)$.

Problem 5-10. Excitation $f(t)$ of an LTI circuit is shown in ▶ Figure P5-10; the impulse response $h(t) = e^{-2t}\varepsilon(t)$ is known. Find the zero-state response $y_f(t)$ of the circuit by the method in the frequency domain.

Problem 5-11. A circuit and a voltage source $u_S(t)$ are shown in ▶ Figure P5-11. Find the corresponding zero-state response $u_C(t)$.

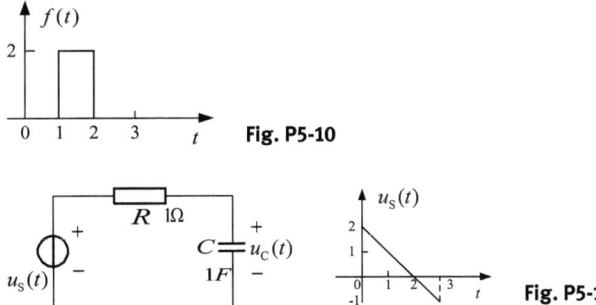

Fig. P5-10

Fig. P5-11

Problem 5-12. A circuit is shown in ▶ Figure P5-12. If $u_S(t)$ are the following signals, find corresponding zero-state responses $i_0(t)$.

(1) $u_S(t) = \delta(t)$ (2) $u_S(t) = \varepsilon(t)$ (3) $u_S(t) = e^{-t}\varepsilon(t)$

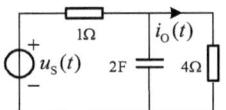

Fig. P5-12

Problem 5-13. A circuit is shown in ▶ Figure P5-13 and $f(t) = 10e^{-t}\varepsilon(t) + 2\varepsilon(t)$. Find the unit impulse response $h(t)$ and the zero-state response $i_f(t)$ to $i(t)$.

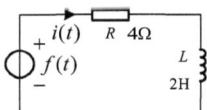

Fig. P5-13

Problem 5-14. A system is shown in ▶ Figure P5-14. Find the system function $H(j\omega)$ and $h(t)$. If $f(t) = te^{-2t}\varepsilon(t)$, find the zero-state response of the system.

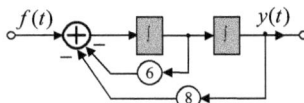

Fig. P5-14

6 Analysis of continuous-time systems in the complex frequency domain

Question: Can we find a method to make up the deficiencies of Fourier analysis?
Solution: Seek a method to make an unbounded signal becomes absolutely integrable → Analyze systems with the aid of the Fourier analysis approach.
Results: Laplace transform, system function, system models in the s domain.

6.1 Concept of the Laplace transform

The Fourier transform is an effective mathematical tool, which can be used to analyze the responses resulting from some nonperiodic and periodic signals of a system in the frequency domain. However, it has the following drawbacks:
(1) It is not applicable to signals whose Fourier transforms do not exist.
(2) It is incapable of giving the complete response of a system.

To compensate for the shortages, another mathematical method called the Laplace transform was introduced.

The Laplace transform is also a kind of integral transform method, which was proposed by the French mathematician Pierre Simon de Laplace (1749–1825) in 1780. It is not only an effective analysis means for LTI systems, but also a widely used method in other technologies.

The Laplace transform can be drawn from the Fourier transform according to the following procedures. For a signal $f(t)$ which is not absolutely integrable, we hope that the product of $f(t)$ and an attenuation factor $e^{-\sigma t}$ can be absolutely integrable. That is, it is possible to find an appropriate value of σ to make $f(t)e^{-\sigma t}$ absolutely integrable, and then obtain the Fourier transform of $f(t)e^{-\sigma t}$.

Assuming $\mathcal{F}[f(t)e^{-\sigma t}] = F_\sigma(j\omega)$, from definition of the Fourier transform, we obtain

$$F_\sigma(j\omega) = \int_{-\infty}^{+\infty} [f(t)e^{-\sigma t}]e^{-j\omega t}dt = \int_{-\infty}^{+\infty} f(t)e^{-(\sigma+j\omega)t}dt \ . \tag{6.1-1}$$

The corresponding inverse Fourier transform is

$$f(t)e^{-\sigma t} = \frac{1}{2\pi} \int_{-\infty}^{+\infty} F_\sigma(j\omega)e^{j\omega t}d\omega \ . \tag{6.1-2}$$

Letting $s = \sigma + j\omega$, $F(s) = F_\sigma(j\omega)$, equation (6.1-1) can be changed into

$$F(s) = \int_{-\infty}^{+\infty} f(t)e^{-st}dt \ . \tag{6.1-3}$$

https://doi.org/10.1515/9783110419535-006

Similarly to the Fourier transform, $F(s)$ is defined as the Laplace transform of $f(t)$, which is also called the image function of $f(t)$. Like the case when variable ω is considered as the *real frequency* in the Fourier transform, the complex variable s in equation (6.1-3) is regarded as the *complex frequency*.

If both sides of equation (6.1-2) are multiplied by $e^{\sigma t}$, and $F_\sigma(j\omega)$ is replaced by $F(s)$,

$$f(t) = \frac{1}{2\pi} \int_{-\infty}^{+\infty} F_\sigma(j\omega) e^{(\sigma+j\omega)t} d\omega \overset{s=\sigma+j\omega}{=} \frac{1}{2\pi j} \int_{\sigma-j\infty}^{\sigma+j\infty} F(s) e^{st} ds . \tag{6.1-4}$$

Equation (6.1-4) is called the inverse Laplace transform of $F(s)$, where $f(t)$ is called the original function of $F(s)$.

Equations (6.1-3) and (6.1-4) are called the bilateral Laplace transform pair. Symbols "$\mathcal{L}$" and "$\mathcal{L}^{-1}$" represent the Laplace transform and the inverse transform operations, respectively, and they can be related by

$$F(s) = \mathcal{L}[f(t)] ,$$
$$f(t) = \mathcal{L}^{-1}[F(s)] ,$$
$$f(t) \overset{\mathcal{L}}{\longleftrightarrow} F(s) .$$

Obviously, the Laplace transform can convert a time function $f(t)$ into a complex function $F(s)$. Because the variable s is called the complex frequency, the Laplace transform is considered to provide a new path for the analysis of the response of the system to a nonperiodic signal in the complex frequency domain, which refers to the conception of the Fourier transform.

Comparing equations (6.1-3) with (5.1-3), we find that $F(s)$ is similar to $F(j\omega)$ in form, and equation (5.1-3) can be changed into equation (6.1-3) only if $j\omega$ is replaced by s, but it does not mean $s=j\omega$. Because $s = \sigma+j\omega$, the above result virtually illustrates that if $\sigma = 0$, which is the real part of the variable s, the Laplace transform should become the Fourier transform. This means that the Fourier transform is a special case of the Laplace transform, whereas the Laplace transform is a generalization of the Fourier transform.

Comparing the functions of these two transforms, we reach the following conclusions:

The relationship between a signal in the time and frequency domains is established by the Fourier transform, whereas the relationship between a signal in the time domain and the complex frequency domain is established by the Laplace transform. The Fourier transform decomposes a signal $f(t)$ into a continuous sum of imaginary exponential signals like $e^{j\omega t}$, whereas the Laplace transform can express a signal $f(t)$ as a continuous sum of complex exponential signals like e^{st}. Clearly, they are different approaches but give equally satisfactory results.

Compared with the Fourier transform, the Laplace transform is only a mathematical tool in theory without definite physical meaning. However, it has the operating

Tab. 6.1: Relationship between bilateral and unilateral Laplace transform and Fourier transform.

Transform	Bilateral Laplace transform	Unilateral Laplace transform ($f(t) = 0, t < 0$)	Fourier transform ($\sigma = 0$)
Variable properties	$s = \sigma + j\omega$, $-\infty < t < +\infty$	$s = \sigma + j\omega$, $0 < t < +\infty$	$s = j\omega$, $-\infty < t < +\infty$

capability that the Fourier transform has not and it can accomplish some tasks that cannot be finished by the Fourier transform.

In practice, signals are all causal (unilateral) signals with a starting moment, so if the starting time is set as the origin of time, we have

$$f(t) = 0 \quad (t < 0).$$

As a result, equation (6.1-3) will be changed into

$$F(s) = \int_{0_-}^{+\infty} f(t)e^{-st}dt \tag{6.1-5}$$

Equation (6.1-5) is called the unilateral Laplace transform or the image function of $f(t)$. Because a signal may have a step at moment $t = 0$, the lower limit of the integral of the unilateral Laplace transform is stipulated as 0_-, equation (6.1-4) is changed into

$$f(t) = \begin{cases} 0 & (t < 0) \\ \frac{1}{2\pi j} \int_{\sigma-j\infty}^{\sigma+j\infty} F(s)e^{st}ds & (t > 0) \end{cases} \tag{6.1-6}$$

Equations (6.1-5) and (6.1-6) are known as the unilateral Laplace transform pair.

Table 6.1 shows the relationships between the bilateral Laplace transform, the unilateral Laplace transform and Fourier transform.

Here, we only discuss the unilateral Laplace transform pair, because it is much closer to the real situation, and in subsequent chapters the term Laplace transform is just the unilateral Laplace transform without explanation. Thus, for $t < 0$, the values of $f(t)$ are irrelevant to the result of the Laplace transform. For example, the three signals shown in ▶ Figure 6.1 are different, but their unilateral Laplace transforms are all $\frac{1}{s+a}$.

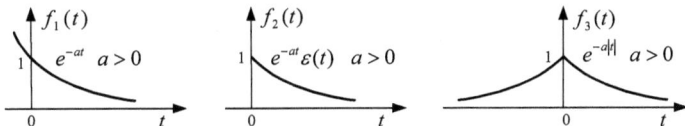

Fig. 6.1: Three signals with same unilateral Laplace transform.

As stated with the introduction of the Laplace transform, it is possible to find an appropriate value of σ to make $f(t)e^{-\sigma t}$ absolutely integrable. Thus, the Fourier transform of $f(t)e^{-\sigma t}$ should exist, and the Laplace transform of $f(t)$ can be obtained. However, how can we determine an appropriate value of σ? If the range of σ is considered as the ROC (region of convergence), how can we determine the ROC? The ROC concept of the Laplace transform is given by the following example. Note that σ is also expressed as $Re(s)$.

Example 6.1-1. Find the image function $F(s)$ of an exponential $f(t) = e^{at}\varepsilon(t)$, $(a > 0)$.

Solution. According to the definition,

$$F(s) = \int_{0_-}^{\infty} e^{at}e^{-st}dt = \int_{0_-}^{\infty} e^{-(s-a)t}dt = \frac{e^{-(s-a)t}}{-(s-a)}\bigg|_{0_-}^{\infty} = \frac{1}{(s-a)}\left[1 - \lim_{t\to\infty} e^{-(s-a)t}\right]. \quad (6.1\text{-}7)$$

Because $s = \sigma + j\omega$, the second term in equation (6.1-7) can be written as

$$\lim_{t\to\infty} e^{-(s-a)t} = \lim_{t\to\infty} e^{-(\sigma-a)t}e^{-j\omega t}. \quad (6.1\text{-}8)$$

If we choose $\sigma > a$, the $e^{-(\sigma-a)t}$ will be attenuated with the increase of time t, so,

$$\lim_{t\to\infty} e^{-(s-a)t} = 0.$$

Thereby, equation (6.1-7) will be convergent. The image function of $f(t)$ is

$$F(s) = \frac{1}{s-a}$$

If $\sigma < a$, the $e^{-(\sigma-a)t}$ will increase with time t. When $t \to \infty$, equation (6.1-8) will tend to infinity, thus equation (6.1-7) is not convergent (divergent), and the image function of $f(t)$ does not exist.

From the above discussions it can be seen that whether $f(t)e^{-\sigma t}$ holds the absolutely integrable condition depends on the properties of $f(t)$ and the relative relationship between $f(t)$ and σ. In general, if the value of $\lim_{t\to\infty} f(t)e^{-\sigma t}$ is zero for $\sigma > \sigma_0$, the function $f(t)e^{-\sigma t}$ is convergent for the range of $\sigma > \sigma_0$, and its integral exists, so the Laplace transform of it also exists.

In the complex plane (s plane) with σ as the horizontal axis and $j\omega$ as the vertical axis, and σ_0 is called the convergence coordinate, the vertical line crossing point σ_0 is the boundary of the ROC and is known as the convergence axis. This convergence axis divides the s plane into two regions, that is, $\sigma > \sigma_0$ is the ROC of $F(s)$, but $\sigma < \sigma_0$ is the divergent region, which are shown in ► Figure 6.2a. The Laplace transform of $f(t)$ only exists in its ROC, thus equation (6.1-5) should be written as

$$F(s) = \int_{0_-}^{+\infty} f(t)e^{-st}dt \quad (\sigma > \sigma_0). \quad (6.1\text{-}9)$$

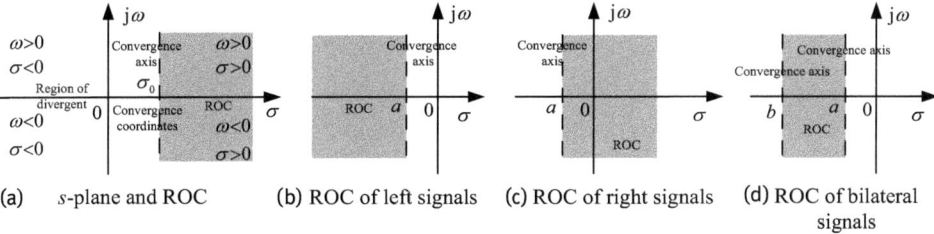

(a) s-plane and ROC (b) ROC of left signals (c) ROC of right signals (d) ROC of bilateral signals

Fig. 6.2: s plane and ROC.

In Example 6.1-1, because $\sigma_0 = a$, the complete answer should be written as

$$F(s) = \frac{1}{s-a} \quad \text{ROC: } \sigma > a \quad \text{or} \quad Re(s) > a .$$

This means that $f(t)e^{-\sigma t}$ is convergent for the range $\sigma > a$, and its Laplace transform exists.

By further analysis, we reach the following points with respect to the ROC:

(1) For a left-sided signal $f(t)$, the ROC lies to the left of the convergent axis $\sigma = a$ shown in ▶ Figure 6.2b.

(2) For a right-sided signal $f(t)$, the ROC lies to the right of the convergent axis $\sigma = a$ shown in ▶ Figure 6.2c.

(3) For a bilateral signal $f(t)$, the ROC is a strip region between two convergent axes $\sigma = a$ and $\sigma = b$ shown in ▶ Figure 6.2d.

(4) The ROC of a right-sided signal must be the right half-plane to the right of the rightmost pole of $H(s)$; the ROC of a left-sided signal must be the left half-plane to the left of the leftmost pole of $H(s)$.

(5) If $f(t)$ is time limited and absolutely integrable, its ROC is the whole s plane.

(6) The ROC does not contain the convergence axis or its boundary, namely, the ROC is an open set.

(7) The ROC does not contain any pole of $F(s)$.

(8) Because the original function and the image function in the unilateral Laplace transform are one to one correspondence, for convenience, the ROC of a unilateral Laplace transform of a signal usually does not need to be labeled or emphasized.

(9) The original function and the image function in the bilateral Laplace transform are not on a one-to-one corresponding relationship, that is, different original functions can have the same image function. However, there is a one-to-one corresponding relationship between the original function and the image function with its ROC, thus it is necessary to label the ROC. For example, both bilateral Laplace transforms of the causal signal $f_1(t) = e^{-at}\varepsilon(t)$ and the noncausal signal $f_2(t) = -e^{-at}\varepsilon(-t)$ are $\frac{1}{s+a}$, but the ROCs of $F_1(s)$ and $F_2(s)$ are, respectively, $Re(s) > -a$ and $Re(s) < -a$.

6.2 Laplace transforms of common signals

Laplace transforms of two core signals will be discussed first, and then others are listed at end of the section.

1. Unit step signal $\varepsilon(t)$

Because the unit step signal has a step at moment $t = 0$, which means $f(0_-) = 0$ and $f(0_+) = 1$, its image function is

$$F(s) = \mathcal{L}\left[\varepsilon(t)\right] = \int_{0_-}^{\infty} e^{-st}dt = -\frac{e^{-st}}{s}\bigg|_{0_-}^{\infty} = \frac{1}{s},$$

namely,

$$\varepsilon(t) \overset{\mathcal{L}}{\longleftrightarrow} \frac{1}{s}.\tag{6.2-1}$$

2. Unit impulse signal $\delta(t)$

According to the definition, the Laplace transform of $\delta(t)$ is

$$F(s) = \mathcal{L}[\delta(t)] = \int_{0_-}^{\infty} \delta(t)e^{-st}dt = 1,$$

namely,

$$\delta(t) \overset{\mathcal{L}}{\longleftrightarrow} 1.\tag{6.2-2}$$

The Laplace transforms of some common signals are listed in Table 6.2.

6.3 Laplace transforms of periodic signals

If there is a periodic signal $f_T(t)$ with the period T, and its first cycle waveform beginning at $t \geq 0$ is $f_1(t)$, its unilateral Laplace transform $F_T(s)$ equals the product of the Laplace transform $F_1(s)$ of $f_1(t)$ and a factor $\frac{1}{1-e^{-Ts}}$, that is,

$$\mathcal{L}[f_T(t)] = F_T(s) = \frac{1}{1 - e^{-Ts}}F_1(s).\tag{6.3-1}$$

Readers can decompose $f_T(t)$ into an algebraic sum of infinite periodic waveforms $(f_T(t) = \sum_{n=0}^{+\infty} f_1(t - nT))$ and then use the time shifting property of the Laplace transform and the summation of a series (Referring to Example 6.4-10) to prove equation (6.3-1). Note that the periodic signal here is a unilateral signal.

Tab. 6.2: Laplace transforms of common signals.

No.	$f(t), t > 0, \alpha > 0$	$F(s)$	ROC
1	$\delta(t)$	1	$Re(s) > -\infty$
2	$\delta^{(n)}(t)$	s^n	$Re(s) > -\infty$
3	$\varepsilon(t)$	$\frac{1}{s}$	$Re(s) > 0$
4	$e^{-\alpha t}$	$\frac{1}{s+\alpha}$	$Re(s) > -\alpha$
5	t^n (n Positive integer)	$\frac{n!}{s^{n+1}}$	$Re(s) > 0$
6	$te^{-\alpha t}$	$\frac{1}{(s+\alpha)^2}$	$Re(s) > -\alpha$
7	$t^n e^{-\alpha t}$	$\frac{n!}{(s+\alpha)^{n+1}}$	$Re(s) > -\alpha$
8	$\sin \omega_0 t$	$\frac{\omega_0}{s^2+\omega_0^2}$	$Re(s) > 0$
9	$\cos \omega_0 t$	$\frac{s}{s^2+\omega_0^2}$	$Re(s) > 0$
10	$e^{-\alpha t} \sin \omega_0 t$	$\frac{\omega_0}{(s+\alpha)^2+\omega_0^2}$	$Re(s) > -\alpha$
11	$e^{-\alpha t} \cos \omega_0 t$	$\frac{s+\alpha}{(s+\alpha)^2+\omega_0^2}$	$Re(s) > -\alpha$
12	$t \sin \omega_0 t$	$\frac{2\omega_0 s}{(s^2+\omega_0^2)^2}$	$Re(s) > 0$
13	$t \cos \omega_0 t$	$\frac{s^2-\omega_0^2}{(s^2+\omega_0^2)^2}$	$Re(s) > 0$
14	$sh\alpha t$	$\frac{\alpha}{s^2-\alpha^2}$	$Re(s) > \alpha$
15	$ch\alpha t$	$\frac{s}{s^2-\alpha^2}$	$Re(s) > \alpha$

6.4 Properties of the Laplace transform

6.4.1 Linearity

If

$$\mathcal{L}[f_1(t)] = F_1(s) \quad \text{and} \quad \mathcal{L}[f_2(t)] = F_2(s) ,$$

and then

$$\mathcal{L}[af_1(t) + bf_2(t)] = aF_1(s) + bF_2(s) . \qquad (6.4\text{-}1)$$

Proof.

$$\mathcal{L}[af_1(t) + bf_2(t)] = \int_{0_-}^{\infty} [af_1(t) + bf_2(t)]e^{-st}\,dt$$

$$= a \int_{0_-}^{\infty} f_1(t)e^{-st}\,dt + b \int_{0_-}^{\infty} f_2(t)e^{-st}\,dt$$

$$= aF_1(s) + bF_2(s)$$

$\square$

Example 6.4-1. Find the Laplace transform $F(s)$ of $f(t) = \cos \omega_0 t$.

Solution.

$$F(s) = \mathcal{L}\,[\cos \omega_0 t]$$

$$= \mathcal{L}\left[\frac{e^{-j\omega_0 t}}{2} + \frac{e^{j\omega_0 t}}{2}\right] = \mathcal{L}\left[\frac{e^{-j\omega_0 t}}{2}\right] + \mathcal{L}\left[\frac{e^{j\omega_0 t}}{2}\right]$$

$$= \frac{1}{2}\left(\frac{1}{s + j\omega_0} + \frac{1}{s - j\omega_0}\right) = \frac{s}{s^2 + \omega_0^2}$$

With the same method, the image function of $\sin \omega_0 t$ is

$$F(s) = \mathcal{L}\,[\sin \omega_0 t] = \frac{\omega_0}{s^2 + \omega_0^2}\,.$$

Example 6.4-2. Find the Laplace transform $F(s)$ of the hyperbolic sine $sh\alpha t$ and the hyperbolic cosine $ch\alpha t$.

Solution. Because

$$sh\alpha t = \frac{e^{\alpha t} - e^{-\alpha t}}{2} \quad \text{and} \quad ch\alpha t = \frac{e^{\alpha t} + e^{-\alpha t}}{2}\,,$$

according to linearity properties, we can obtain the image function of $sh\alpha t$

$$F(s) = \mathcal{L}[sh\alpha t] = \mathcal{L}\left[\frac{e^{\alpha t}}{2} - \frac{e^{-\alpha t}}{2}\right]$$

$$= \frac{1}{2}\mathcal{L}\left[e^{\alpha t}\right] - \frac{1}{2}\mathcal{L}\left[e^{-\alpha t}\right] = \frac{1}{2}\left(\frac{1}{s - \alpha} - \frac{1}{s + \alpha}\right) = \frac{\alpha}{s^2 - \alpha^2}\,.$$

Similarly, we also obtain the image function of $ch\alpha t$

$$F(s) = \mathcal{L}[ch\alpha t] = \mathcal{L}\left[\frac{e^{\alpha t}}{2} + \frac{e^{-\alpha t}}{2}\right] = \frac{1}{2}\left(\frac{1}{s - \alpha} + \frac{1}{s + \alpha}\right) = \frac{s}{s^2 - \alpha^2}\,.$$

6.4.2 Time shifting

If

$$\mathcal{L}[f(t)] = F(s)\,,$$

then

$$\mathcal{L}\,[f(t - t_0)\,\varepsilon\,(t - t_0)] = e^{-st_0}F(s)\,. \tag{6.4-2}$$

Proof.

$$\mathcal{L}\,[f(t - t_0)\,\varepsilon\,(t - t_0)] \quad = \quad \int_{0_-}^{\infty} f(t - t_0)\,\varepsilon\,(t - t_0)\,e^{-st}dt = \int_{t_0}^{\infty} f(t - t_0)\,e^{-st}dt$$

$$\stackrel{t - t_0 = x}{=} \int_{0}^{\infty} f(x)e^{-s(x + t_0)}dx = e^{-st_0} \int_{0}^{\infty} f(x)e^{-sx}dx = e^{-st_0}F(s)$$

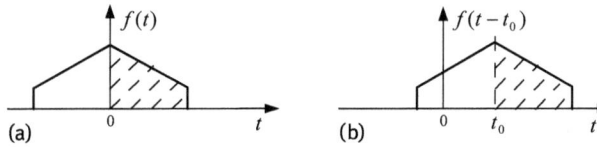

Fig. 6.3: Time shifting properties diagram.

Note that the original function in equation (6.4-2) is $f(t - t_0)\varepsilon(t - t_0)$ rather than $f(t - t_0)$. Waveforms of $f(t)$ and $f(t - t_0)$ are, respectively, shown in ▶ Figure 6.3a and b. Obviously, the effective (shaded) parts of the unilateral Laplace transforms of $f(t - t_0)$ and $f(t - t_0)\varepsilon(t - t_0)$ are different; the time shifting property cannot be used by the former. □

Example 6.4-3. The signals $f_1(t)$ and $f_2(t)$ are plotted in ▶ Figure 6.4; the image function of $f_1(t)$ is $F_1(s)$. Please find the image function $F_2(s)$ of $f_2(t)$.

Solution. $f_1(t)$ and $f_2(t)$ can be related by

$$f_2(t) = f_1(t) - f_1(t - 1) .$$

According to the linearity and the time shifting of the Laplace transform, we have

$$F_2(s) = F_1(s) - e^{-s}F_1(s) = \left(1 - e^{-s}\right) F_1(s) .$$

Example 6.4-4. Find the Laplace transform of signal $f(t) = t^2\varepsilon(t - 1)$.

Solution. The expression of the signal can be transformed as

$$f(t) = (t - 1)^2\varepsilon(t - 1) + 2(t - 1)\varepsilon(t - 1) + \varepsilon(t - 1) .$$

According to the time shifting, we have

$$\mathcal{L}[\varepsilon(t - 1)] = \frac{1}{s}e^{-s} ,$$

$$\mathcal{L}[2(t - 1)\varepsilon(t - 1)] = \frac{2}{s^2}e^{-s} ,$$

$$\mathcal{L}[(t - 1)^2\varepsilon(t - 1)] = \frac{2}{s^3}e^{-s} .$$

According to the linearity property, we have

$$\mathcal{L}[f(t)] = \left(\frac{2}{s^3} + \frac{2}{s^2} + \frac{1}{s} \right) e^{-s} .$$

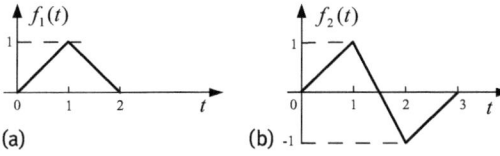

(a) (b) **Fig. 6.4:** E6.4-3.

6.4.3 Complex frequency shifting

If

$$\mathcal{L}[f(t)] = F(s) \, ,$$

then

$$\mathcal{L}\left[f(t)e^{s_0 t}\right] = F(s - s_0) \, . \tag{6.4-3}$$

Proof.

$$\mathcal{L}\left[f(t)e^{s_0 t}\right] = \int_{0_-}^{\infty} \left[f(t)e^{s_0 t}\right]e^{-st}\mathrm{d}t = \int_{0_-}^{\infty} f(t)e^{-(s-s_0)t}\mathrm{d}t = F(s - s_0)$$

□

Example 6.4-5. Find the Laplace transforms of the attenuation sine $e^{-at}\sin\beta t$ and the attenuation cosine $e^{-at}\cos\beta t$, where $a > 0$.

Solution. We know that

$$\mathcal{L}[\sin\beta t] = \frac{\beta}{s^2 + \beta^2} \, .$$

According to the complex frequency shifting property, we obtain

$$\mathcal{L}\left[e^{-at}\sin\beta t\right] = \frac{\beta}{(s + a)^2 + \beta^2} \, .$$

Similarly,

$$\mathcal{L}[\cos\beta t] = \frac{s}{s^2 + \beta^2} \, ,$$

so,

$$\mathcal{L}\left[e^{-at}\cos\beta t\right] = \frac{s + a}{(s + a)^2 + \beta^2} \, .$$

6.4.4 Time scaling

If

$$\mathcal{L}[f(t)] = F(s) \, ,$$

then

$$\mathcal{L}[f(at)] = \frac{1}{a}F\left(\frac{s}{a}\right) \quad (a > 0) \, . \tag{6.4-4}$$

Proof.

$$\mathcal{L}[f(at)] = \int_{0_-}^{\infty} f(at)e^{-st}\mathrm{d}t \overset{\tau=at}{=\!=} \int_{0_-}^{\infty} f(\tau)e^{-(\frac{\tau}{a})s}\mathrm{d}\left(\frac{\tau}{a}\right) = \frac{1}{a}\int_{0_-}^{\infty} f(\tau)e^{-(\frac{s}{a})\tau}\mathrm{d}\tau = \frac{1}{a}F\left(\frac{s}{a}\right) \, .$$

□

Example 6.4-6. If $\mathcal{L}[f(t)] = F(s)$ is known, find $\mathcal{L}[f(at - b)\varepsilon(at - b)](a, b > 0)$.

Solution. From the time shifting, we have

$$\mathcal{L}[f(t - b)\varepsilon(t - b)] = e^{-bs}F(s) \, .$$

According to the time scaling property, we obtain

$$\mathcal{L}[f(at - b)\varepsilon(at - b)] = \frac{1}{a}e^{-\frac{b}{a}s}F\left(\frac{s}{a}\right) \, .$$

6.4.5 Differentiation in the time domain

If

$$\mathcal{L}[f(t)] = F(s) \, ,$$

then

$$\mathcal{L}\left[\frac{df(t)}{dt}\right] = sF(s) - f(0_-) \, , \tag{6.4-5}$$

where $f(0_-)$ is the starting value of $f(t)$ at moment $t = 0$.

Proof.

$$\mathcal{L}\left[\frac{df(t)}{dt}\right] = \int_{0_-}^{\infty} f'(t)e^{-st}dt = f(t)e^{-st}\Big|_{0_-}^{\infty} + s\int_{0_-}^{\infty} f(t)e^{-st}dt = sF(s) - f(0_-)$$

From the first-order derivative, the second-order and nth-order derivatives can be deduced as

$$\mathcal{L}\left[\frac{d^2 f(t)}{dt^2}\right] = s^2 F(s) - sf(0_-) - f'(0_-) \, , \tag{6.4-6}$$

$$\mathcal{L}\left[\frac{d^n f(t)}{dt^n}\right] = s^n F(s) - s^{n-1}f(0_-) - s^{n-2}f'(0_-) - \cdots - f^{(n-1)}(0_-) \, . \tag{6.4-7}$$

Because the causal signal can satisfy the following relationship:

$$f(0_-) = f'(0_-) = \cdots = f^{(n-1)}(0_-) = 0 \, ,$$

Equations (6.4-5)–(6.4-7) can be simplified as

$$\mathcal{L}\left[\frac{df(t)}{dt}\right] = sF(s) \, , \tag{6.4-8}$$

$$\mathcal{L}\left[\frac{d^2 f(t)}{dt^2}\right] = s^2 F(s) \, , \tag{6.4-9}$$

$$\mathcal{L}\left[\frac{d^n f(t)}{dt^n}\right] = s^n F(s) \, . \tag{6.4-10}$$

Note: This property is unsuitable for all-order derivatives of the unit step signal.   □

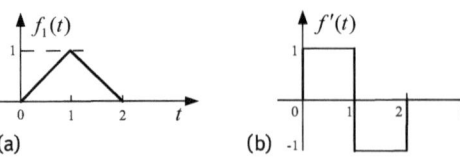

(a)   (b) -1   **Fig. 6.5:** E6.4-8.

Example 6.4-7. Find the image function of $\delta'(t)$ which is called the impulse couple.

Solution. Knowing $\mathcal{L}[\delta(t)] = 1$, and according to the differential property in the time domain, we have

$$\mathcal{L}\left[\delta'(t)\right] = s .$$

Example 6.4-8. Find the Laplace transform of $f(t)$ shown in ▶ Figure 6.5.

Solution. The first-order derivative of $f(t)$ is

$$f'(t) = \varepsilon(t) - 2\varepsilon(t-1) + \varepsilon(t-2) ,$$

and it can be illustrated in ▶ Figure 6.5b, so,

$$\mathcal{L}\left[f'(t)\right] = \frac{1}{s} - \frac{2}{s}e^{-s} + \frac{1}{s}e^{-2s} = \frac{1}{s}(1 - e^{-s})^2 ,$$

from the differential property in the time domain we have

$$\mathcal{L}\left[f'(t)\right] = s\mathcal{L}[f(t)] ,$$

so,

$$\mathcal{L}[f(t)] = \frac{1}{s^2}(1 - e^{-s})^2 .$$

6.4.6 Integration in the time domain

If

$$\mathcal{L}[f(t)] = F(s) ,$$

then

$$\mathcal{L}\left[\int_{-\infty}^{t} f(\tau)d\tau\right] = \frac{1}{s}F(s) + \frac{1}{s}\int_{-\infty}^{0_-} f(\tau)d\tau . \qquad (6.4\text{-}11)$$

Proof. Since

$$\mathcal{L}\left[\int_{-\infty}^{t} f(\tau)d\tau\right] = \mathcal{L}\left[\int_{-\infty}^{0_-} f(\tau)d\tau\right] + \mathcal{L}\left[\int_{0_-}^{t} f(\tau)d\tau\right] ,$$

so, the first term $\int_{-\infty}^{0_-} f(\tau)d\tau$ in the above equation is a constant, that is,

$$\mathcal{L}\left[\int_{-\infty}^{0_-} f(\tau)d\tau\right] = \frac{1}{s}\int_{-\infty}^{0_-} f(\tau)d\tau .$$

Moreover,

$$\mathcal{L}\left[\int_{0_-}^{t} f(\tau)\mathrm{d}\tau\right] = \int_{0_-}^{\infty}\left[\int_{0_-}^{t} f(\tau)\mathrm{d}\tau\right]e^{-st}\mathrm{d}t = -\frac{e^{-st}}{s}\int_{0_-}^{t} f(\tau)\mathrm{d}\tau\bigg|_{0_-}^{\infty} + \frac{1}{s}\int_{0_-}^{\infty} e^{-st}f(t)\mathrm{d}t ,$$

the first term on the right is equal to 0 for $t = 0_-$ and $t \rightarrow \infty$, so,

$$\mathcal{L}\left[\int_{0_-}^{t} f(\tau)\mathrm{d}\tau\right] = \frac{1}{s}F(s) ,$$

and then

$$\mathcal{L}\left[\int_{-\infty}^{t} f(\tau)\mathrm{d}\tau\right] = \frac{1}{s}F(s) + \frac{1}{s}\int_{-\infty}^{0_-} f(\tau)\mathrm{d}\tau .$$

□

Example 6.4-9. Find the image function of $t^n\varepsilon(t)$.

Solution. Since

$$\int_{0}^{t} \varepsilon(\tau)\mathrm{d}\tau = t\varepsilon(t) ,$$

and according to the integral property in the time domain, we have

$$\mathcal{L}[t\varepsilon(t)] = \mathcal{L}\left[\int_{0}^{t} \varepsilon(\tau)\mathrm{d}\tau\right] = \frac{1}{s}\mathcal{L}[\varepsilon(t)] = \frac{1}{s^2} .$$

Because

$$\int_{0}^{t} \tau\varepsilon(\tau)\mathrm{d}\tau = \frac{1}{2}t^2\varepsilon(t) ,$$

we have

$$\mathcal{L}\left[t^2\varepsilon(t)\right] = 2\mathcal{L}\left[\int_{0}^{t} \tau\varepsilon(\tau)\mathrm{d}\tau\right] = \frac{2}{s}\mathcal{L}[t\varepsilon(t)] = \frac{2}{s^3} ,$$

and then

$$\mathcal{L}[t^n] = \frac{n!}{s^{n+1}} .$$

6.4.7 Convolution theorem

If

$$\mathcal{L}[f_1(t)] = F_1(s) \quad \text{and} \quad \mathcal{L}[f_2(t)] = F_2(s) ,$$

then

$$\mathcal{L}[f_1(t) * f_2(t)] = F_1(s)F_2(s) . \tag{6.4-12}$$

This states that the product of the Laplace transforms of two signals is equal to the Laplace transform of convolution of the two signals.

Proof. Since

$$\mathcal{L}\left[f_1(t) * f_2(t)\right] = \mathcal{L}\left[\int_0^\infty f_1(\tau)f_2(t-\tau)d\tau\right] = \int_0^\infty\left[\int_0^\infty f_1(\tau)f_2(t-\tau)d\tau\right]e^{-st}dt\,,$$

exchanging the integral order of the equation, we obtain

$$\mathcal{L}\left[f_1(t) * f_2(t)\right] = \int_0^\infty f_1(\tau)\left[\int_0^\infty f_2(t-\tau)e^{-st}dt\right]d\tau = \int_0^\infty f_1(\tau)e^{-s\tau}F_2(s)d\tau$$

$$= F_2(s)\int_0^\infty f_1(\tau)e^{-s\tau}d\tau = F_1(s)F_2(s)$$

□

Example 6.4-10. Find the image function of the periodic signal $f(t)$ shown in ▶ Figure 6.6a.

Solution. Since the original signal can be expressed as a convolution of its first cycle waveform and a unilateral unit comb or a unilateral impulse train, that is,

$$f(t) = f_1(t) * f_2(t)\,,$$

based on the results of Example 6.4-8, we have

$$\mathcal{L}\left[f_1(t)\right] = \frac{1}{s^2}\left(1-e^{-s}\right)^2\,.$$

Because a unilateral impulse train is of the from

$$f_2(t) = \delta(t) + \delta(t-2) + \delta(t-4) + \cdots = \sum_{n=0}^\infty \delta(t-2n)\,,$$

and

$$\mathcal{L}\left[f_2(t)\right] = 1 + e^{-2s} + e^{-4s} + \cdots = \sum_{n=0}^\infty e^{-2ns} = \frac{1}{1-e^{-2s}}\,,$$

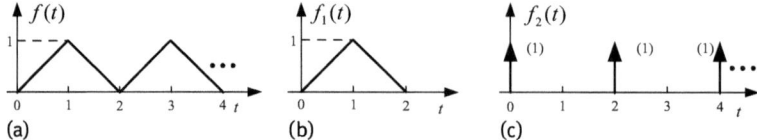

(a) (b) (c)

Fig. 6.6: E6.4-10.

from the convolution theorem, we obtain

$$\mathcal{L}[f(t)] = \mathcal{L}[f_1(t) * f_2(t)] = \frac{1}{s^2}(1 - e^{-s})^2 \frac{1}{1 - e^{-2s}} = \frac{1 - e^{-s}}{s^2(1 + e^{-s})} .$$

The answer to this question can also be directly written by using equation (6.3-1) and the results of Example 6.4-8.

6.4.8 Initial value theorem

If the Laplace transforms of signal $f(t)$ and its derivative $\frac{df(t)}{dt}$ exist, and the Laplace transform of $f(t)$ is $F(s)$, we obtain

$$f(0_+) = \lim_{t \to 0_+} f(t) = \lim_{s \to \infty} sF(s) . \tag{6.4-13}$$

Proof. According to the differential property in the time domain,

$$sF(s) - f(0_-) = \mathcal{L}\left[\frac{df(t)}{dt}\right] = \int_{0_-}^{\infty} \frac{df(t)}{dt} e^{-st} dt = \int_{0_-}^{0_+} \frac{df(t)}{dt} e^{-st} dt + \int_{0_+}^{\infty} \frac{df(t)}{dt} e^{-st} dt$$

$$= f(0_+) - f(0_-) + \int_{0_+}^{\infty} \frac{df(t)}{dt} e^{-st} dt$$

so,

$$sF(s) = f(0_+) + \int_{0_+}^{\infty} \frac{df(t)}{dt} e^{-st} dt . \tag{6.4-14}$$

If $s \to \infty$, the limit of the second term on the right side of equation (6.4-14) should be

$$\lim_{s \to \infty} \left\{ \int_{0_+}^{\infty} \frac{df(t)}{dt} e^{-st} dt \right\} = \int_{0_+}^{\infty} \frac{df(t)}{dt} [\lim_{s \to \infty} e^{-st}] dt = 0 ,$$

therefore, calculating the limit of equation (6.4-14) when $s \to \infty$, we have

$$f(0_+) = \lim_{s \to \infty} sF(s) .$$

$\square$

6.4.9 Final value theorem

If the Laplace transform of $f(t)$ and its derivative $\frac{df(t)}{dt}$ exist, and the Laplace transform $f(t)$ is $F(s)$, there should be

$$f(\infty) = \lim_{t \to \infty} f(t) = \lim_{s \to 0} sF(s) . \tag{6.4-15}$$

Proof. Calculate the limit of equation (6.4-14) when $s \to 0$,

$$\lim_{s \to 0} sF(s) = f(0_+) + \int_{0_+}^{\infty} \frac{df(t)}{dt} e^{-st} dt = f(0_+) + \lim_{s \to 0} \int_{0_+}^{\infty} \frac{df(t)}{dt} e^{-st} dt$$

$$= f(0_+) + \lim_{t \to \infty} f(t) - f(0_+)$$

thus,

$$\lim_{t \to \infty} f(t) = \lim_{s \to 0} sF(s) .$$

The initial and final value theorems are usually used in the cases that $f(0_+)$ and $f(\infty)$ can be calculated by $F(s)$, and $f(t)$ does not have to be found. However, please note the application conditions for the initial and final value theorems in the following.

(1) When we want to find the initial value of a signal by equation (6.4-13), if $F(s)$ is a rational algebraic expression, it should be a real fraction. This means the order of the numerator polynomial should be lower than that of the denominator in $F(s)$. However, if $F(s)$ is not a real fraction, it should be operated by long division to extract the real fraction terms from $F(s)$ denoted by $F_0(s)$. It can be proved that $f(0_+)$ equals $f_0(0_+)$ which is the initial value of $f_0(t)$ from the inverse Laplace transform of $F_0(s)$:

$$f(0_+) = f_0(0_+) = \lim_{s \to \infty} sF_0(s) . \tag{6.4-16}$$

(2) The final value of $f(t)$ can be calculated by equation (6.4-15) only if the final value of $f(t)$ exists, otherwise, a false conclusion is obtained. The judgment conditions about it is that when the poles of $F(s)$ distribute on the left half of the s plane or $F(s)$ only has a first-order pole at the origin, the final value theorem can be used. For example, when $a > 0$, the value of $\lim_{t \to \infty} e^{at}$ does not exist, but the wrong conclusion of $\lim_{t \to \infty} e^{at} = \lim_{s \to 0} sF(s) = 0$ will arise if the final value theorem is used directly. The reason is that the root $s = a$ of the denominator in $\mathcal{L}\left[e^{at}\right] = \frac{1}{s-a}$ is on the real axis of right half-plane, so the final value theorem cannot be used in this case.

□

Example 6.4-11. Solve the initial and final values of the inverse transform for the following expressions.

(1) $F(s) = \frac{s^3+s^2+2s+1}{s^2+2s+1}$

(2) $F(s) = \frac{s^2+2s+3}{(s^2+w_0^2)(s+1)}$

Solution. (1) Because it is an improper fraction, $F(s)$ should be dealt with by long division in order to obtain real fraction terms,

$$F(s) = s - 1 + \frac{3s + 2}{s^2 + 2s + 1} ,$$

so,

$$f(0_+) = \lim_{s\to\infty} s\frac{3s + 2}{s^2 + 2s + 1} = 3 .$$

The poles of $F(s)$ are on the left half-plane of the s plane, so the final value is

$$f(\infty) = \lim_{s\to 0} sF(s) = 0 .$$

(2)

$$f(0_+) = \lim_{s\to\infty} s\frac{s^2 + 2s + 3}{\left(s^2 + w_0^2\right)(s + 1)} = 1 .$$

Considering the pair of conjugate poles $s = \pm jw_0$ of $F(s)$ on the imaginary axis, the final value of $f(t)$ does not exist.

6.4.10 Differentiation in the s domain

If

$$\mathcal{L}[f(t)] = F(s) ,$$

then

$$\mathcal{L}\left[(-t)^n f(t)\right] = \frac{d^n F(s)}{ds^n} . \tag{6.4-17}$$

Proof. Because

$$F(s) = \int_{0_-}^{\infty} f(t)e^{-st}dt ,$$

we have

$$\frac{dF(s)}{ds} = \frac{d}{ds}\int_{0_-}^{\infty} f(t)e^{-st}dt .$$

Exchanging order of the integral and differential, we have

$$\frac{dF(s)}{ds} = \int_{0_-}^{\infty} f(t)\frac{d}{ds}e^{-st}dt = \int_{0_-}^{\infty} (-t)f(t)e^{-st}dt = \mathcal{L}[(-t)f(t)] ,$$

so,

$$\mathcal{L}[(-t)f(t)] = \frac{dF(s)}{ds} .$$

Repeating this, we have

$$\mathcal{L}\left[(-t)^n f(t)\right] = \frac{d^n F(s)}{ds^n} .$$

□

6.4.11 Integration in the s domain

If

$$\mathcal{L}[f(t)] = F(s) ,$$

then

$$\mathcal{L}\left[\frac{f(t)}{t}\right] = \int\limits_{s}^{\infty} F(\eta)d\eta . \tag{6.4-18}$$

Proof. Substituting $F(s) = \int_{0_-}^{\infty} f(t)e^{-st}dt$ into the integral $\int_{s}^{\infty} F(\eta)d\eta$, and exchanging the integral order, we have

$$\int\limits_{s}^{\infty} F(\eta)d\eta = \int\limits_{s}^{\infty}\int\limits_{0_-}^{\infty} f(t)e^{-\eta t}dtd\eta = \int\limits_{0_-}^{\infty} f(t)\left[\int\limits_{s}^{\infty} e^{-\eta t}d\eta\right]dt$$

$$= \int\limits_{0_-}^{\infty} \frac{f(t)}{t}e^{-st}dt = \mathcal{L}\left[\frac{f(t)}{t}\right] .$$

□

Example 6.4-12. Find the Laplace image function of $t^2 e^{-at}\varepsilon(t)$.

Solution. Because

$$\mathcal{L}\left[e^{-at}\varepsilon(t)\right] = \frac{1}{s+a} ,$$

according to the differential property in the s domain, we obtain

$$\mathcal{L}\left[(-t)^2 e^{-at}\varepsilon(t)\right] = \frac{d^2}{ds^2}\left(\frac{1}{s+a}\right) = \frac{2}{(s+a)^3}$$

or

$$\mathcal{L}\left[t^2 e^{-at}\varepsilon(t)\right] = \frac{2}{(s+a)^3} .$$

Example 6.4-13. Find the Laplace image function of $\frac{\sin t}{t}\varepsilon(t)$.

Solution. Because

$$\mathcal{L}[\sin t \cdot \varepsilon(t)] = \frac{1}{s^2+1} ,$$

according to the differential property in the s domain, we obtain

$$\mathcal{L}\left[\frac{\sin t}{t}\varepsilon(t)\right] = \int\limits_{s}^{\infty} \frac{1}{\eta^2+1}d\eta = \frac{\pi}{2} - \arctan(s) = \arctan\frac{1}{s} .$$

The main properties of the Laplace transform are listed in Table 6.3 for reference.

Tab. 6.3: The main properties of Laplace transform.

No.	Name	Time domain $f(t)\varepsilon(t)$	Complex frequency domain $F(s)$
1	Linearity	$af_1(t) + bf_2(t)$	$aF_1(s) + bF_2(s)$
2	Time shifting	$f(t - t_0)\,\varepsilon\,(t - t_0)$	$e^{-st_0}F(s)$
3	Complex frequency shifting	$f(t)e^{s_0 t}$	$F(s - s_0)$
4	Scaling	$f(at)(a > 0)$	$\frac{1}{a}F\left(\frac{s}{a}\right)$
5	Time differential	$\frac{df(t)}{dt}$	$sF(s) - f(0_-)$
		$\frac{d^n f(t)}{dt^n}$	$s^n F(s) - s^{n-1}f(0_-) - s^{n-2}f'(0_-) - \cdots - f^{(n-1)}(0_-)$
6	Time integral	$\int_{-\infty}^{t} f(\tau)d\tau$	$\frac{1}{s}F(s) + \frac{1}{s}\int_{-\infty}^{0_-} f(\tau)d\tau$
7	Time convolution	$f_1(t) * f_2(t)$	$F_1(s)F_2(s)$
8	Convolution in s domain	$f_1(t) \cdot f_2(t)$	$\frac{1}{2\pi j}\int_{\sigma-j\infty}^{\sigma+j\infty} F_1(z)F_2(s - z)dz$
9	Initial value theorem	$f(0_+) = \lim_{t\to 0_+} f(t) = \lim_{s\to\infty} sF(s)$	
10	Final value theorem	$f(\infty) = \lim_{t\to\infty} f(t) = \lim_{s\to 0} sF(s)$	
11	Differential in s domain	$(-t)^n f(t)$	$\frac{d^n F(s)}{ds^n}$
12	Integral in s domain	$\frac{f(t)}{t}$	$\int_s^{\infty} F(\eta)d\eta$
13	Time transformation	$f(at-b)\varepsilon(at-b)$ $a > 0, b \geq 0$	$\frac{e^{-\frac{bs}{a}}}{a}F\left(\frac{s}{a}\right)$

6.5 Solutions for the inverse Laplace transform

For the unilateral Laplace transform, the inverse transform of $F(s)$ can be written as

$$\mathcal{L}^{-1}[F(s)] = \begin{cases} 0 & t < 0 \\ \frac{1}{2\pi j}\int_{\sigma-j\infty}^{\sigma+j\infty} F(s)e^{st}ds & t > 0 \end{cases}$$

Generally, the above integral should be solved by the residue theorem in the theory of complex functions. However, it is commonly seen that the image function is a rational fraction with s in the form, that is, it can be written as a ratio of two polynomials in the s domain,

$$F(s) = \frac{B(s)}{A(s)} = \frac{b_m s^m + b_{m-1}s^{m-1} + b_1 s + b_0}{s^n + a_{n-1}s^{n-1} + a_1 s + a_0}, \tag{6.5-1}$$

where a_i and b_j are real constants. Because the orders of the numerator and denominator are arbitrary, this image function does not appear in standard tables of Laplace transforms, such as Table 6.2. However, in common cases, it can be expressed as a sum of functions that do appear in standard tables, using a technique called partial fraction expansion, so its inverse Laplace transform can be obtained.

If $m \geq n$, $F(s)$ can be divided as a sum of a rational polynomial and rational proper fractions by long division, for example,

$$F(s) = \frac{2s^3 + s^2 - 1}{s^2 + 3s + 2} = 2s - 5 + \frac{11s + 9}{s^2 + 3s + 2} = \text{polynomial} + \text{proper fraction} ,$$

where the inverse Laplace transforms of the rational polynomial can be easily obtained according to the Laplace transform of a typical signal $\delta(t)$ and the properties of the Laplace transform, such as $\mathcal{L}^{-1}[2s] = 2\delta'(t)$ and $\mathcal{L}^{-1}[5] = 5\delta(t)$. Therefore, we should only discuss the inverse transform problem about the rational real fraction, which is equation (6.5-1) for $m < n$.

First, let us introduce the concept about zero and the pole. In equation (6.5-1), the roots $p_1, p_2, \ldots, p_n$, which satisfy $A(s) = 0$, are called poles, and the roots $z_1, z_2, \ldots, z_m$, which satisfy $B(s) = 0$, are called zeros. Then, equation (6.5-1) can be rewritten as

$$F(s) = \frac{B(s)}{A(s)} = \frac{b_m (s - z_1)(s - z_2) \ldots (s - z_m)}{(s - p_1)(s - p_2) \ldots (s - p_n)} . \tag{6.5-2}$$

According to different types of poles, the partial fraction expansion method can be divided into the following situations for discussion.

1. There are only non-repeated real poles in $F(s)$

The so-called non-repeated real poles mean that $p_1, p_2, \ldots, p_n$ are real numbers and are different from each other. In this simplest case, n coefficients, such as $k_1, k_2, \ldots, k_n$, can be found and $F(s)$ can be written in partial fraction form

$$F(s) = \frac{B(s)}{A(s)} = \frac{k_1}{s - p_1} + \frac{k_2}{s - p_2} + \cdots + \frac{k_n}{s - p_n} = \sum_{i=1}^{n} \frac{k_i}{s - p_i} . \tag{6.5-3}$$

Therefore, according to the Laplace transform of the exponential signal, the original function $f(t)$ of the image function $F(s)$ could be written as

$$f(t) = \left(k_1 e^{p_1 t} + k_2 e^{p_2 t} + \cdots + k_n e^{p_n t} \right) \varepsilon(t) = \sum_{i=1}^{n} k_i e^{p_i t} \varepsilon(t) . \tag{6.5-4}$$

Now, how can we determine $k_1, k_2, \ldots, k_n$ in equation (6.5-4)? Let us multiply both sides of equation (6.5-3) by $(s - p_1)$,

$$\frac{B(s)}{(s - p_2)(s - p_3) \ldots (s - p_n)} = k_1 + \frac{(s - p_1) k_2}{s - p_2} + \cdots + \frac{(s - p_1) k_n}{s - p_n} .$$

Letting $s = p_1$,

$$k_1 = \frac{B(s)}{(s - p_2)(s - p_3) \ldots (s - p_n)} \bigg|_{s = p_1} .$$

We can use the same technique to find the other k_i,

$$k_i = (s - p_i) \frac{B(s)}{A(s)} \bigg|_{s = p_i} . \tag{6.5-5}$$

Now, the task of finding the inverse Laplace transform of a real fraction with non-repeated real poles has been fulfilled, that is, the original function $f(t)$ will be expressed as an algebraic sum of exponential signals like equation (6.5-4) in form.

Example 6.5-1. Find the original function $f(t)$ of $F(s) = \frac{2s+1}{s^2+8s+15}$.

Solution.

$$F(s) = \frac{2s + 1}{s^2 + 8s + 15} = \frac{2s + 1}{(s + 3)(s + 5)} = \frac{k_1}{s + 3} + \frac{k_2}{s + 5} ,$$

and according to equation (6.5-5), the coefficients are

$$k_1 = \left.\frac{2s + 1}{s + 5}\right|_{s=-3} = -\frac{5}{2}; \quad k_2 = \left.\frac{2s + 1}{s + 3}\right|_{s=-5} = \frac{9}{2}$$

Thus, the original function $f(t)$ is

$$f(t) = \left(\frac{9}{2}e^{-5t} - \frac{5}{2}e^{-3t}\right)\varepsilon(t) .$$

2. There are conjugate complex poles in $F(s)$

If $A(s) = 0$ has a pair of conjugate roots, such as $s = \alpha \pm j\beta$, $F(s)$ can be written as

$$F(s) = \frac{B(s)}{(s - \alpha^2) + \beta^2} = \frac{B(s)}{(s - \alpha - j\beta)(s - \alpha + j\beta)} = \frac{k_1}{s - \alpha - j\beta} + \frac{k_2}{s - \alpha + j\beta} .$$

According to equation (6.5-5), we can obtain k_1 and k_2,

$$k_1 = (s - \alpha - j\beta)\left.\frac{B(s)}{A(s)}\right|_{s=\alpha+j\beta} = \frac{B(\alpha + j\beta)}{2j\beta} , \tag{6.5-6}$$

$$k_2 = (s - \alpha + j\beta)\left.\frac{B(s)}{A(s)}\right|_{s=\alpha-j\beta} = \frac{B(\alpha - j\beta)}{-2j\beta} . \tag{6.5-7}$$

Since $B(\alpha + j\beta)$ and $B(\alpha - j\beta)$ are conjugate, k_1 and k_2 are related in conjugation. If $k_1 = A + jB$ and $k_2 = A - jB$ or $k_2 = k_1^*$, the original function $f(t)$ will be

$$f(t) = \mathcal{L}^{-1}\left[\frac{k_1}{s - \alpha - j\beta} + \frac{k_2}{s - \alpha + j\beta}\right]$$
$$= \left[k_1 e^{(\alpha+j\beta)t} + k_2 e^{(\alpha-j\beta)t}\right]\varepsilon(t) \tag{6.5-8}$$
$$= e^{\alpha t}\left(k_1 e^{j\beta t} + k_1^* e^{-j\beta t}\right)\varepsilon(t)$$
$$= 2e^{\alpha t}(A\cos\beta t - B\sin\beta t)\varepsilon(t) .$$

Example 6.5-2. Find the original function $f(t)$ of $F(s) = \frac{s+3}{s^2+2s+5}$.

Solution.

$$F(s) = \frac{s + 3}{(s + 1 - 2j)(s + 1 + 2j)} = \frac{k_1}{s + 1 - 2j} + \frac{k_2}{s + 1 + 2j} .$$

According to equations (6.5-6) and (6.5-7), we can obtain k_1, k_2,

$$k_1 = \left.\frac{s + 3}{s + 1 + 2j}\right|_{s=-1+2j} = \frac{1 - j}{2}$$

$$k_2 = \left.\frac{s + 3}{s + 1 - 2j}\right|_{s=-1-2j} = \frac{1 + j}{2}$$

According to equation (6.5-8), this yields

$$f(t) = 2e^{-t}\left(\frac{1}{2}\cos 2t + \frac{1}{2}\sin 2t\right)\varepsilon(t) = e^{-t}(\cos 2t + \sin 2t)\varepsilon(t) .$$

3. There are repeated poles in $F(s)$

This means there are two or more poles are equal in $p_1, p_2, \ldots, p_n$.

Suppose that $F(s)$ only has one m repeated root p_1, then $F(s)$ can be written as

$$F(s) = \frac{B(s)}{A(s)} = \frac{B(s)}{(s - p_1)^m} = \frac{k_{1m}}{(s - p_1)^m} + \frac{k_{1,m-1}}{(s - p_1)^{m-1}} + \cdots + \frac{k_{11}}{s - p_1} . \qquad (6.5\text{-}9)$$

Using $\mathcal{L}\left[t^n\varepsilon(t)\right] = \frac{n!}{s^{n+1}}$ and the frequency shifting property, the original function of $\frac{k_{1m}}{(s-p_1)^m}$ can be obtained by

$$\mathcal{L}^{-1}\left[\frac{k_{1m}}{(s - p_1)^m}\right] = \frac{k_{1m}}{(m - 1)!}t^{m-1}e^{p_1 t}\varepsilon(t) .$$

Now, $k_{1m}, k_{1,m-1}, \ldots, k_{11}$ can be determined. Multiplying both sides of equation 6.5-9 by $(s - p_1)^m$

$$B(s) = k_{1m} + (s - p_1)k_{1,m-1} + \cdots + (s - p_1)^{m-1}k_{11} , \qquad (6.5\text{-}10)$$

and setting $s = p_1$, we obtain

$$k_{1m} = \left.(s - p_1)^m \frac{B(s)}{A(s)}\right|_{s=p_1} . \qquad (6.5\text{-}11)$$

However, when we use the same method to find other $k_{1,m-1}, k_{1,m-2}, \ldots, k_{11}$, we will see that the denominator in equation 6.5-11 will be zero, so we can take the derivative with respect to s on both sides of equation (6.5-10),

$$\frac{d}{ds}\left[(s - p_1)^m F(s)\right] = k_{1,m-1} + 2(s - p_1)k_{1,m-2} + \cdots + (m - 1)(s - p_1)^{m-2}k_{11} ,$$

and setting $s = p_1$, we obtain

$$k_{1,m-1} = \left.\frac{d}{ds}\left[(s - p_1)^m F(s)\right]\right|_{s=p_1} .$$

Similarly, we obtain

$$k_{1,m-2} = \left.\frac{1}{2}\frac{d^2}{ds^2}\left[(s - p_1)^m F(s)\right]\right|_{s=p_1} .$$

Thus, we obtain the general form of k_{1i},

$$k_{1i} = \frac{1}{(m-i)!} \cdot \frac{d^{m-i}}{ds^{m-i}} [(s-p_1)^m F(s)] \Big|_{s=p_1} . \qquad (6.5\text{-}12)$$

Example 6.5-3. Find the original function $f(t)$ of $F(s) = \frac{s+2}{s^2+6s+9}$.

Solution. Since

$$F(s) = \frac{s+2}{(s+3)^2} = \frac{k_1}{(s+3)^2} + \frac{k_2}{s+3}$$

according to equation (6.5-11), we can obtain k_1, k_2,

$$k_1 = (s+2)|_{s=-3} = -1 \quad \text{and} \quad k_2 = \frac{d}{ds}(s+2)\Big|_{s=-3} = 1 .$$

So,

$$F(s) = -\frac{1}{(s+3)^2} + \frac{1}{s+3}$$

and

$$\mathcal{L}^{-1}\left[\frac{1}{(s+3)^2}\right] = te^{-3t}\varepsilon(t) ,$$

$$\mathcal{L}^{-1}\left[\frac{1}{s+3}\right] = e^{-3t}\varepsilon(t) .$$

Thus, the original function $f(t)$ is

$$f(t) = -te^{-3t}\varepsilon(t) + e^{-3t}\varepsilon(t) = (1-t)e^{-3t}\varepsilon(t) .$$

Example 6.5-4. Find the original function $f(t)$ of $F(s) = \frac{s^4+2}{s^3+4s^2+4s}$.

Solution. $F(s)$ can be rewritten in the form

$$F(s) = \frac{s^4+2}{s^3+4s^2+4s} = s-4+\frac{12s^2+16s+2}{s(s+2)^2} = s-4+\frac{k_1}{s}+\frac{k_{22}}{(s+2)^2}+\frac{k_{21}}{s+2} .$$

Since $F(s)$ has one simple and one pair of twice-repeated poles, the corresponding coefficients k_1, k_{21}, k_{22} can be obtained by the solution methods for simple poles and two-repeated poles, respectively.

$$k_1 = \frac{12s^2+16s+2}{(s+2)^2}\Big|_{s=0} = \frac{1}{2}$$

$$k_{22} = \frac{12s^2+16s+2}{s}\Big|_{s=-2} = -9$$

Moreover,

$$k_{21} = \frac{d}{ds}\left(\frac{12s^2 + 16s + 2}{s}\right)\Bigg|_{s=-2} = \frac{23}{2}$$

$$\mathcal{L}^{-1}[s - 4] = \delta'(t) - 4\delta(t)$$

$$\mathcal{L}^{-1}\left[\frac{1}{2s}\right] = \frac{1}{2}\varepsilon(t)$$

$$\mathcal{L}^{-1}\left[\frac{9}{(s+2)^2}\right] = 9te^{-2t}\varepsilon(t)$$

$$\mathcal{L}^{-1}\left[\frac{23}{2(s+2)}\right] = \frac{23}{2}e^{-2t}\varepsilon(t)$$

So, the original function is

$$f(t) = \delta'(t) - 4\delta(t) + \frac{1}{2}\varepsilon(t) - 9te^{-2t}\varepsilon(t) + \frac{23}{2}e^{-2t}\varepsilon(t) .$$

The above examples state that the premise to find the inverse Laplace transform of an image function is that the function is expanded in the partial fractions, and the key is that we must master the properties of the Laplace transform and the Laplace transforms of basic signals.

6.6 Analysis method of the system function in the *s* domain

6.6.1 System function

If the Laplace transforms of the zero-state response and the excitation are, respectively, set as $Y_f(s) = \mathcal{L}\,[y_f(t)]$ and $F(s) = \mathcal{L}[f(t)]$, the system function $H(s)$ is defined by

$$H(s) \stackrel{\text{def}}{=} \frac{Y_f(s)}{F(s)} . \qquad (6.6\text{-}1)$$

In system analysis, the excitation and response may be a voltage or a current, therefore, the system function may be the impedance (the ratio of voltage and current), the admittance (the ratio of current and voltage), or the constant (the ratio of two currents or the ratio of two voltages). In addition, if the excitation and the response are in the same port of a system, the system function is called the driving point function, but if they are not in the same port, it is called the transfer function. For example, in ▶ Figure 6.7, the ratio of $U_1(s)$ and $I_1(s)$, and the ratio of $I_2(s)$ and $U_2(s)$ are both

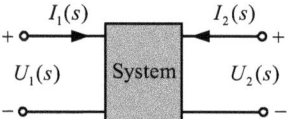

Fig. 6.7: Graph for defining the driving point functions and transfer functions.

Tab. 6.4: Names of system functions.

Position of excitation and response	Excitation	Response	System function
At the same port	Current $I_1(s)$	Voltage $U_1(s)$	Driving point impedance $H(s) = \frac{U_1(s)}{I_1(s)}$
	Voltage $U_1(s)$	Current $I_1(s)$	Driving point admittance $H(s) = \frac{I_1(s)}{U_1(s)}$
At different ports	Current $I_1(s)$	Voltage $U_2(s)$	Transform impedance $H(s) = \frac{U_2(s)}{I_1(s)}$
	Voltage $U_1(s)$	Current $I_2(s)$	Transform admittance $H(s) = \frac{I_2(s)}{U_1(s)}$
	Voltage $U_1(s)$	Voltage $U_2(s)$	Transform voltage ratio $H(s) = \frac{U_2(s)}{U_1(s)}$
	Current $I_1(s)$	Current $I_2(s)$	Transform current ratio $H(s) = \frac{I_2(s)}{I_1(s)}$

driving point functions, but the ratio of $U_1(s)$ and $U_2(s)$, and the ratio of $U_1(s)$ and $I_2(s)$ are both transfer functions. (Related content can refer to the two-port networks in *Circuits Analysis*, Chinese edition, written by Weigang Zhang, Tsinghua University Press, 2015.1). Obviously, the driving point function can only be the impedance or the admittance function, but the transfer function can be the impedance function, the admittance function or the numerical ratio. These system functions are listed in Table 6.4. In system analysis, they are collectively referred to as system function, network function or transfer function, rather than their distinguishing attributes in general, and they are denoted by $H(s)$.

After the definition of system function is known, we will study separately the relationship between the system function and the impulse response, and that between the system function and the system differential equation.

According to equation (6.6-1), we have

$$Y_f(s) = F(s)H(s) . \tag{6.6-2}$$

At the same time,

$$y_f(t) = f(t) * h(t) .$$

With the Laplace transform on both sides of the expression and using convolution theorem, we obtain

$$\mathcal{L}\left[y_f(t)\right] = \mathcal{L}[f(t)] \cdot \mathcal{L}[h(t)]$$

or

$$Y_f(s) = F(s) \cdot \mathcal{L}[h(t)] .$$

Comparing this equation with equation (6.6-2), we will obtain the relationship between the system function and the impulse response, that is, they are a Laplace transform pair.

$$H(s) = \mathcal{L}[h(t)] , \tag{6.6-3}$$

$$h(t) = \mathcal{L}^{-1}[H(s)] . \tag{6.6-4}$$

We know that the relationship between the input and the output of an nth-order LTI system is described by a differential equation,

$$\sum_{i=0}^{n} a_i y^{(i)}(t) = \sum_{j=0}^{m} b_j f^{(j)}(t) . \tag{6.6-5}$$

With the Laplace transform on both sides of equation (6.6-5), the image function of zero-state response $y_f(t)$ can be deduced as

$$Y_f(s) = \frac{\sum_{j=0}^{m} b_j s^j}{\sum_{i=0}^{n} a_i s^i} F(s) = \frac{B(s)}{A(s)} F(s) = H(s)F(s) . \tag{6.6-6}$$

Since

$$H(s) = \frac{\sum_{j=0}^{m} b_j s^j}{\sum_{i=0}^{n} a_i s^i} = \frac{b_m s^m + b_{m-1} s^{m-1} + \cdots + b_1 s + b_0}{a_n s^n + a_{n-1} s^{n-1} + \cdots + a_1 s + a_0} , \tag{6.6-7}$$

the relationship between the function system and the coefficients of a differential equation are:
(1) The coefficient of term s^j in the numerator polynomial of the system function corresponds to the coefficient b_j belonging to the j-th order derivative of the input $f(t)$.
(2) The coefficient of term s^i in the denominator polynomial corresponds to the coefficient a_i of the i-th order derivative of the output $y(t)$.

Therefore, the system function of an LTI system can be obtained from the system differential equation. In turn, the differential equation of the system can be also determined by the system function.

Comparing equations (6.6-2) and (6.6-7) with the related equations of $H(p)$ and $H(j\omega)$ in Chapters 3–5, the definitions and the forms of $H(j\omega)$ and $H(s)$ are similar, hence, they can be considered as two expressions of the system function, which are used in the frequency domain and the complex frequency domain respectively. Additionally, the operator P represents a kind of operation instead of a variable, however, $H(p)$ is also quite similar to the system function in form, so the expression $H(s) = H(p)|_{p=s}$ can be also used as another way of finding the system function.

The conclusions for this section are as follows:
(1) The $H(p)$, $H(j\omega)$ and $H(s)$ can all express the relationship between the excitation and the response, or, they are all mathematical models used to describe characteristics of a system, and are determined by the system structure and the component parameters.
(2) The $H(p)$ reveals the differential relationship between the excitation and the response in the time domain, but $H(j\omega)$ and $H(s)$ show virtually their algebraic relationships in the frequency domain and in the complex frequency domain.
(3) Although $H(p)$, $H(j\omega)$ and $H(s)$ have different meanings, they have the same structure, and all can be transformed into each other by variable substitutions.

Example 6.6-1. The step response is $g(t) = (1 - e^{-2t})\varepsilon(t)$. Find the corresponding excitation $f(t)$ to produce a zero-state response as $y_f(t) = (1 - e^{-2t} - te^{-2t})\varepsilon(t)$.

Solution. This kind of problem can be solved by means of $H(s)$. Considering that $g(t)$ is the zero-state response with the action of $\varepsilon(t)$,

$$H(s) = \frac{Y_f(s)}{F(s)} = \frac{\frac{1}{s} - \frac{1}{s+2}}{\frac{1}{s}} = \frac{2}{s+2} \ .$$

Since

$$F(s) = \frac{Y_f(s)}{H(s)} = \frac{\frac{1}{s} - \frac{1}{s+2} - \frac{1}{(s+2)^2}}{\frac{2}{s+2}} = \frac{1}{s} - \frac{1}{2(s+2)} \ ,$$

the corresponding excitation $f(t)$ is

$$f(t) = \mathcal{L}^{-1}[F(s)] = \mathcal{L}^{-1}\left[\frac{1}{s} - \frac{1}{2(s+2)}\right] = \left(1 - \frac{1}{2}e^{-2t}\right)\varepsilon(t) \ .$$

Example 6.6-2. Find the system function of the LTI system described by the following differential equation:

$$\frac{d^2}{dt^2}y(t) + 3\frac{d}{dt}y(t) + 2y(t) = 2\frac{d}{dt}f(t) - 3f(t)$$

Solution. According to equation (6.6-7), we obtain

$$H(s) = \frac{2s - 3}{s^2 + 3s + 2} \ .$$

6.6.2 Analysis method with the system function

According to the definition of the system function, we have

$$Y_f(s) = F(s)H(s) \ . \tag{6.6-8}$$

The zero-state response of a system should be

$$y_f(t) = \mathcal{L}^{-1}[Y_f(s)] = \mathcal{L}^{-1}[F(s)H(s)] \ . \tag{6.6-9}$$

Equation (6.6-9) provides a new method to solve the zero-state response, which is called the analysis method based on the system function in the complex frequency domain.

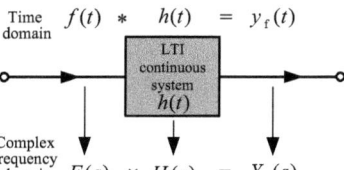

Fig. 6.8: Complex frequency domain analysis method.

The above analysis process for an LTI system by using Laplace transform is plotted in ▸ Figure 6.8.

Example 6.6-3. The excitation of an LTI circuit is a rectangular pulse with amplitude 1 and width 1; the impulse response of the circuit is $h(t) = e^{-at}\varepsilon(t)$. Find the zero-state response $y_f(t)$.

Solution. From the given conditions, we have

$$\mathcal{L}[f(t)] = \frac{1 - e^{-s}}{s} ,$$

so, the system function of the circuit is

$$H(s) = \mathcal{L}[h(t)] = \frac{1}{s + a} .$$

According to equation (6.6-8), the image function of the zero-state response is

$$Y_f(s) = H(s) \cdot F(s) = \frac{1 - e^{-s}}{s(s + a)} = \frac{1}{s(s + a)} - \frac{e^{-s}}{s(s + a)} . \tag{6.6-10}$$

The first term on the right side of the upper equation can be decomposed into

$$\frac{1}{s(s + a)} = \frac{1}{a}\left(\frac{1}{s} - \frac{1}{s + a}\right) ,$$

so,

$$\mathcal{L}^{-1}\left[\frac{1}{a}\left(\frac{1}{s} - \frac{1}{s + a}\right)\right] = \frac{1}{a}\varepsilon(t) - \frac{1}{a}e^{-at}\varepsilon(t) = \frac{1}{a}\left(1 - e^{-at}\right)\varepsilon(t) .$$

The original function of the second term on the right side of equation (6.6-10) is

$$\mathcal{L}^{-1}\left[\frac{e^{-s}}{s(s + a)}\right] = \frac{1}{a}\left[1 - e^{-a(t-1)}\right]\varepsilon(t - 1) ,$$

thus, the zero-state response $y_f(t)$ of this circuit is

$$y_f(t) = \frac{1}{a}\left(1 - e^{-at}\right)\varepsilon(t) - \frac{1}{a}\left[1 - e^{-a(t-1)}\right]\varepsilon(t - 1) .$$

From the above, it can be seen that the zero-state response can be found by means of the formula $y_f(t) = \mathcal{F}^{-1}[Y_f(j\omega)] = \mathcal{F}^{-1}[F(j\omega)H(j\omega)]$ in the Fourier transform method and by $y_f(t) = \mathcal{L}^{-1}[Y_f(s)] = \mathcal{L}^{-1}[F(s)H(s)]$ in the Laplace transform method, and there seems to be no essential difference between them. In the formulas, the Laplace transform method is just the Fourier transform method when the variable $j\omega$ is replaced by s. Therefore, once an excitation signal satisfies the Dirichlet conditions, the zero-state response of a system can both be obtained by the Fourier transform method and the Laplace transform method. The advantage of the Laplace transform method is only that it can help to find the zero-state responses to some signals that cannot meet the Dirichlet conditions, so, it is just a supplement to the Fourier transform method rather than the core value of the Laplace transform approach. Then, what is the most important characteristic of the Laplace transform method?

6.7 Analysis methods with system models in the *s* domain

6.7.1 Analysis with mathematic models

As we know, the mathematical model of an LTI system is a differential equation, and the differential operation can be transformed into multiplication by means of the differential property of the Laplace transform in the frequency domain. Therefore, differential equations that are difficult to solve can be converted into a simple algebraic equation. Simultaneously, the boundary conditions of the system are also contained in the algebraic equation, so the full solution of the equation can be obtained in one stroke. This is the ace in the hole of the Laplace transform analysis method.

The steps to solve the full response of system using the Laplace transform are as follows:

Step 1: According to circuit laws (KVL, KCL, etc.), build the system's mathematical model, namely get a linear differential equation with constant coefficients (if there are integral terms in it, we should perform the differential operation on the equation to remove the integral operations, and then a differential equation with higher one order than the original one can be obtained).

Step 2: Change this differential equation of the system in the time domain into the algebraic equation in the complex frequency domain using the differential property of the Laplace transform, and substitute the boundary conditions into it.

Step 3: Find the solution of the algebraic equation in the complex frequency domain.

Step 4: Restore the solution in the time domain from the one in the complex frequency domain using the inverse Laplace transform.

This system analysis method is also called the transform method, in which the Laplace transform is a kind of transferring machine; when we put a differential equation in the time domain into this machine, it can produce the algebraic equation in the complex frequency domain as a product that is easy to handle, just like peanuts are squeezed into peanut oil by a machine. The schematic diagram is shown in ▶ Figure 6.9.

Example 6.7-1. In the circuit shown in ▶ Figure 6.10, we know $u_S(t) = \frac{3}{5}e^{-2t}\varepsilon(t)$ and the starting condition $u_C(0_-) = -2$ V, $RC = 0.2$. Solve the voltage across the capacitor.

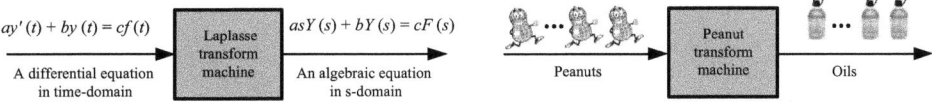

Fig. 6.9: The machines of Laplace transform and peanuts transform.

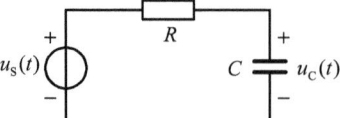

Fig. 6.10: E6.7-1.

Solution. According to Kirchhoff's law, the differential equation of the system is

$$\frac{du_C(t)}{dt} + \frac{1}{RC}u_C(t) = \frac{1}{RC}u_S(t) \ .$$

Appling Laplace transform and the differential property to both sides of the upper equation, we obtain

$$sU_C(s) - u_C(0_-) + 5U_C(s) = 5U_S(s) \ ,$$

namely,

$$U_C(s) = \frac{1}{s+5}[5U_S(s) + u_C(0_-)] \ .$$

Because

$$U_S(s) = \mathcal{L}^{-1}[u_S(t)] = \frac{3}{5} \cdot \frac{1}{s+2}, \quad u_C(0_-) = -2 \ ,$$

we have

$$U_C(s) = \frac{3}{(s+2)(s+5)} - \frac{2}{s+5} = \frac{1}{s+2} - \frac{1}{s+5} - \frac{2}{s+5} = \frac{1}{s+2} - \frac{3}{s+5}$$

The capacitor voltage $u_C(t)$ is

$$u_C(t) = e^{-2t}\varepsilon(t) - 3e^{-5t}\varepsilon(t) \ .$$

From this example, we can see that the zero-state response and the zero-input response of system can be found by the Laplace transform at the same time, which is impossible to do with the Fourier transform method. In other words, the most obvious characteristic of the Laplace transform is that the complete response of system can be directly obtained by solving the system mathematical model at one time. The reason for this advantage is that the differential property of Laplace transform can contain the starting values of a signal.

Now, calculation for the complete response is given. If the differential equation model of an LTI system is

$$y^{(n)}(t) + a_{n-1}y^{(n-1)}(t) + \cdots + a_1y'(t) + a_0y(t)$$
$$= b_mf^{(m)}(t) + b_{m-1}f^{(m-1)}(t) + \cdots + b_1f'(t) + b_0f(t) \ , \quad (6.7\text{-}1)$$

letting $\mathcal{L}[y(t)] = Y(s)$, $\mathcal{L}[f(t)] = F(s)$, using the differential property of Laplace transform in the time domain, the Laplace transform of $y^{(i)}(t)$ which is i-th order derivative of $y(t)$ is

$$\mathcal{L}\left[y^{(i)}(t)\right] = s^iY(s) - P_i(s) \ ,$$

where

$$P_i(s) = s^{i-1}y(0_-) + s^{i-2}y'(0_-) + \cdots + sy^{(i-2)}(0_-) + y^{(i-1)}(0_-) \,.$$

If $f(t)$ is a causal signal, its value is zero at time $t = 0_-$,

$$f(0_-) = f'(0_-) = \cdots = f^{(m-1)}(0_-) = 0 \,,$$

and then $\mathcal{L}\left[f^{(j)}(t)\right] = s^j F(s)$. With Laplace transform on both sides of equation (6.7-1), we obtain

$$A(s)Y(s) - C(s) = B(s)F(s) \,, \tag{6.7-2}$$

and

$$A(s) = s^n + a_{n-1}s^{n-1} + \cdots + a_1 s + a_0 \,,$$
$$C(s) = P_n(s) + a_{n-1}P_{n-1}(s) + \cdots + a_1 P_1(s) \,,$$
$$B(s) = b_m s^m + b_{m-1}s^{m-1} + \cdots + b_1 s + b_0 \,.$$

From equation (6.7-2), we have

$$Y(s) = \frac{C(s)}{A(s)} + \frac{B(s)}{A(s)}F(s) \,, \tag{6.7-3}$$

where the first term $\frac{C(s)}{A(s)}$ is only related to the response $y(t)$ and its all-order derivatives at $t = 0_-$; and the second term $\frac{B(s)}{A(s)}F(s)$ is only related to the excitation $f(t)$. Therefore, $\frac{C(s)}{A(s)}$ is the image function $Y_x(s)$ of the zero-input response, but $\frac{B(s)}{A(s)}F(s)$ is the image function $Y_f(s)$ of the zero-state response. That is,

$$Y(s) = \mathcal{L}\left[y_x(t)\right] + \mathcal{L}\left[y_f(t)\right] = Y_x(s) + Y_f(s) \,. \tag{6.7-4}$$

Example 6.7-2. The differential equation of a system is

$$y''(t) + 5y'(t) + 6y(t) = f'(t) + 6f(t) \,,$$

the starting conditions are $y(0_-) = 1$, $y'(0_-) = 2$, and the excitation is $\varepsilon(t)$. Find the zero-input response and the zero-state response.

Solution. With the Laplace transform on both sides of the equation at the same time, we obtain

$$s^2 Y(s) - sy(0_-) - y'(0_-) + 5sY(s) - 5y(0_-) + 6Y(s) = sF(s) - f(0_-) + 6F(s) \,.$$

Because

$$f(0_-) = 0 \,,$$

we have

$$\left(s^2 + 5s + 6\right) Y(s) - sy(0_-) - y'(0_-) - 5y(0_-) = (s + 6)F(s) \,,$$

and the solution is

$$Y(s) = \frac{s+6}{s^2+5s+6}F(s) + \frac{sy(0_-) + y'(0_-) + 5y(0_-)}{s^2+5s+6}.$$

The first term is only related to the excitation, which is the image function of the zero-state response. The second term is only related to the starting conditions, which is the image function $Y_x(s)$ of the zero-input response. Substituting $F(s) = \frac{1}{s}$ into the first term, we obtain

$$Y_f(s) = \frac{s+6}{s(s+2)(s+3)} = \frac{1}{s} - \frac{2}{s+2} + \frac{1}{s+3}.$$

Thus, the zero-state response is

$$y_f(t) = \varepsilon(t) - 2e^{-2t}\varepsilon(t) + e^{-3t}\varepsilon(t).$$

According to the starting conditions $y(0_-) = 1$, $y'(0_-) = 2$, we obtain

$$Y_x(s) = \frac{s+7}{(s+2)(s+3)} = \frac{5}{s+2} - \frac{4}{s+3},$$

and the zero-input response is

$$y_x(t) = 5e^{-2t}\varepsilon(t) - 4e^{-3t}\varepsilon(t).$$

The complete response is

$$y(t) = y_f(t) + y_x(t) = \varepsilon(t) + 3e^{-2t}\varepsilon(t) - 3e^{-3t}\varepsilon(t).$$

So far, two purposes of the introduction of the Laplace transform have been achieved. One was to remedy the shortcoming or deficiency of the Fourier transform, which can solve the problem that Fourier transforms of some signals are nonexistent. The other was to obtain the zero-state and the zero-input responses, which form the complete response, at one time. The key to solve the first problem is the introduction of the attenuation factor $e^{-\sigma t}$, whereas the fundamental reason for solving the second one lies in the differential property of the Laplace transform.

The free response is constituted by the superposition of all components related to the poles of $H(s)$ in the complete response, and the remaining part of the full response is the forced response.

6.7.2 Analysis with a circuit model

As stated in the above, if we want to analyze a system by means of the Laplace transform, the first thing is to establish the system's mathematical model, that is, the differential equation in the time domain. This is not difficult for general simple circuits, but for some complex circuits, their differential equations may be too complicated to build. So, we must try to find a new way for this problem.

We know that the zero-input response roots in the energy storage elements (dynamic elements) in a system, such as inductors and capacitors. That is, the memory (storage) property of the dynamic elements results in the zero-input response. Therefore, understanding the circuit models of dynamic elements in the s domain is the basis or premise for finding the full response from the circuit models.

As we know, a circuit system is generally composed of resistors, inductors and capacitors. If circuit models of the components can be drawn in the s domain, then the mathematic model of the system in the s domain can also be written directly from KCL and KVL laws, etc., so that this method omits the complicated process where the differential equation of the system in the time domain must be established first and then be transformed to the algebraic equation in Section 6.7.1; as a result, the purpose of simplifying the solution steps of the equation has been achieved. So, first we will introduce the s domain mathematic models of Kirchhoff's laws.

In the time domain, the expression of the KCL is $\sum i(t) = 0$. The Laplace transform on both sides of the equation, yields

$$\sum I(s) = 0 . \tag{6.7-5}$$

Equation (6.7-5) is the representation of the KCL in the s domain, where $I(s)$ is the image function of the corresponding branch current $i(t)$. Equation (6.7-5) states that the algebraic sum of the outflow and the inflow image currents at any a node in a given circuit is zero.

Similarly, the representation of the KVL in the s domain is

$$\sum U(s) = 0 . \tag{6.7-6}$$

That means that for any a closed loop in a circuit, the algebra sum of image voltages on all components equals zero.

Next, we will discuss the circuit models for the elements R, L and C in the s domain.

1. Resistor model

At any moment, voltage $u_R(t)$ and current $i_R(t)$ for an LTI resistor R are restrained by

$$u_R(t) = Ri_R(t) .$$

With the Laplace transform on both sides of the equation and setting $\mathcal{L}\,[u_R(t)] = U_R(s)$ and $\mathcal{L}\,[i_R(t)] = I_R(s)$, we obtain

$$U_R(s) = RI_R(s) . \tag{6.7-7}$$

Equation (6.7-7) shows the relationship between the current and the voltage of R in the s domain, which is called the s domain mathematical model of R. The circuit models of a resistor R in the time domain and the s domain are illustrated in ▶ Figure 6.11.

(a) *Time*-domain model   (b) s-domain model

Fig. 6.11: Time domain and s domain circuit models of a resistor.

2. Inductor model

At any moment, voltage $u_L(t)$ and current $i_L(t)$ of an LTI inductor L are restrained by

$$u_L(t) = L\frac{di_L(t)}{dt} \; .$$

With the Laplace transform on both sides of the equation and letting $\mathcal{L}\,[u_L(t)] = U_L(s)$, $\mathcal{L}\,[i_L(t)] = I_L(s)$, we obtain the s domain mathematical model of an inductor

$$U_L(s) = sLI_L(s) - Li_L(0_-) \; . \tag{6.7-8}$$

In equation (6.7-8), sL is called the inductive reactance in the s domain. It can be seen that the image function $U_L(s)$ of the voltage on an inductor consists of two parts; one is the product of the inductive reactance sL and the image current $I_L(s)$, the other is the equivalent voltage source $Li_L(0_-)$ in the s domain. As a result, the inductor L in the s domain can be seen as a series circuit with an inductive reactance sL and a voltage source $Li_L(0_-)$. So, the circuit models of an inductor in the time and s domains are shown in ▸ Figure 6.12.

3. Capacitor model

At any moment, voltage $u_C(t)$ and current $i_C(t)$ of an LTI capacitor C are restrained by

$$u_C(t) = \frac{1}{C} \int_{-\infty}^{t} i_C(\tau)d\tau \; .$$

With the Laplace transform on both sides of the equation and setting $\mathcal{L}\,[u_C(t)] = U_C(s)$, $\mathcal{L}\,[i_C(t)] = I_C(s)$, we can get the s domain mathematical model of a capacitor

$$U_C(s) = \frac{1}{sC}I_C(s) + \frac{1}{sC}\int_{-\infty}^{0_-} i_C(\tau)d\tau = \frac{1}{sC}I_C(s) + \frac{1}{s}u_C(0_-) \; , \tag{6.7-9}$$

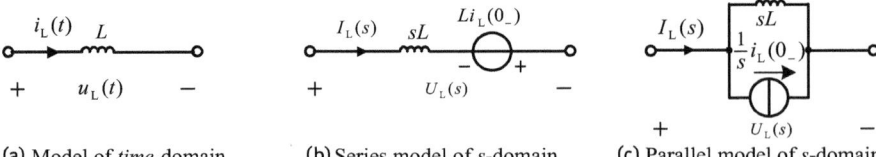

(a) Model of *time*-domain   (b) Series model of s-domain   (c) Parallel model of s-domain

Fig. 6.12: Time domain and s domain circuit models of a inductor.

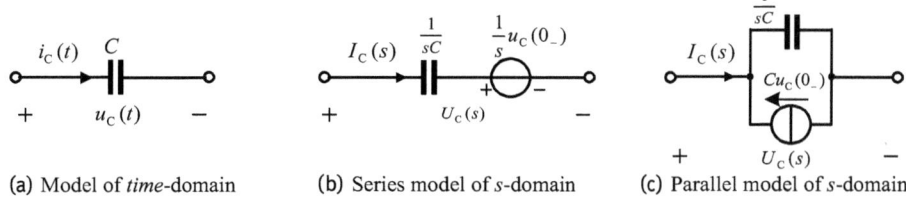

Fig. 6.13: Time domain and s domain circuit models of a capacitor.

where $U_C(s)$ and $I_C(s)$ are the image voltage and image current, and $\frac{1}{sC}$ is the capacitive reactance in the s domain. It can be seen that the image voltage across a capacitor is composed of two parts; the first term is the product of the capacitive reactance and the image current, the second is equivalent to a voltage source in the s domain, which can be called the internal image voltage source. Based on the s domain KVL model, the relation between $U_C(s)$ and $I_C(s)$ can be considered as a series circuit with a capacitive reactance $\frac{1}{sC}$ and an image voltage source $\frac{1}{s}u_C(0_-)$. The circuit models of a capacitor in the time and s domains are shown in ▶ Figure 6.13.

The component circuit models in equations (6.7-8) and (6.7-9) are called the series circuit models. If the two formulas are rewritten as

$$I_L(s) = \frac{1}{sL}U_L(s) + \frac{1}{s}i_C(0_-) \tag{6.7-10}$$

and

$$I_C(s) = sCU_C(s) - Cu_C(0_-) , \tag{6.7-11}$$

the s domain parallel circuit models of two elements can be obtained, which are illustrated in ▶ Figure 6.12c and ▶ Figure 6.13c, respectively.

In practice, which model should be adopted by us depends on the specific conditions. If we need to write out voltage equations, the series model is the best choice, and the parallel model is more suitable for current equations.

It can be seen from equations (6.7-8) and (6.7-9) that on their right-hand sides, besides responses $sLI_L(s)$ and $\frac{1}{sC}I_C(s)$ which can reflect the effect of the excitation signal, there are also terms such as $Li_L(0_-)$ and $\frac{1}{s}u_C(0_-)$, which can reflect the effect from the state of the system, and the existence of the two terms just is the root cause to obtain the complete response in one stroke by the Laplace transform. So, the essential difference between the Laplace and Fourier transform analysis methods lies in their different differential properties.

Example 6.7-3. In the circuit shown in ▶ Figure 6.14a, $f(t) = 3e^{-10t}\varepsilon(t)$, $u_C(0_-) = 5$ V are known. Find the voltage $y(t)$ on a resistor.

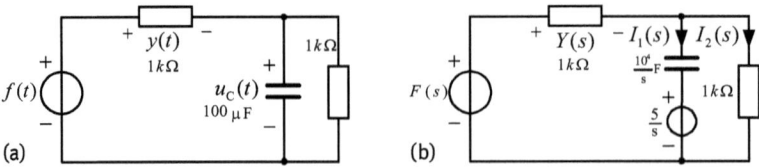

Fig. 6.14: E6.7-3.

Solution. The s domain model of the circuit is shown in ▶ Figure 6.14b. From Kirchhoff's laws, we have

$$Y(s) = 1,000\,[I_1(s) + I_2(s)]$$

$$F(s) = Y(s) + \frac{10^4}{s}I_1(s) + \frac{5}{s}$$

$$F(s) = Y(s) + 1,000 I_2(s)$$

Solving the three simultaneous equations yields

$$Y(s) = \frac{s+10}{s+20}F(s) - \frac{5}{s+20}\,.$$

Substituting $F(s) = \frac{3}{s+10}$ into the upper equation, we obtain

$$Y(s) = -\frac{2}{s+20}\,,$$

so,

$$y(t) = -2e^{-20t}\varepsilon(t)\,\text{V}\,.$$

Example 6.7-4. For the circuit shown in ▶ Figure 6.15a, the excitation is $u(t)$ and the response is $i(t)$. Solve the impulse response and step response of the circuit.

Solution. According to the s domain model of zero-state response [as shown in ▶ Figure 6.15b], we obtain the system function

$$H(s) = \frac{I(s)}{U(s)} = \frac{1}{sL_1 + \frac{R_1(sL_2+R_2)}{sL_2+R_1+R_2}} = \frac{1}{s + \frac{2(s+3)}{s+5}} = \frac{s+5}{s^2+7s+6}\,.$$

Then,

$$h(t) = \mathcal{L}^{-1}[H(s)] = \mathcal{L}^{-1}\left[\frac{s+5}{s^2+7s+6}\right] = \mathcal{L}^{-1}\left[\frac{4}{5} \times \frac{1}{s+1} + \frac{1}{5} \times \frac{1}{s+6}\right]$$

$$= \left(\frac{4}{5}e^{-t} + \frac{1}{5}e^{-6t}\right)\varepsilon(t)\,.$$

The image function $G(s)$ of the step response is

$$G(s) = H(s) \cdot \mathcal{L}\,[\varepsilon(t)] = \frac{s+5}{s(s^2+7s+6)} = \frac{5}{6} \times \frac{1}{s} - \frac{4}{5} \times \frac{1}{s+1} - \frac{1}{30} \times \frac{1}{s+6}\,,$$

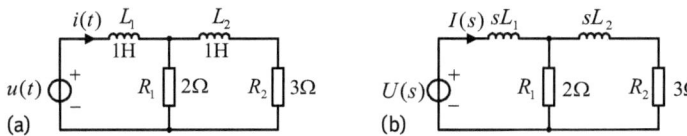

Fig. 6.15: E6.7-4.

so, the step response is

$$g(t) = \mathcal{L}^{-1}[G(s)] = \left(\frac{5}{6} - \frac{4}{5}e^{-t} - \frac{1}{30}e^{-6t} \right) \varepsilon(t) .$$

Example 6.7-5. A circuit is shown in ► Figure 6.16a. When $t < 0$, the switch S turns off, and then the circuit is in steady state; when $t = 0$, the switch S turns on. Find the output voltage $u(t)$ when $t \geq 0$.

Solution. First, find the starting values of the capacitor voltage and inductor current,

$$u_C(0_-) = 12 \, \text{V}, \quad i_L(0_-) = \frac{u_S}{R_2} = 3 \, \text{A} .$$

According to the s domain model of the circuit shown in ► Figure 6.16b, we can obtain the algebraic equation of the s domain model of the circuit, that is,

$$\left[\frac{1}{R_3} + \frac{1}{R_1 + \frac{1}{sC}} + \frac{1}{R_2 + sL} \right] U(s) = \frac{U_S(s)}{R_3} + \frac{\frac{u_C(0_-)}{s}}{R_1 + \frac{1}{sC}} - \frac{Li_L(0_-)}{R_2 + sL}$$

Substituting the element parameters and initial conditions into the upper equation, we obtain

$$\left[\frac{1}{2} + \frac{1}{2 + \frac{2}{s}} + \frac{1}{s + 4} \right] U(s) = \frac{6}{s} + \frac{\frac{12}{s}}{2 + \frac{2}{s}} - \frac{3}{s + 4} ,$$

so,

$$U(s) = \frac{9s^2 + 51s + 24}{s \left(s^2 + 5.5s + 3 \right)} = \frac{9s^2 + 51s + 24}{s(s + 4.89)(s + 0.615)} .$$

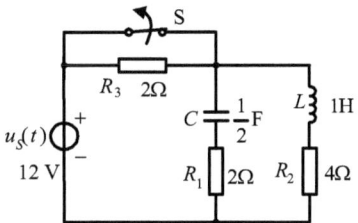

Fig. 6.16: E6.7-5.

We have the partial fraction expansion of the above equation,

$$U(s) = \frac{7.98}{s} - \frac{0.493}{s + 4.89} + \frac{1.5}{s + 0.615} .$$

Thus, the output voltage $u(t)$ is

$$u(t) = \mathcal{L}^{-1}[U(s)] = \left(7.98 - 0.493e^{-4.89t} + 1.5e^{-0.615t}\right)\varepsilon(t) .$$

6.8 Analysis method from signal decomposition in the s domain

According to the contents of Section 5.6.3, there is a similar derivation process for the signal decomposition analysis method in complex frequency domain.

From equation (6.1-4), an aperiodic signal $f(t)$ can be represented as the linear combination of infinite terms of complex exponential signals just like e^{st}. Therefore, the zero-state response $y_{f1}(t)$ to the basic signal e^{st} must be solved first. We have

$$y_{f1}(t) = e^{st} * h(t) = \int_{-\infty}^{\infty} h(\tau)e^{s(t-\tau)}d\tau = e^{st}\int_{-\infty}^{\infty} h(\tau)e^{-s\tau}d\tau .$$

The term $\int_{-\infty}^{\infty} h(\tau)e^{-s\tau}d\tau = \int_{-\infty}^{\infty} h(t)e^{-st}dt$ is just the Laplace transform $H(s)$ of $h(t)$, and, therefore,

$$y_{f1}(t) = H(s)e^{st} . \tag{6.8-1}$$

Equation (6.8-1) states that the zero-state response to the basic signal e^{st} is the product of the signal itself and a constant coefficient, which is independent of time t. This coefficient is the Laplace transform $H(s)$ of the impulse response $h(t)$. Thus, we reach the same conclusion as equation (6.6-8).

The Laplace transform $Y_f(s)$ of the zero-state response of a system to any aperiodic signal $f(t)$ is equal to the product of the Laplace transform $F(s)$ of this signal and the system function $H(s)$. That is,

$$Y_f(s) = F(s)H(s) .$$

The derivation process is similar to that shown in ▶ Figure 5.26.

From the above introduction and Chapters 3–5, we find that if the continuous system analysis is regarded as a big stage for performance, then the system function is a leading role. It holds the frequency domain with the left hand and the complex frequency domain with the right hand, steps on the time domain operator, wears the impulse response as a vest, and frequently appears on the stage to show and attract the attention of audiences with the zero-state response as a spotlight. Moreover, in the subsequent contents of the discrete system analysis, audiences can still see his charming figure; the difference is just that it is on another stage. Therefore, it is no exaggeration to say that the system function is a key for system analysis. Readers are advised to understand its concept and meaning fully in order to judge the whole from one sample and to draw parallels from its inference.

6.9 Relationship among the time, frequency and complex frequency domain methods

From Chapters 3–6, we have learned some basic analysis methods for an LTI system, and have built the following three basic mathematical models for solving the zero-state response.

The model in the time domain is

$$y_f(t) = f(t) * h(t) . \tag{6.9-1}$$

The model in the frequency domain is

$$Y_f(j\omega) = F(j\omega)H(j\omega) . \tag{6.9-2}$$

The model in the complex frequency domain is

$$Y_f(s) = F(s)H(s) . \tag{6.9-3}$$

Equations (6.9-2) and (6.9-3) are, respectively, the transform forms of equation (6.9-1) in the frequency and the complex frequency domains. Their main function is to transform the integral operation in the time domain into the algebraic operation in transform domain, and thus can greatly simplify the system analysis process.

The Laplace transform extends the application range of the Fourier transform and is widely used in system analysis, but because
(1) the Laplace transform has no obvious physical meaning;
(2) the result of the analysis of a physical system has to be implemented on the time and frequency characteristics of the system in the end;

So, we find that the system analysis methods are essentially divided into only two types: time domain and frequency domain. The method in the complex frequency domain is not only the extension of one in frequency domain (the Laplace transform can be regarded as the generalized Fourier transform), but also the link between methods in the time and frequency domains. The relationship between them is shown conceptually in ▶ Figure 6.17.

Note that ▶ Figure 6.17 emphasizes the relationship between the three methods in concept, but in the real analysis, the Fourier transform of a signal could not be obtained directly by replacing s with $j\omega$ in the Laplace transform only when the Laplace transform ROC contains a $j\omega$ axis. If the ROC does not contain the $j\omega$ axis, then the signal does not have the corresponding Fourier transform. If the boundary of the ROC is located on the $j\omega$ axis, the Fourier transform of a signal could not be obtained by replacing s with $j\omega$ directly, even though the Fourier transform of the signal exists, but it could be obtained by

$$F(j\omega) = F(s)\big|_{s=j\omega} + \pi \sum_{i=1}^{N} K_i \delta(\omega - \omega_i) , \tag{6.9-4}$$

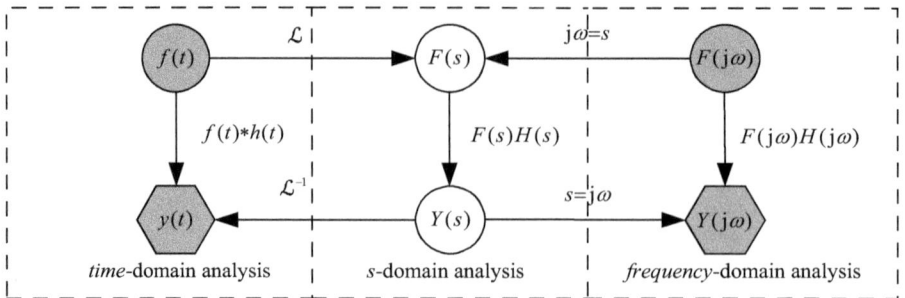

Fig. 6.17: The bridgefunction of the Laplace transform in system analysis.

where the second term on the right of the expression is the sum of the impulse signals located on the poles of $F(s)$, such as $j\omega_1, j\omega_2, \ldots, j\omega_N$. The K_i represents all coefficients of the partial fractions decomposed from $F(s)$.

Example 6.9-1. Using $F(j\omega) = F(s)\big|_{s=j\omega} + \pi \sum_{i=1}^{N} K_i \delta(\omega - \omega_i)$ solve the Fourier transform of signal $f(t) = \cos \omega_0 t \varepsilon(t)$.

Solution. According to Table 6.2, we obtain

$$F(s) = \frac{s}{s^2 + \omega_0^2}, \tag{6.9-5}$$

This can be expanded as the partial fractions

$$F(s) = \frac{s}{s^2 + \omega_0^2} = \frac{1}{2}\left(\frac{1}{s + j\omega_0}\right) + \frac{1}{2}\left(\frac{1}{s - j\omega_0}\right). \tag{6.9-6}$$

The poles of the first formula are $j\omega_0$ and $-j\omega_0$, and we have

$$K_1 = \frac{1}{2}, \quad K_2 = \frac{1}{2}.$$

So,

$$F(j\omega) = F(s)\big|_{s=j\omega} + \pi \sum_{i=1}^{N} K_i \delta(\omega - \omega_i) = \frac{j\omega}{\omega_0^2 - \omega^2} + \frac{\pi}{2}\left[\delta(\omega - \omega_0) + \delta(\omega + \omega_0)\right].$$

We should explain that because $h(t)$, $H(j\omega)$ and $H(s)$ are all defined under the zero-state condition, when we employ above models to find the system response we can directly write the results as "system response is $y(t)$" and do not need to consider whether the response is the zero-state, zero-input or full response.

6.10 Solved questions

Question 6-1. Find the inverse Laplace transform of $\frac{s+3}{s^2+2s+2}e^{-s}$.

Solution. Using partial fraction expansion, we have

$$\frac{s+3}{s^2+2s+2} = \frac{s+1}{(s+1)^2+1} + \frac{2}{(s+1)^2+1}.$$

So, the original function of $\frac{s+3}{s^2+2s+2}$ is $e^{-t}[\cos t + 2\sin t]\varepsilon(t)$, and the original function of $\frac{s+3}{s^2+2s+2}e^{-s}$ is $e^{-(t-1)}[\cos(t-1) + 2\sin(t-1)]\varepsilon(t-1)$.

Question 6-2. For the circuit in ▸ Figure Q6-2-1 and ▸ Figure Q6-2-2, $i_L(0_-) = 1\,\text{A}$, $u_C(0_-) = 1\,\text{V}$ are known. Plot the equivalent circuit in its s domain and calculate $i_R(t)$.

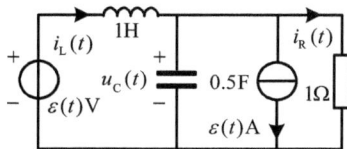

Fig. Q6-2-1

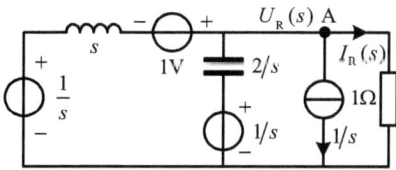

Fig. Q6-2-2

Solution. The equivalent circuit of ▸ Figure Q6-2-1 s domain is shown in ▸ Figure Q6-2-2.

Let the node voltage at point A be $U_R(s)$. Using node voltage method, the node voltage equation of point A can be written as

$$\left(1 + \frac{s}{2} + \frac{1}{s}\right)U_R(s) = -\frac{1}{s} + \frac{\frac{1}{s}}{\frac{2}{s}} + \frac{\frac{1}{s}+1}{s}.$$

So,

$$U_R(s) = \frac{s^2+2}{s(s^2+2s+2)} = \frac{1}{s} + \frac{-2}{(s+2)^2+1}.$$

Using the inverse Laplace transform, we obtain $u_R(t) = 1 - 2e^{-t}\sin t$, $t > 0$. So,

$$i_R(t) = \frac{u_R(t)}{1} = u_R(t).$$

Question 6-3. The block diagram of the LTI system in ▶ Figure Q6-3 is known, and when $f(t) = 3(1 + e^{-t})\varepsilon(t)$, the total response of the system is $y(t) = (4e^{-2t} + 3e^{-3t} + 1)\varepsilon(t)$.

(1) Find input–output equation of the system.
(2) Find the zero-input response $y_x(t)$ of the system.
(3) Find the starting states $y(0_-)$ and $y'(0_-)$ of the system.

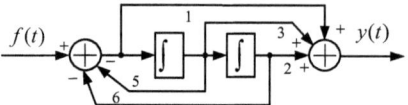

Fig. Q6-3

Solution. (1) From the diagram, the differential equation of the system is

$$y''(t) + 5y'(t) + 6y(t) = f''(t) + 3f'(t) + 2f(t) . \tag{6.10-1}$$

(2) With the Laplace transform in the equation (6.10-1), we have

$$Y(s) = \frac{sy(0_-) + y'(0_-) + 5y(0_-)}{s^2 + 5s + 6} + \frac{s^2 + 3s + 2}{s^2 + 5s + 6}F(s) = Y_x(s) + \frac{s+1}{s+3}F(s) . \tag{6.10-2}$$

Finding Laplace transforms of the known excitation and response and putting them into the equation (6.10-2),

$$Y_x(s) = \frac{4}{s+2} + \frac{3}{s+3} + \frac{1}{s} - \frac{s+1}{s+3}\left(\frac{1}{s} + \frac{3}{s+1}\right) = \frac{2s+8}{(s+2)(s+3)}$$
$$= \frac{4}{s+2} - \frac{2}{s+3} . \tag{6.10-3}$$

Then, calculating the inverse Laplace transform of the equation (6.10-3), yields

$$y_x(t) = 2\left(2e^{-2t} - e^{-3t}\right)\varepsilon(t) . \tag{6.10-4}$$

(3) From equation (6.10-2), we have

$$Y_x(s) = \frac{sy(0_-) + y'(0_-) + 5y(0_-)}{s^2 + 5s + 6} . \tag{6.10-5}$$

From equation (6.10-3), we have

$$Y_x(s) = \frac{2s+8}{(s+2)(s+3)} . \tag{6.10-6}$$

Comparing equations (6.10-5) and (6.10-6), we obtain

$$\begin{cases} y(0_-) = 2 \\ y'(0_-) + 5y(0_-) = 8 \end{cases} \rightarrow \begin{cases} y(0_-) = 2 \\ y'(0_-) = -2 \end{cases} .$$

Question 6-4. The differential equation of an LTI causal system is

$$y''(t) + 3y'(t) + 2y(t) = 5f'(t) + 4f(t) \quad t > 0 \,.$$

The input $f(t) = e^{-3t}\varepsilon(t)$ and the starting states are $y(0_-) = 2$, $y'(0_-) = 1$. In the s domain find

(1) the zero-input response $y_x(t)$ and the zero-state response $y_f(t)$ of the system;
(2) the system function $H(s)$, the impulse response $h(t)$, and judge whether the system is stable; and
(3) if the input is $f(t) = e^{-3t}\varepsilon(t-2)$, calculate (1) and (2) again.

Solution. (1) With the Laplace transform on both sides of the differential equation, the input–output expression in the s domain can be obtained,

$$s^2 Y(s) - sy(0_-) - y'(0_-) + 3\,[sY(s) - y(0_-)] + 2Y(s) = (5s+4)F(s) \,.$$

So, the total response of the system in the s domain is

$$Y(s) = \frac{sy(0_-) + y'(0_-) + 3y(0_-)}{s^2 + 3s + 2} + \frac{5s+4}{s^2 + 3s + 2}F(s) \,.$$

The starting conditions are $y(0_-) = 2$ and $y'(0_-) = 1$, and the total response is

$$Y(s) = \frac{2s+7}{s^2 + 3s + 2} + \frac{5s+4}{s^2 + 3s + 2}F(s) = Y_x(s) + Y_f(s) \,.$$

With the inverse Laplace transform in $Y_x(s)$ and zero-input response can be obtained by

$$y_x(t) = 5e^{-t} - 3e^{-t} \quad (t \ge 0) \,.$$

Because the image function of $f(t) = e^{-3t}\varepsilon(t)$ is $F(s) = \frac{1}{s+3}$, the zero-state response is

$$y_f(t) = \left(6e^{-2t} - 0.5e^{-t} - 5.5e^{-3t}\right)\varepsilon(t) \,.$$

(2) According to (1), the expression of the zero-state response of the system in the s domain is

$$Y_f(s) = \frac{5s+4}{s^2 + 3s + 2}F(s) \,.$$

So,

$$H(s) = \frac{Y_f(s)}{F(s)} = \frac{5s+4}{s^2 + 3s + 2} = -\frac{1}{s+1} + \frac{6}{s+2} \,.$$

Finding the inverse Laplace transform of the equation, we obtain the unit impulse response

$$h(t) = \left(6e^{-2t} - e^{-t}\right)\varepsilon(t) \,.$$

There are a zero $z = -\frac{4}{5}$ and two poles $p_1 = -1$, $p_2 = -2$ for the system function; obviously, both poles are on the left half-plane in the s domain, so the system is stable.

(3) When input is $f(t) = e^{-3t}\varepsilon(t-2)$, the system function is changeless, so the system is still stable, and the $h(t)$ and the $y_x(t)$ of the system are also changeless. Because

$$f(t) = e^{-3t}\varepsilon(t-2) \overset{\mathcal{L}}{\longleftrightarrow} F(s) = \frac{e^{-6}}{s+3}e^{-2s},$$

we obtain

$$Y_f(s) = \frac{5s+4}{s^2+3s+2}F(s) = \frac{5s+4}{s^2+3s+2}\frac{e^{-6}}{s+3}e^{-2s}$$

$$= e^{-6}\left[6\frac{1}{s+2} - \frac{1}{2}\frac{1}{s+1} - \frac{11}{2}\frac{1}{s+3}\right]e^{-2s}.$$

Using the inverse Laplace transform in the above equation, we obtain the zero-state response

$$y_f(t) = e^{-6}\left[6e^{-2(t-2)} - 0.5e^{-(t-2)} - 5.5e^{-3(t-2)}\right]\varepsilon(t-2).$$

6.11 Learning tips

The complex frequency domain system analysis is the complement and promotion of that of frequency domain, so we suggest that readers pay attention to the following points.
(1) The Fourier transform is a special case of the Laplace transform.
(2) The system function and impulse response are the Laplace transform pair.
(3) The Laplace transform of the zero-state response of a system is the product of the Laplace transform of excitation and the system function.
(4) The s domain model of a circuit is actually a diagram of the Laplace transform applied on a differential equation of the system.

6.12 Problems

Problem 6-1. Find the unilateral Laplace transforms of the following signals:
(1) $(3\sin 2t + 2\cos 3t)\varepsilon(t)$,
(2) $2\delta(t) - e^{-t}\varepsilon(t)$,
(3) $\cos^2 2t\varepsilon(t)$
(4) $e^{-(t+a)}\cos \omega t\varepsilon(t)$,
(5) $e^{-(t-1)}\varepsilon(t-1)$,
(6) $e^{-(t-1)}\varepsilon(t)$

(7) $t[\varepsilon(t) - \varepsilon(t-1)]$,
(8) $(t+1)\varepsilon(t+1)$,
(9) $\sin \pi t[\varepsilon(t) - \varepsilon(t-2)]$
(10) $te^{-(t-1)}\varepsilon(t-2)$,
(11) $t^2\varepsilon(t-1)$,
(12) $t^2\cos t$

Problem 6-2. The unilateral Laplace transform $F(s) = \frac{1}{s^2+3s-5}$ of a causal signal $f(t)$ is known. Find the unilateral Laplace transforms of following signals:
(1) $e^{-t}f(4t)$
(2) $f(2t-4)$

(3) $tf''(t)$
(4) $f(t)\sin 2t$

(5) $\int_0^{t-2} f(\tau)e^{\tau}d\tau$
(6) $\frac{1}{t}f(t)$

Problem 6-3. Find the unilateral Laplace transforms of the signals shown in ▶ Figure P6-3.

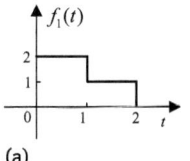
(a)

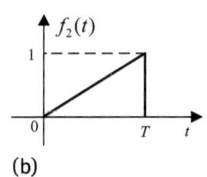
(b)

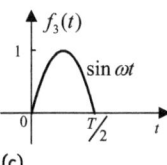
(c)

Fig. P6-3

Problem 6-4. A causal periodic signal $f(t) = f(t)\varepsilon(t)$ with periodic T is known, the image function is $F(s)$, and the image function of the signal during the first periodic $f_1(t) = f(t)[\varepsilon(t) - \varepsilon(t - T)]$ is $F_1(s)$. Prove $F(s) = \frac{F_1(s)}{1 - e^{-sT}}$.

Problem 6-5. Using the conclusions of Problem 6-4, find the image functions of the following causal periodic signals shown in ▶ Figure P6-5.

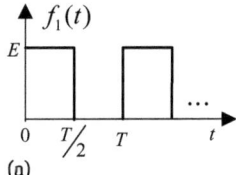
(a)

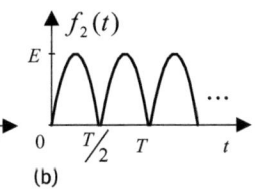
(b)

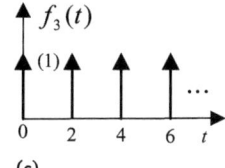
(c)

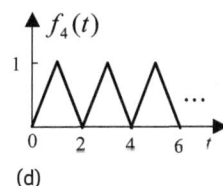
(d)

Fig. P6-5

Problem 6-6. Find the initial values and final values of the following inverse Laplace transforms:

(1) $F(s) = \frac{s+3}{(s+1)(s+2)^2}$;

(2) $F(s) = \frac{s^3+6s^2+6s}{s^2+6s+8}$;

(3) $F(s) = \frac{s^2+2s+3}{(s+1)(s^2+4)}$;

(4) $F(s) = \frac{2s+1}{s^3+3s^2+2s}$;

(5) $F(s) = \frac{1-e^{-2s}}{s(s^2+4)}$;

(6) $F(s) = \frac{1}{s(1+e^{-s})}$

Problem 6-7. Find the original functions of the following image functions:

(1) $\frac{4s+2}{s^2+8s+15}$

(2) $\frac{s^3+s-3}{s^2+3s+2}$

(3) $\frac{3s+4}{s^3+5s^2+8s+4}$

(4) $\frac{s+5}{s(s^2+2s+5)}$

(5) $\frac{4s^2+6}{s^3+s^2-2}$

(6) $\frac{s+3}{(s+1)^3(s+2)}$

Problem 6-8. Find the inverse Laplace transforms of following functions:

(1) $\frac{2-e^{-3s}}{s+2}$

(2) $\frac{s(1+e^{-sT})}{s^2+\pi^2}$

(3) $\left(\frac{1+e^{-2s}}{s}\right)^2$

(4) $\frac{1-e^{-s}}{4s(s^2+1)}$

(5) $\frac{s+6e^{-s}}{s^2+9}$

(6) $\frac{1}{1+e^{-s}}$

(7) $\frac{1}{s(1+e^{-s})}$

(9) $\frac{1}{s(1-e^{-s})}$

(8) $\frac{1}{1-e^{-s}}$

(10) $\ln\frac{s}{s+9}$

Problem 6-9. The excitation is $f(t) = e^{-t}\varepsilon(t)$, and the corresponding zero-state response is $y_f(t) = \left(\frac{1}{2}e^{-t} - e^{-2t} + \frac{1}{2}e^{-3t}\right)\varepsilon(t)$. Find
(1) the impulse response $h(t)$;   (2) the input–output equation.

Problem 6-10. The impulse response and the zero-state response of a circuit are, respectively, $h(t) = \delta(t) - 11e^{-10t}\varepsilon(t)$ and $y_f(t) = (1 - 11t)e^{-10t}\varepsilon(t)$. Find the excitation $f(t)$.

Problem 6-11. The system function is $H(s) = \frac{s}{s^2+3s+2}$. Find the system response to the following excitation $f(t)$ and point out the free response and the forced response:
(1) $f(t) = 10\varepsilon(t)$ (2) $f(t) = 10\sin(t)\varepsilon(t)$

Problem 6-12. In the circuit shown in ▶ Figure P6-12, $i_L(0_-) = 2\,\text{A}$, $i_S(t) = 5t\varepsilon(t)\,\text{A}$. Find $u_L(t)$.

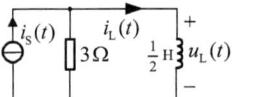

Fig. P6-12

Problem 6-13. The circuit shown in ▶ Figure P6-13 is in steady state when $t < 0$, and when $t = 0$ the switch S turns on. Find the capacitor voltage $u_C(t)$ when $t \geq 0$.

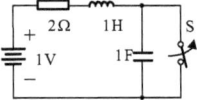

Fig. P6-13

Problem 6-14. A circuit is shown in ▶ Figure P6-14. It is known that $C = 1\,\text{F}$, $L = \frac{1}{2}\,\text{H}$, $R_1 = \frac{1}{5}\,\Omega$, $R_2 = 1\,\Omega$, $u_C(0_-) = 5\,\text{V}$, $i_L(0_-) = 4\,\text{A}$ and $u_S(t) = 10\varepsilon(t)$. Find the full response $i_1(t)$ for $t \geq 0$.

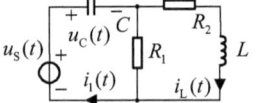

Fig. P6-14

Problem 6-15. The starting state of an LTI system is given, an input is $f_1(t) = \delta(t)$ and the corresponding full response is $y_1(t) = -3e^{-t}(t \geq 0)$; another input is $f_2(t) = \varepsilon(t)$ and the full response is $y_2(t) = 1 - 5e^{-t}(t \geq 0)$. Find the full response to the input $f(t) = t\varepsilon(t)$.

Problem 6-16. The starting states of an LTI system stay the same, when the input $f_1(t) = \delta(t)$, the full response $y_1(t) = \delta(t) + e^{-t}\varepsilon(t)$, and when the input $f_2(t) = \varepsilon(t)$, the full response $y_2(t) = 3e^{-t}\varepsilon(t)$. Find the response of the system $y_3(t)$ when

$$f_3(t) = \begin{cases} 0 & t < 0 \\ t & 0 < t < 1 \\ 1 & t > 1 \end{cases} .$$

Problem 6-17. The system function of a second-order LTI system $H(s) = \frac{s+3}{s^2+3s+2}$. The input signal $f(t) = e^{-3t}\varepsilon(t)$ and the starting conditions $y(0_-) = 1, y'(0_-) = 2$. Find the full response $y(t)$, $y_x(t)$ and $y_f(t)$, and determine the free response and the forced response.

Problem 6-18. The differential equation of a system $y'(t) + 2y(t) = f(t)$, the starting condition $y(0_-) = 1$, and $f(t) = \sin(2t) \cdot \varepsilon(t)$. Find the full response.

Problem 6-19. The differential equation of a system $y''(t) + 4y'(t) + 3y(t) = 3f(t)$, the excitation $f(t) = \varepsilon(t)$, and the starting conditions $y(0_-) = 2, y'(0_-) = -1$. Using Laplace transform, find the full response of the system.

Problem 6-20. Input–output equation of an LTI system $y''(t) + 5y'(t) + 6y(t) = 6f(t)$. Find the zero-input and zero-state responses $y_x(t)$ and $y_f(t)$ of the system when $f(t) = 2e^{-t}\varepsilon(t), y(0_-) = 0$, and $y'(0_-) = 1$.

Problem 6-21. The differential equation of a system $y''(t) + 2y'(t) + y(t) = f'(t)$, the starting conditions $y(0_-) = 1, y'(0_-) = 2$, and the input signal. Find the zero-input, zero-state response and the full responses of the system and point out the free and the forced responses.

7 Simulation and stability analysis of continuous-time systems

Questions: Graphs are a very useful tool to analyze and solve. Can we, then, use a graph instead of a mathematical model to represent a system? Further, how can we test the stability of a feedback system?

Solution: Use several basic graphs to represent basic mathematical operations → simulate the system structure with these basic graphic elements → find the relationship between system function and system stability.

Results: System representation with the block diagram and the flow graphs, and pole-zero analysis of the system function.

7.1 System simulation

A simple LTI system (single-input/single-output) can be represented by its mathematical model, namely the linear constant coefficient differential equation. In reality, people would like to use a graph as an auxiliary tool or even as a replacement for the mathematical model in order to make a problem clearer and more intuitive. The graphical representation (simulation) concept of systems was introduced in Chapter 2, and the related details will be discussed in this chapter.

7.1.1 Basic arithmetic units

The block diagrams of some basic operators in the time domain were given in Chapter 2, and ▶ Figure 7.1a, b, respectively, show the symbols and the operational relationships of an adder/summer and a number multiplier in the time and s domains. The symbols of two units are same in the time and s domains, but the representation of an integrator is a little complex. We know that the relationship between the output

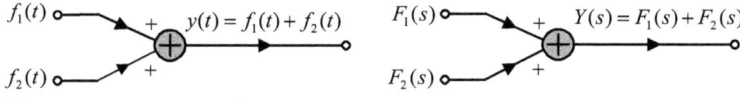

$f_1(t)$　$y(t) = f_1(t) + f_2(t)$　$F_1(s)$　$Y(s) = F_1(s) + F_2(s)$

$f_2(t)$　　　　　　　　$F_2(s)$

(a)　　Summer models in time-domain and complex frequency-domain

$f(t)$　$y(t) = af(t)$　$F(s)$　$Y(s) = aF(s)$

(b)　　Multiplier models in time-domain and complex frequency-domain

Fig. 7.1: Summer and multiplier models in time domain and complex frequency domain.

https://doi.org/10.1515/9783110419535-007

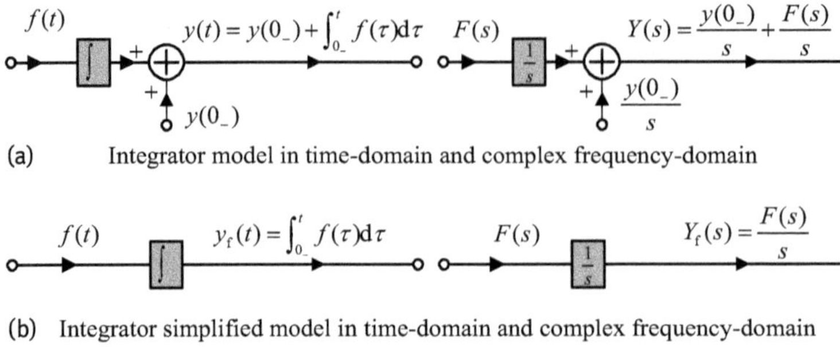

(a)   Integrator model in time-domain and complex frequency-domain

(b)   Integrator simplified model in time-domain and complex frequency-domain

Fig. 7.2: Integrator models in time domain and complex frequency domain.

and the input signal of an integrator in the time domain is

$$y(t) = \int_{-\infty}^{t} f(\tau)d\tau = \int_{-\infty}^{0_-} f(\tau)d\tau + \int_{0_-}^{t} f(\tau)d\tau = y(0_-) + \int_{0_-}^{t} f(\tau)d\tau, \qquad (7.1\text{-}1)$$

and the Laplace transform is

$$Y(s) = \frac{y(0_-)}{s} + \frac{F(s)}{s}. \qquad (7.1\text{-}2)$$

Therefore, the time and s domain block diagrams of an integrator are, respectively, shown in ▶ Figure 7.2a. If the starting condition is $y(0_-) = 0$, the simplified model of an integrator is as illustrated in ▶ Figure 7.2b.

7.1.2 Simulating system with block diagrams

We can say that:
(1) *The process using basic operation units (integrator, scalar multiplier and adder) to describe the system function (equation model) of a system is called system simulation.*
(2) *The simulation based on block diagrams of operation units is called block diagram simulation, while with flow graphs of operation units it is called flow graph simulation.*

The following examples will illustrate processes or methods of block diagram simulation.

Example 7.1-1. Draw the block diagram of the RC circuit shown in ▶ Figure 7.3, where $u_i(t)$ and $u_o(t)$ are, respectively, the excitation and the response.

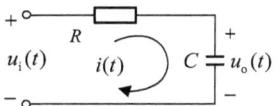

Fig. 7.3: E7.1-1.

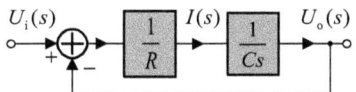

Fig. 7.4: E7.1-1 result.

Solution. The system equations in the complex frequency domain are

$$I(s) = \frac{1}{R}[U_i(s) - U_o(s)] , \qquad (7.1\text{-}3)$$

$$U_o(s) = \frac{1}{Cs}I(s) . \qquad (7.1\text{-}4)$$

Equation (7.1-3) shows that there should be a summer to add $-U_o(s)$ and $U_i(s)$ in the block diagram; the output $U_o(s)$ and the input $U_i(s)$ can merge together to form an intermediate function $I(s)$. Then $I(s)$ will be transformed into the output $U_o(s)$ according to equation (7.1-4), so the final result can be shown in ▶ Figure 7.4.

Example 7.1-2. The input–output equation of a system is given by

$$y''(t) + 3y'(t) + 2y(t) = f'(t) + f(t) .$$

Plot the system simulation block diagram.

Solution. The zero-state algebraic equation of the system in the s domain is

$$s^2 Y(s) + 3sY(s) + 2Y(s) = sF(s) + F(s) . \qquad (7.1\text{-}5)$$

Equation (7.1-5) can be rewritten as

$$Y(s) = \frac{s+1}{s^2+3s+2}F(s) = \frac{\frac{1}{s}+\frac{1}{s^2}}{1+\frac{3}{s}+\frac{2}{s^2}}F(s) = \frac{\frac{1}{s}}{1+\frac{3}{s}+\frac{2}{s^2}}F(s) + \frac{\frac{1}{s^2}}{1+\frac{3}{s}+\frac{2}{s^2}}F(s) . \quad (7.1\text{-}6)$$

Suppose that

$$X(s) = \frac{1}{1+\frac{3}{s}+\frac{2}{s^2}}F(s) , \qquad (7.1\text{-}7)$$

then equation (7.1-6) can be changed into

$$Y(s) = \frac{1}{s}X(s) + \frac{1}{s^2}X(s) .$$

This means that $Y(s)$ can be seen as an output of a summing point or a sum of which the two inputs are, respectively, $\frac{1}{s}X(s)$ and $\frac{1}{s^2}X(s)$, so equation (7.1-7) can be written as

$$X(s) = F(s) - \frac{3}{s}X(s) - \frac{2}{s^2}X(s) .$$

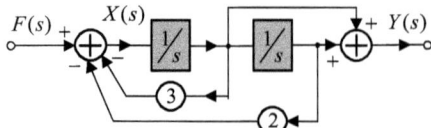

Fig. 7.5: E7.1-2.

This means that $X(s)$ can be seen as the output of a summing point of which the three inputs are, respectively, $F(s)$, $-\frac{3}{s}X(s)$, and $-\frac{2}{s^2}X(s)$; the simulation block diagram of the system is shown in ▸ Figure 7.5.

The expressing form of a system model can be changed, which means a math expression may have different transformation forms, so, the form of system simulation with a block diagram should be nonunique. Similarly to the transformation of a mathematical expression, we show some common transformation forms of block diagrams in Table 7.1, which can improve the understanding of the concept of simulation and simplify the structures of block diagrams.

Tab. 7.1: Common transformations of block diagrams.

Original	Equivalent	Remark
$F(s) \rightarrow \boxed{G_1(s)} \rightarrow \boxed{G_1(s)} \rightarrow Y(s)$	$F(s) \rightarrow \boxed{G_1 G_2} \rightarrow Y(s)$	1. Merge two cascaded boxes $F \cdot G_1 \cdot G_2 = F(G_1 \cdot G_2)$
$F(s) \rightarrow \Sigma \rightarrow \boxed{G(s)} \rightarrow Y(s)$, $X(s) +(-)$	$F(s) \rightarrow \boxed{G} \rightarrow \pm \Sigma \rightarrow Y(s)$, $+(-) \boxed{G} \leftarrow X(s)$	2. Move adder to the back of the box $(F \pm X)G = FG \pm XG$
$F(s) \rightarrow \boxed{G(s)} \rightarrow \pm \Sigma \rightarrow Y(s)$, $X(s) +(-)$	$F(s) \rightarrow \Sigma \rightarrow \boxed{G} \rightarrow Y(s)$, $+(-) \boxed{1/G} \leftarrow X(s)$	3. Move adder to the front of the box $F \cdot G \pm X = \left(F \pm \frac{1}{G} \cdot X\right) G$
$F(s) \rightarrow \boxed{G(s)} \rightarrow Y(s)$, $F(s)$	$F(s) \rightarrow \boxed{G} \rightarrow Y(s)$, $\rightarrow \boxed{1/G} F(s)$	4. Move bifurcation to the back of the box
$F(s) \rightarrow \boxed{G} \rightarrow Y(s)$, $Y(s)$	$F(s) \rightarrow \boxed{G} \rightarrow Y(s)$, $\rightarrow \boxed{G} Y(s)$	5. Move bifurcation to the front of the box
$F(s) \rightarrow \Sigma \rightarrow \boxed{G(s)} \rightarrow Y(s)$, $+(-) \boxed{H(s)} \leftarrow$	$F(s) \rightarrow \boxed{\dfrac{G}{1 \mp GH}} \rightarrow Y(s)$	6. Eliminate feedback loop
$F(s) \rightarrow \Sigma \rightarrow \boxed{G(s)} \rightarrow Y(s)$, $Y(s) +(-)$	$F(s) \rightarrow \boxed{\dfrac{G}{1 \mp G}} \rightarrow Y(s)$	7. Special case of 6 ($H = 1$)

7.1.3 Simulating systems with flow graphs

Although block diagram simulation of systems has been of great convenience in system analysis, people are still eager to find easier solutions for system simulation. This is true of flow graph simulation.

1. Concept of the signal flow graph

The flow graph is a short form of the signal flow graph, and its essence is not different from the block diagram. Thus, it is actually the simplified form of a block diagram and is basically similar to a block diagram in the way it is expressed. However, it is characterized by blocks being replaced by directed segments and adders being omitted.

▸ Figure 7.6 shows the block diagram and flow graph of a system with $H(s)$ as the transfer function. The basic method of flow graphs is to draw a directed segment from one node to another according to the flow direction of signals, and mark the transfer function next to the segment.

A figure consisting of directed segments with side-noted transfer functions (indicating transmission direction and the corresponding processing of a signal) and nodes (the system variables, indicating the start and end points of a signal), is called a signal flow graph.

In ▸ Figure 7.6, the branch with $H(s)$ starts from the node representing an excitation signal $F(s)$ and ends the node representing a response signal $Y(s)$. This shows that the signal at the end point of this branch is equal to the product of the signal at the starting point and the transfer function of the branch, namely, $Y(s) = H(s)F(s)$. Note that $H(s)$ is also called the gain of a system.

A node can have many forward (input) and backward (output) branches, so there is a rule for flow graphs: the signal (variable) on any a node is only equal to the sum of all signals from the forward branches. In ▸ Figure 7.7a, for example, the node signal variables such as x_4, x_5 and x_6 can be given by the following equations:

$$x_4 = H_{14}x_1 + H_{24}x_2 + H_{34}x_3 \, ,$$

$$x_5 = H_{45}x_4 \, ,$$

$$x_6 = H_{46}x_4 \, .$$

Note that the node plays the role of a adder and can add all signals brought by the forward branches. If a signal needs to be subtracted, we can set a negative sign in front of the transfer function of the branch with this signal; this it is shown in ▸ Figure 7.7b,

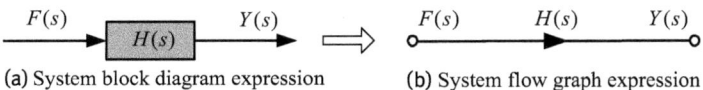

(a) System block diagram expression   (b) System flow graph expression

Fig. 7.6: System block diagram and flow graph expression.

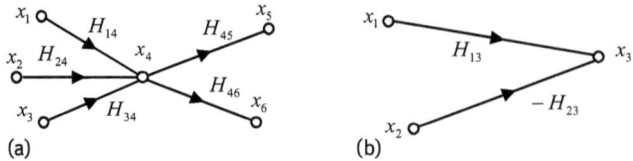

Fig. 7.7: Flow graph of the nodes and branchs.

and then

$$x_3 = H_{13}x_1 - H_{23}x_2 .$$

It is obvious that the flow graphs can simplify system representation. In a block diagram, the interconnection form of subsystems and transfer functions are important information, and the flow graphs only keep these necessary messages and abandons other unnecessary things such as boxes and summation symbols, etc., so, it enables us to focus on the essentials of a system.

2. Mason's formula

The transfer function of a system can be obtained easily from flow graphs by using Mason's formula. This makes it an important tool for system analysis. Moreover, it can also be used as a mathematical model for system simulation.

First, we will introduce several important buzzwords in Mason's formula

(1) source point (excitation point) – the node with only output branches;

(2) subordinate point (response point) – the node with at least one input branch;

(3) branch – the directed segment that points directly from one node to another;

(4) opened path – if a signal passes with directed branches from node A to B, and on the way no one node is met again, the path passed by this signal is called an opened path from A to B;

(5) loop – if a signal passes with directed branches from node A back to A, and on the way, no one node is met again except A, this path passed by the signal is called a loop;

(6) Opened path transfer function T – the product of the transfer functions of all branches located on an opened path;

(7) loop transfer function L – the product of the transfer functions of all branches located on a loop;

(8) flow graph determinant Δ –

$$\Delta = 1 - \sum_i L_i + \sum_{i,j} L_i L_j - \sum_{i,j,k} L_i L_j L_k + \dots \tag{7.1-8}$$

In equation (7.1-8), there are

$$\sum_i L_i - \text{the sum of all the loop transfer functions,} \qquad (7.1\text{-}9)$$

$$\sum_{i,j} L_i L_j - \begin{array}{l}\text{the sum of the products of transfer functions of two}\\ \text{loops that do not touch each other,}\end{array} \qquad (7.1\text{-}10)$$

$$\sum_{i,j,k} L_i L_j L_k - \begin{array}{l}\text{the sum of the products of transfer functions of each}\\ \text{three loops that do not touch each other.}\end{array} \qquad (7.1\text{-}11)$$

"No touching" means that there are neither common nodes nor branches among different loops.

With the above concept, we can give the form of Mason's formula.

The system transfer function T_{FY} from excitation point F to response point Y can be expressed in the form

$$T_{FY} = \frac{\sum_{N=1}^{M} T_N \Delta_N}{\Delta} ,$$

or

$$H(s) = \frac{\sum_{N=1}^{M} T_N \Delta_N}{\Delta} , \qquad (7.1\text{-}12)$$

where M represents the number of opened paths from F to Y. The T_N represents the opened path transfer function of the Nth opened path from F to Y. The Δ_N represents the flow graph determinant of the rest flow graphs after the Nth open path is removed.

Mason's formula is another expression means of a system's mathematical model and is also a shortcut to achieve flow graphs. Next, some examples are given to illustrate its application.

Example 7.1-3. Find the transfer function of the flow diagram shown in ► Figure 7.8.

Solution. The graph only has one loop, whose transfer function is $-H_2 H_3 H_4$. Therefore, the flow graph determinant Δ is

$$\Delta = 1 - \sum_i L_i = 1 + H_2 H_3 H_4 .$$

There is only one opened path from the input to the output, whose transfer function is $H_1 H_2 H_3$. Once rid of this open path, the loop is also interrupted. Therefore,

$$\Delta_1 = 1 .$$

The transfer function of the flow graph is

$$H(s) = \frac{T_1 \Delta_1}{\Delta} = \frac{H_1 H_2 H_3}{1 + H_2 H_3 H_4} .$$

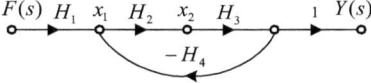

Fig. 7.8: E7.1-3.

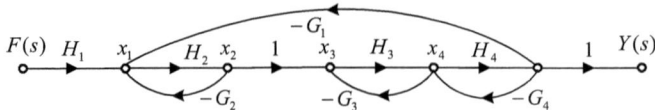

Fig. 7.9: E7.1-4.

Example 7.1-4. Find the transfer function of the flow graph shown in ▶ Figure 7.9.

Solution. The graph has four loops

$$(a)\ x_1 - x_2 - x_1, \quad (b)\ x_3 - x_4 - x_3,$$
$$(c)\ x_4 - Y - x_4, \quad (d)\ x_1 - x_2 - x_3 - x_4 - Y - x_1 .$$

The transfer function of each loop is

$$(a)\ - H_2 G_2, \quad (b)\ - H_3 G_3, \quad (c)\ - H_4 G_4, \quad (d)\ - H_2 H_3 H_4 G_1 .$$

So,

$$\sum_i L_i = -(H_2 G_2 + H_3 G_3 + H_4 G_4 + H_2 H_3 H_4 G_1) .$$

There are two pairs of nontouching loops, such as $x_1 - x_2 - x_1$ and $x_3 - x_4 - x_3$, $x_1 - x_2 - x_1$ and $x_4 - Y - x_4$, so,

$$\sum_{ij} L_i L_j = H_2 G_2 H_3 G_3 + H_2 G_2 H_4 G_4 .$$

There are no three or more loops that do not touch each other, so,

$$\Delta = 1 - \sum_i L_i + \sum_{ij} L_i L_j$$

$$= 1 + (H_2 G_2 + H_3 G_3 + H_4 G_4 + H_2 H_3 H_4 G_1 + H_2 G_2 H_3 G_3 + H_2 G_2 H_4 G_4) .$$

There is only an opened path from the input to the output, and the transfer function is $H_1 H_2 H_3 H_4$, that is,

$$T_1 = H_1 H_2 H_3 H_4 .$$

All loops are in contact with the opened path, and they will break off if the opened path is taken out. So, we have

$$\Delta_1 = 1 - 0 + 0 - \cdots = 1 .$$

According to Mason's formula, the system transfer function is

$$H(s) = \frac{T_1 \Delta_1}{\Delta}$$

$$= \frac{H_1 H_2 H_3 H_4}{1 + H_2 G_2 + H_3 G_3 + H_4 G_4 + H_2 H_3 H_4 G_1 + H_2 G_2 H_3 G_3 + H_2 G_2 H_4 G_4} .$$

3. Flow graph simulation

It has been seen that the system function can be derived from flow graphs by using Mason's formula, so we will introduce how to build flow graphs based on the system function $H(s)$, which is called the flow graph simulation.

Flow graph simulation methods can be divided into three types: the direct, parallel and series simulations. Obviously, considering the relationship between flow graphs and block diagrams, block diagram simulation methods also have these three forms.

1.) Direct simulation. Flow graphs can be directly built based on the general form of the system function $H(s)$ in this method.

Example 7.1-5. If the system function of a system is $H(s) = \frac{1}{s^3+a_2s^2+a_1s+a_0}$, simulate this system with flow graphs.

Solution. The system function is rewritten as $H(s) = \frac{\frac{1}{s^3}}{1-\frac{-a_2}{s}-\frac{-a_1}{s^2}-\frac{-a_0}{s^3}}$, and a comparison with Mason's formula, leads to the following conclusions.

From the numerator, there is only one opened path, and the flow graph determinant without the opened path is 1; this opened path is composed of three integrators.

From the denominator, only three loops touch each other. Their starting points or ends are, respectively, the output terminals or input terminals of three integrators.

Accordingly, the direct form 1 is depicted in ▸ Figure 7.10a, and the corresponding block diagram in ▸ Figure 7.10b. If ▸ Figure 7.10a is transposed, that is, the signal transmission directions of all branches are reversed, and the source point and the response point are exchanged, then the direct form 2 can be shown as in ▸ Figure 7.11a; ▸ Figure 7.11b is the corresponding block diagram of the system.

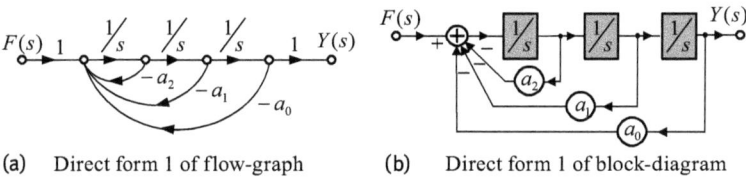

(a)   Direct form 1 of flow-graph   (b)   Direct form 1 of block-diagram

Fig. 7.10: Direct form 1 of simulation diagrams of E7.1-5.

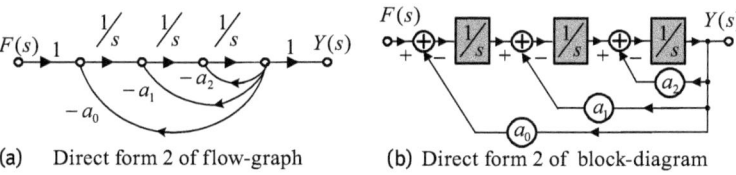

(a)   Direct form 2 of flow-graph   (b) Direct form 2 of block-diagram

Fig. 7.11: Direct form 2 of simulation diagrams of E7.1-5.

Clearly, the system function will remain the same after the flow graph is transposed, so the direct simulation can be divided into two forms. Let us see another example in the following.

Example 7.1-6. The system function of a system is $H(s) = \frac{b_2s^2+b_1s+b_0}{s^3+a_2s^2+a_1s+a_0}$. Simulate this system with a flow graph.

Solution. The main difference between this example and the above lies in their numerators. This system function can be rewritten as

$$H(s) = \frac{\dfrac{b_2}{s} + \dfrac{b_1}{s^2} + \dfrac{b_0}{s^3}}{1 - \dfrac{-a_2}{s} - \dfrac{-a_1}{s^2} - \dfrac{-a_0}{s^3}} \ .$$

Comparing it with Mason's formula, we obtain the following points.

(1) From the numerator, there are three opened paths, and the flow graph determinants after each opened path is removed are all 1. These three opened paths which, respectively, point to the subordinate point from the output terminals of the three integrators or from the excitation point to their input terminals.
(2) From the denominator, only three loops touch each other. The starting points or ends of the three loops are, respectively, the output terminals or input terminals of the three integrators. Accordingly, the flow graph and block diagram in direct forms can be depicted as in ▶ Figure 7.12 and ▶ Figure 7.13.

We find some simulating rules from the above examples and give the general forms of the direct simulation after they have been collated and summed up.

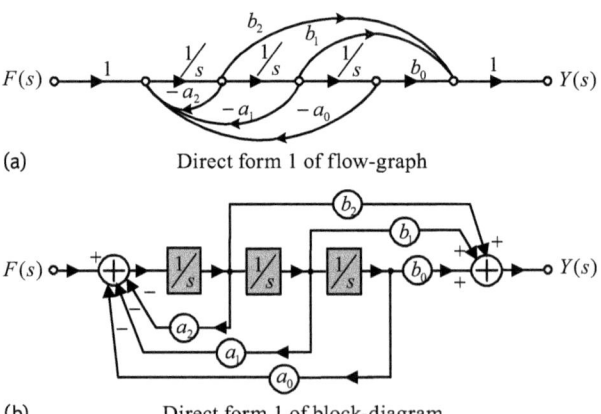

(a)   Direct form 1 of flow-graph

(b)   Direct form 1 of block-diagram

Fig. 7.12: Direct form 1 of simulation diagrams of E7.1-6.

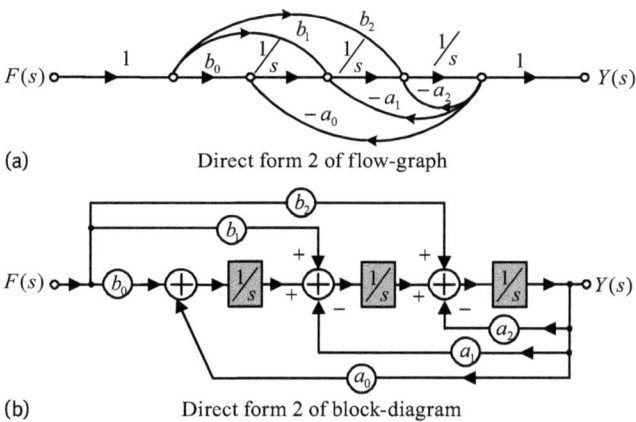

(a)   Direct form 2 of flow-graph

(b)   Direct form 2 of block-diagram

Fig. 7.13: Direct form 2 of simulation diagrams of E7.1-6.

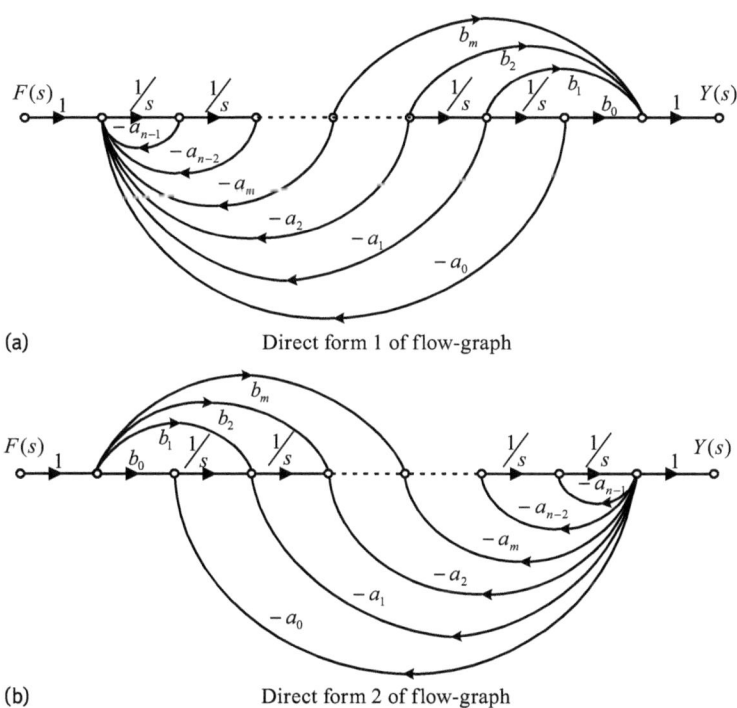

(a)   Direct form 1 of flow-graph

(b)   Direct form 2 of flow-graph

Fig. 7.14: General forms of flow graph.

For the systems with the following system function

$$H(s) = \frac{b_m s^m + b_{m-1} s^{m-1} + \cdots + b_1 s + b_0}{s^n + a_{n-1} s^{n-1} + \cdots + a_1 s + a_0}$$

$$= \frac{b_m s^{-(n-m)} + b_{m-1} s^{-(n-m+1)} + \cdots + b_1 s^{-(n-1)} + b_0 s^{-n}}{1 + a_{n-1} s^{-1} + \cdots + a_1 s^{-(n-1)} + a_0 s^{-n}};$$

(7.1-13)

the denominator is a flow graph determinant consisting of n loops, and each loop touches the others. The numerator can be regarded as the sum of transfer functions of $(m + 1)$ opened paths, and each path touches the others. Therefore, the two direct forms of flow graph simulation can be plotted as in ▶ Figure 7.14.

2.) Series simulation. The general form of a system function $H(s)$ is always a fraction, but it can be decomposed into the product of several subfractions. Each subfraction can be directly simulated into a subflow graph, and finally, these subflow graphs can be connected in series and form a series form of simulation. Note that in some books, the series form is also called the cascade form.

For example, the system function of a system is $H(s) = \frac{5s+5}{s^3+7s^2+10s}$; now the system function will be transformed as follows:

$$H(s) = \frac{5s + 5}{s^3 + 7s^2 + 10s} = \frac{5(s + 1)}{s(s + 2)(s + 5)}$$

$$= \frac{5}{s + 2} \cdot \frac{s + 1}{s + 5} \cdot \frac{1}{s}.$$

It can be seen that $H(s)$ can become the product of three subsystem functions, which can be simulated directly, as shown in ▶ Figure 7.15a–c. Then, they can constitute the series form of the simulation diagram of the whole system shown in ▶ Figure 7.15d and e.

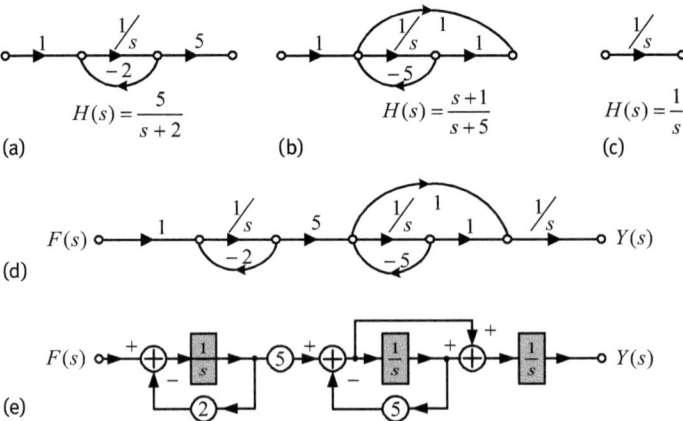

Fig. 7.15: System series simulation schematic diagram.

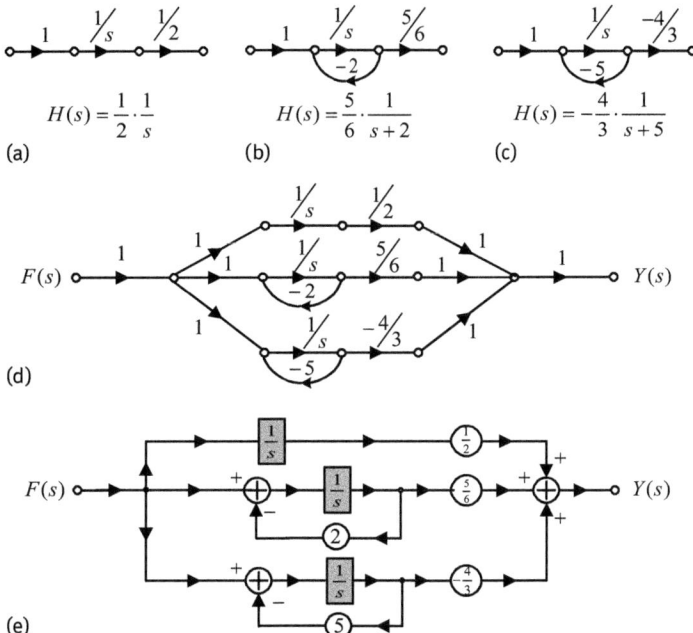

Fig. 7.16: System parallel simulation schematic diagram.

3.) Parallel simulation. $H(s)$ is also transformed into the sum of several fractions, then each of them is directly simulated into a subdiagram. Finally, these subdiagrams are connected in parallel to form the parallel form of simulation.

If $H(s)$ is decomposed as follows:

$$H(s) = \frac{5s + 5}{s^3 + 7s^2 + 10s} = \frac{5(s + 1)}{s(s + 2)(s + 5)} = \frac{1}{2} \cdot \frac{1}{s} + \frac{5}{6} \cdot \frac{1}{s + 2} - \frac{4}{3} \cdot \frac{1}{s + 5}$$

It can be seen that $H(s)$ has become a sum of three subsystem functions. After each subsystem is simulated in direct forms, the parallel form of the diagram of the whole system can be constituted as shown in ▶ Figure 7.16.

Example 7.1-7. The system function of a system is $H(s) = \frac{2s+3}{s(s+3)(s+2)^2}$. Simulate this system with a block diagram and a flow graph in direct, series and parallel forms.

Solution. (1) $H(s)$ is rewritten as $H(s) = \frac{2s+3}{s^4+7s^3+16s^2+12s}$; its direct form diagrams are shown in ▶ Figure 7.17.

(2) $H(s)$ is rewritten as $H(s) = \frac{1}{s} \cdot \frac{1}{s+2} \cdot \frac{2s+3}{s+2} \cdot \frac{1}{s+3}$; its series form diagrams are shown in ▶ Figure 7.18.

(3) $H(s)$ is rewritten as $H(s) = \frac{1}{4} \cdot \frac{1}{s} - \frac{5}{4} \cdot \frac{1}{s+2} + \frac{1}{2} \cdot \frac{1}{(s+2)^2} + \frac{1}{s+3}$; its parallel form diagrams are shown in ▶ Figure 7.19.

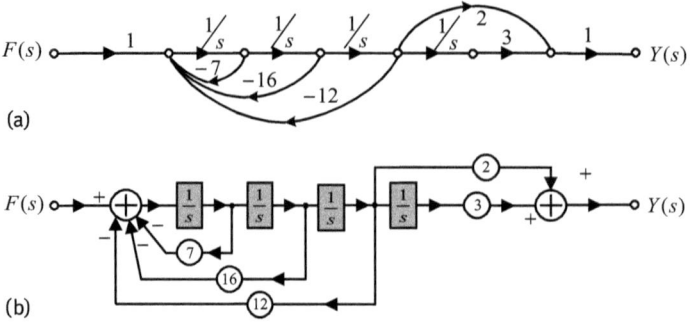

(a)

(b)

Fig. 7.17: Direct simulation diagram of E7.1-7.

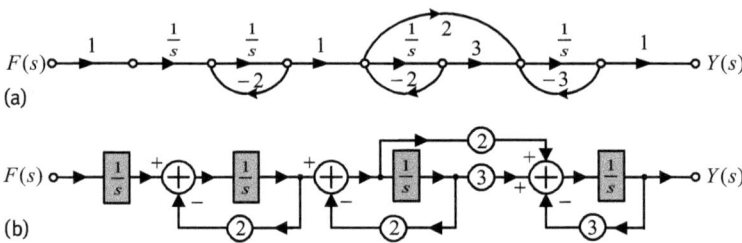

(a)

(b)

Fig. 7.18: Series simulation diagram of E7.1-7.

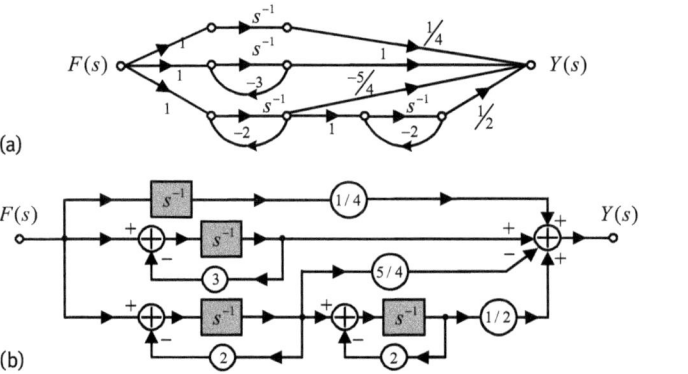

(a)

(b)

Fig. 7.19: Parallel simulation diagram of E7.1-7.

Note: From the above examples, we can see that although Mason's formula is derived from the flow graph, it is also suitable in system simulation with block diagrams. Obviously, the block diagram can be replaced completely by the flow graph in view of its functionality.

7.2 System stability analysis

System stability as an important conception in the system analysis process is discussed next.

7.2.1 System stability

We first describe stability with the simple physical system in ▸ Figure 7.20, which shows the movement of a small ball located at different positions when it is forced. When the ball is on point A, it will leave the original position and roll to point B or C if it is forced to the left or the right. This means that A is an instable point and the small ball located on A is in an unstable state. Suppose that we put the ball on B or C and apply a leftward or rightward force to move it; the ball will be back in the original position after the force is canceled. This states that B or C is the stable point and the ball on it is in a stable state. A cone on the desktop can be used as a similar example, as shown in ▸ Figure 7.21. If the cone can go back to the original state after the touch force is canceled, the cone is in a stable state. Conversely, if it has been inverted, even a slight disturbance can change its original state, so the upside down cone is in unstable state.

If a linear system is disturbed, no matter how tiny the disturbance is, there will be a response that grows with time (even after the disturbance is removed), so this system is unstable. If the response is limited, the system is boundary stable. If the response eventually becomes zero, the system is stable.

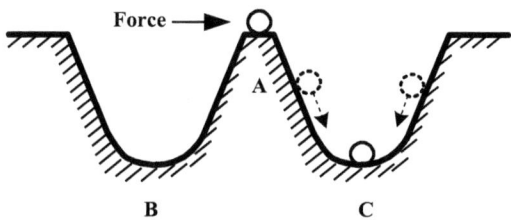

Fig. 7.20: Ball stability schematic diagram.

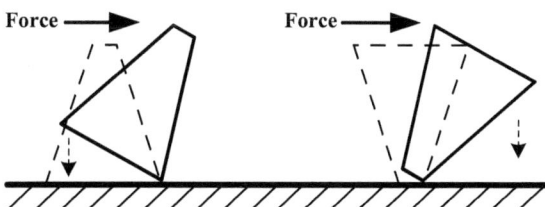

Fig. 7.21: Cone stable and unstable state diagram.

Tab. 7.2: The zero-input response stability criterion.

Zero-input response	Stability
Increases with time	Instability
Maintained within a certain limit	Boundary stability or critical stability
Eventually becomes zero	Asymptotic stability

As we know, the whole response of system can be divided into zero-input response and zero-state response, so the system stability problem can be approached from two aspects, that is, zero-input response stability and zero-state response stability.

1. Zero-input response stability

The zero-input response is stable, which is also known as asymptotic stability or internal stability, which means that the response of system caused by the initial energy storage is gradually reduced to zero with time. That is,

$$\lim_{t \to \infty} y_x(t) = 0 . \tag{7.2-1}$$

Under any a set of initial conditions, if we have

$$\lim_{t \to \infty} |y_x(t)| \le M , \tag{7.2-2}$$

where M is a bounded real constant; and then the system is critical or boundary stable.

If under some certain initial conditions,

$$\lim_{t \to \infty} |y_x(t)| \to \infty , \tag{7.2-3}$$

the system is unstable.

The stability criterion of the zero-input response of a system is listed in Table 7.2.

2. Zero-state response stability

That zero-state response is stable or externally stable refers to that if a system without initial energy storage is driven by a bounded signal, then the output (zero-state response) is also bounded. This phenomenon is known as BIBO (bounded-input bounded-output), and the mathematical description is of the form

$$|f(t)| < M \to |y_f(t)| < \infty . \tag{7.2-4}$$

The BIBO stability of a system can be defined as:

If the response of a system to a bounded excitation is also bounded, the system is a BIBO system, otherwise, is an unstable one.

It can be proved that the necessary and sufficient condition testing the stability of a CT system is that the $h(t)$ is absolutely integrable,

$$\int_{-\infty}^{\infty} |h(t)| dt < \infty .\qquad(7.2\text{-}5)$$

The necessary condition is

$$\lim_{t \to \infty} h(t) = 0 .\qquad(7.2\text{-}6)$$

Usually if there are no common factors in the numerator and denominator of a system function $H(s)$, the characteristics of an LTI system are completely described by the system function. The asymptotic stability of system is equivalent to BIBO stability. If there are common factors, the zeros and poles can be canceled, and the system may be not asymptotically stable but rather BIBO stable. For example, if the system function of a system is $H(s) = \frac{s-1}{s^2+s-2} = \frac{s-1}{(s-1)(s+2)}$, clearly, there is a pole $s = 1$ on the right half-plane, so this system is not asymptotically stable (see Section 7.2.2 for details.). However, if the impulse response is solved by

$$H(s) = \frac{s-1}{s^2+s-2} = \frac{s-1}{(s-1)(s+2)} = \frac{1}{s+2} \to h(t) = e^{-2t}\varepsilon(t) \to \int_{-\infty}^{\infty} |h(t)| \, dt = \frac{1}{2} < \infty ,$$

then, a contradictory conclusion that this system is BIBO stable is results. This phenomenon indicates that the cancellation of zeros and poles covers the essence of system inherent instability, so the system function cannot describe fully and correctly the system characteristics here. Of course, if a system is inherently stable, it still remains stable after the zeros and poles of $H(s)$ offset each other.

Analysis has shown that:
(1) The asymptotically stable system must be a BIBO stable system.
(2) The asymptotically unstable system must be a BIBO unstable system.
(3) A BIBO stable system is not always an asymptotically stable system.
(4) If the order of the denominator is no less than the order of the numerator, and there is no cancellation of the zeros and the poles in $H(s)$, the asymptotical stability is equivalent to BIBO stability.
(5) A real system (except signal generators) must be asymptotically stable and BIBO stable at the same time.

For convenience, stability in the following discussion will be zero-state response stability unless stated otherwise.

7.2.2 Pole-zero analysis of the system function $H(s)$

We know that the stability of a system depends on the characteristics of $h(t)$, but because $h(t)$ is a function of time, so that finding and analyzing $h(t)$ directly are usually more troublesome. Therefore, we naturally wonder whether we can find any analyses methods in the s domain. The answer is "yes". Since $h(t)$ and $H(s)$ are the Laplace transform pair, some characteristics of $h(t)$ can be obtained by $H(s)$.

If $A(s)$ and $B(s)$, respectively, represent the denominator and the numerator of $H(s)$, and $B(s) = 0$ has m roots such as $\xi_1, \xi_2, \ldots, \xi_m$, and $A(s) = 0$ has n roots $\lambda_1, \lambda_2, \ldots, \lambda_n$, then the general form of the system function can be of the form

$$H(s) = \frac{b_m s^m + b_{m-1} s^{m-1} + \cdots + b_1 s + b_0}{s^n + a_{n-1} s^{n-1} + \cdots + a_1 s + a_0} = \frac{B(s)}{A(s)} = \frac{b_m \prod_{j=1}^{m} (s - \xi_j)}{\prod_{i=1}^{n} (s - \lambda_i)}. \tag{7.2-7}$$

We name $\xi_1, \xi_2, \ldots, \xi_m$ and $\lambda_1, \lambda_2, \ldots, \lambda_n$, respectively, zeros and poles of $H(s)$. According to mathematical knowledge, there are three forms of the zero and the pole, such as real, imaginary and complex numbers. The coefficients a_i and b_j in $H(s)$ are generally real numbers. Therefore, if a pole or a zero is an imaginary number or a complex number, it must appear in a pair with conjugate form.

From the solution method of the inverse Laplace transform, when $H(s)$ is expanded into partial fraction form, each of its poles will determine a corresponding function of time. Therefore, the expression of the impulse response $h(t)$ only depends on the poles of $H(s)$, while the amplitude and phase of $h(t)$ are determined by poles and zeros together. In other words, $h(t)$ is completely determined by the distribution of zeros and poles in the s plane.

The locations of poles in the s plane can be divided into three types, such as in the open left half-plane, on the imaginary axis and in the open right half-plane. The distribution characteristics of first-order and higher-order poles will be introduced in the following.

1. First-order poles
If the poles of $H(s)$ $\lambda_1, \lambda_2, \ldots, \lambda_n$ are all of first order, $H(s)$ will be expanded into partial fraction form as

$$H(s) = \sum_{i=1}^{n} \frac{K_i}{s - \lambda_i}. \tag{7.2-8}$$

If λ_i is a real number, the expansion of $H(s)$ will contain the term $H_i(s) = \frac{b}{s-\alpha}$. The pole $\lambda_i = \alpha$ is on the real axis. From three different cases of $\alpha < 0$, $\alpha = 0$ and $\alpha > 0$, the pole may be on the negative real axis, the origin or the positive real axis, which are shown in ▶ Figure 7.22a (where "×" represents a pole). The corresponding impulse response is of the form

$$h_i(t) = b e^{\alpha t} \varepsilon(t), \tag{7.2-9}$$

and its waveforms are as displayed in ▶ Figure 7.22b.

From equation (7.2-9), when $t \to \infty$, if $\alpha < 0$, then $h_i(t) \to 0$; if $\alpha = 0$, $h_i(t)$ is a finite value; if $\alpha > 0$, $h_i(t) \to \infty$.

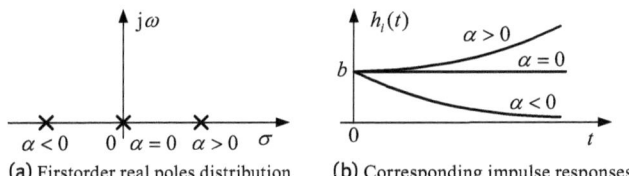

(a) Firstorder real poles distribution   (b) Corresponding impulse responses

Fig. 7.22: First-order real pole distribution and corresponding impulse responses.

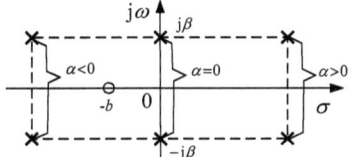

Fig. 7.23: First-order conjugate poles distribution.

If a pole is complex, it must be a pair of conjugate poles $\lambda_{1,2} = \alpha \pm j\beta$. The expansion of $H(s)$ contains the term $H_i(s) = \frac{s+b}{(s-\alpha)^2+\beta^2}$, letting $b > 0$. It has a first-order zero at point $\xi = -b$. According to three situations such as $\alpha < 0$, $\alpha = 0$ and $\alpha > 0$, the pole can be in the open left half-plane, on the imaginary axis or in the open right half-plane, respectively, as shown in ► Figure 7.23, where "∘" denotes the zero. The corresponding impulse response is

$$h_i(t) = \frac{\sqrt{(b+\alpha)^2 + \beta^2}}{\beta} e^{\alpha t} \sin(\beta t + \varphi)\varepsilon(t) , \tag{7.2-10}$$

where

$$\varphi = \arctan \frac{\beta}{b+\alpha} . \tag{7.2-11}$$

If $\alpha < 0$, the pole is in the open left half-plane, and $h_i(t)$ is in the damped oscillation curve. Thus, when $t \to \infty$, $h_i(t) = 0$; if $\alpha = 0$, the pole is on the imaginary axis, and $h_i(t)$ will be a oscillating wave with a constant amplitude; if $\alpha > 0$, the pole is in the open right half-plane, and $h_i(t)$ will be in the form of increased oscillation, so when $t \to \infty$, the amplitude of $h_i(t) \to \infty$. From equations (7.2-10) and (7.2-11), both the amplitude and the phase of $h_i(t)$ are related to the zero positions.

2. Second-order poles

If $H(s)$ has a second-order pole on the real axis as $\lambda_i = \alpha$, the $H_i(s)$ can be written as

$$H_i(s) = \frac{s+b}{(s-\alpha)^2} . \tag{7.2-12}$$

Since there is a first-order zero at $\xi = -b$, the impulse response is of the from

$$h_i(t) = [(b+\alpha)t + 1]e^{\alpha t}\varepsilon(t) . \tag{7.2-13}$$

If $\alpha < 0$, the pole is on the negative real axis, so when $t \to \infty$, $h_i(t) \to 0$; if $\alpha = 0$ or $\alpha > 0$, the pole is at the origin or on the positive real axis, which can result in an increasing curve of $h_i(t)$ with t, so when $t \to \infty$, $h_i(t) \to \infty$.

If $H(s)$ has second-order conjugate poles $\lambda_{1,2} = \alpha \pm j\beta$, the denominator of $H_i(s)$ should be $[(s - \alpha)^2 + \beta^2]^2$, so its inverse transform is

$$[c_1 t e^{\alpha t} \cos(\beta t + \theta) + c_2 e^{\alpha t} \cos(\beta t + \varphi)]\varepsilon(t) , \qquad (7.2\text{-}14)$$

where c_1, c_2, θ and φ are constants which relate to the locations of zeros and poles. If $\alpha < 0$, the poles are on the open left half-plane, so when $t \to \infty$, $h_i(t) \to 0$. If $\alpha = 0$ or $\alpha > 0$, the poles are on the imaginary axis or the open right half-plane, which can result in an increasing amplitude of $h_i(t)$ with t, so when $t \to \infty$, the amplitude of $h_i(t) \to \infty$.

If there are higher-order poles in $H_i(s)$, the change regulations of corresponding $h_i(t)$ with time t are similar to the cases of second-order poles and will not be explained here.

Based on the above contents, we can sum up the following points.

(1) The function form for each component $h_i(t)$ in $h(t)$ only depends on the corresponding locations of poles of $H(s)$, but its amplitude and phase are codetermined by the positions of the poles and zeros.

(2) Poles of $H(s)$ in the open left half-plane correspond to the transient components $h_i(t)$ in $h(t)$. When $t \to \infty$, the transient component $h_i(t) \to 0$. A first-order pole on the negative real axis corresponds to the exponential decay curve $h_i(t) = e^{-\alpha t}\varepsilon(t)$ $(\alpha > 0)$. A pair of conjugate poles $\lambda_{1,2} = \alpha \pm j\beta$ corresponds to a damped oscillation signal like $h_i(t) = e^{-\alpha t} \sin(\beta t + \varphi)\varepsilon(t)$.

(3) A first-order pole of $H(s)$ at the origin point corresponds to a step signal like $h_i(t) = \varepsilon(t)$. A pair of conjugate poles $\lambda_{1,2} = \pm j\beta$ on the imaginary axis corresponds to a constant amplitude oscillation signal as $h_i(t) = \sin(\beta t + \varphi)\varepsilon(t)$.

(4) For the second-order pole of $H(s)$ at the origin point or on the imaginary axis, or first- and second-order poles in open right half-plane, the corresponding $h_i(t)$ all grow with time.

To make things easier, Table 7.3 shows the typical pole-zero distributions of $H(s)$ and the corresponding waveforms of $h(t)$.

It can be deduced from Table 7.3 that if all the poles of $H(s)$ are located on the open left half-plane, the system must be stable; however, it must be unstable if $H(s)$ has one or more poles on the open right half-plane or the imaginary axis. In addition, if only the first-order pole is on the imaginary axis, the system is often called a boundary stable or oscillation system.

Example 7.2-1. System functions are as follows:. Draw the pole-zero diagrams and waveforms of impulse responses.

(1) $H(s) = \frac{s+1}{(s+1)^2+4}$ 　　　　(2) $H(s) = \frac{s}{(s+1)^2+4}$ 　　　　(3) $H(s) = \frac{(s+1)^2}{(s+1)^2+4}$

Tab. 7.3: Pole-zero distributions of $H(s)$ and the corresponding waveforms of $h(t)$.

No.	$H(s)$	Distribution of poles and zeros in the s plane	$h(t)$	Waveform of $h(t)$
1	$\frac{1}{s}$		$\varepsilon(t)$	
2	$\frac{1}{s-\alpha}$ $(\alpha > 0)$		$e^{\alpha t}\varepsilon(t)$	
3	$\frac{1}{s+\alpha}$ $(\alpha > 0)$		$e^{-\alpha t}\varepsilon(t)$	
4	$\frac{\omega_0}{s^2+\omega_0^2}$		$\sin\omega_0 t\,\varepsilon(t)$	
5	$\frac{s}{s^2+\omega_0^2}$		$\cos\omega_0 t\,\varepsilon(t)$	
6	$\frac{\omega_0}{(s-\alpha)^2+\omega_0^2}$ $\alpha > 0$		$e^{\alpha t}\sin\omega_0 t\,\varepsilon(t)$	
7	$\frac{\omega_0}{(s+\alpha)^2+\omega_0^2}$ $\alpha > 0$		$e^{-\alpha t}\sin\omega_0 t\,\varepsilon(t)$	
8	$\frac{1}{(s+\alpha)^2}$ $(\alpha > 0)$		$te^{-\alpha t}\varepsilon(t)$	

Tab. 7.3 (continued): Pole-zero distributions of $H(s)$ and the corresponding waveforms of $h(t)$.

No.	$H(s)$	Distribution of poles and zeros in the s plane	$h(t)$	Waveform of $h(t)$
9	$\frac{1}{s^2}$		$t\varepsilon(t)$	
10	$\frac{2\omega_0 s}{(s^2+\omega_0^2)^2}$		$t\sin\omega_0 t\varepsilon(t)$	

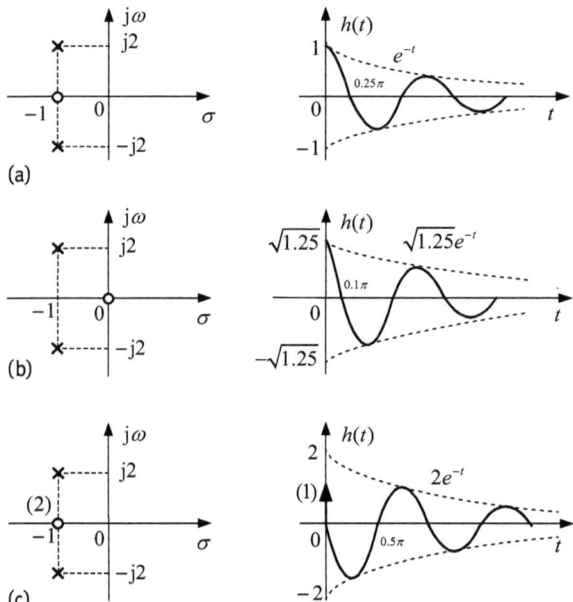

(a)

(b)

(c)

Fig. 7.24: E7.2-1.

Solution. The pole distribution situations of the three system functions are the same, but the zero distribution situations are different. Therefore, impulse response waveforms are also different. The detailed process is as follows.

(1) $h(t) = \mathcal{L}^{-1}\left[\frac{s+1}{(s+1)^2+4}\right] = e^{-t}\cos(2t)\varepsilon(t)$.

The pole-zero diagram of $H(s)$ and impulse response are, respectively, plotted in ▶ Figure 7.24a.

(2)

$$h(t) = \mathcal{L}^{-1}\left[\frac{s}{(s+1)^2+4}\right] = \mathcal{L}^{-1}\left[\frac{s+1}{(s+1)^2+4} - \frac{1}{(s+1)^2+4}\right]$$

$$= e^{-t}[\cos(2t) - \frac{1}{2}\sin(2t)]\varepsilon(t) = \frac{\sqrt{5}}{2}e^{-t}\cos(2t+26.57°)\varepsilon(t)$$

The pole-zero diagram of $H(s)$ and the impulse response are, respectively, plotted in ▶ Figure 7.24b.

(3)

$$h(t) = \mathcal{L}^{-1}\left[\frac{(s+1)^2}{(s+1)^2+4}\right] = \mathcal{L}^{-1}\left[1 - \frac{4}{(s+1)^2+4}\right] = \delta(t) - 2e^{-t}\sin(2t)\varepsilon(t) .$$

The pole-zero diagram of $H(s)$ and the impulse response waveform are, respectively, shown in ▶ Figure 7.24c.

From this example, we can see that the amplitude and phase of the impulse response waveform change when the zero is moved from point −1 to the origin. However, when the order of the zero at point −1 changes from first to second order, not only do the amplitude and phase of the impulse response change, but an impulse signal also appears in the impulse response. Generally, changing the zero position of $H(s)$ can result in changes of the amplitude and phase of $h(t)$ and may also produce an impulse signal in $h(t)$.

Example 7.2-2. $H(s) = \frac{U(s)}{I(s)}$ is the system function of the circuit shown in ▶ Figure 7.25a. Its pole-zero diagram is shown in ▶ Figure 7.25b, and $H(0) = 1$. Solve for the values of R, L and C.

Solution. From ▶ Figure 7.25a and b the system function can be written as

$$H(s) = \frac{U(s)}{I(s)} = \frac{1}{sC + \frac{1}{Ls+R}} = \frac{Ls+R}{LCs^2 + sCR + 1} \tag{7.2-15}$$

and

$$H(s) = \frac{k(s+2)}{(s+1-j\frac{1}{2})(s+1+j\frac{1}{2})} .$$

Since $H(0) = 1$, and setting $s = 0$ for the above equation, we have

$$H(s)|_{s=0} = \frac{2k}{(1-j\frac{1}{2})(1+j\frac{1}{2})} = \frac{8}{5}k = 1 ,$$

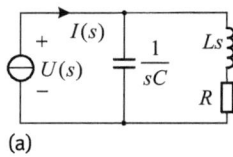

(a)

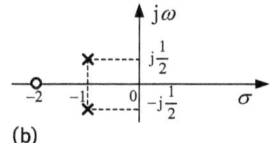

(b)

Fig. 7.25: E7.2-2.

and then

$$k = \frac{5}{8} \, .$$

Therefore, the system function is

$$H(s) = \frac{5}{8} \cdot \frac{s + 2}{(s^2 + 2s + \frac{5}{4})} = \frac{\frac{1}{2}s + 1}{\frac{4}{5}s^2 + \frac{8}{5}s + 1} \, . \tag{7.2-16}$$

Comparing equation (7.2-15) with equation (7.2-16) results in

$$L = \frac{1}{2} \, \text{H}, \quad R = 1 \, \Omega, \quad C = \frac{8}{5} \, \text{F} \, .$$

7.2.3 Relationships between stability and ROC, and poles

1. Relationship between stability and ROC

From Section 7.2.1, an LTI system being stable is the equivalence of $h(t)$ being absolutely integrable; moreover, the Fourier transform of $h(t)$ also exists at this time. This means that the Fourier transform of a signal is equal to the values taken by its Laplace transform along with the $j\omega$ axis. Thus, we reach the following conclusion:

If and only if the ROC of system function $H(s)$ contains $j\omega$ axis, is the LTI system stable.

Note that this system may be not a causal system, that means $h(t)$ could be a left-sided, right-sided or double-sided signal.

2. Relationship between stability and pole positions

From Section 7.2.2, we also know the relationship between stability and pole positions.

If and only if all the poles of system function $H(s)$ lie in the left half of the s plane, is the LTI causal system stable.

In other words, the ROC of $H(s)$ for a causal system is the right half-plane of a certain convergence axis in the s plane.

7.2.4 Stability judgment based on the R–H criterion

It is commonly known that before the stability of a system is determined by means of $H(s)$, we must first find roots of the equation $A(s) = 0$. We can then make conclusions about the stability of the system based on the positions of the roots. Obviously, it is easy to solve $A(s) = 0$ when it is with relative lower power (order), but it is very difficult to solve and determine the stability of a system if it has higher order. Can we find other solution methods for this problem? Fortunately, a relatively simple judging method has been given by the Routh–Hurwitz criterion.

From previous analysis, it is not necessary to know the accurate positions of poles to determine a system's stability. In other words, it is not necessary to calculate the specific roots of $A(s) = 0$; the only thing we should do is to test whether the real part of each pole (root) is greater than, less than or equal to zero. If the system is stable, all the poles are in the open left half-plane (the real parts of roots are all less than zero); as long as there is a pole located in the open right half-plane (the real part is greater than zero), the system will be unstable.

We can define a Hurwitz polynomial as:

A polynomial with negative real roots, which is with real coefficients and all roots in the open left half-plane, is named the Hurwitz polynomial.

Obviously, if $A(s)$ is a Horowitz polynomial, the system is stable.

The necessary condition for which $A(s) = a_n s^n + a_{n-1} s^{n-1} + \cdots + a_1 s + a_0$ is the Hurwitz polynomial means that all its coefficients must be nonzero (there is no term missing) and they all have the same plus or minus sign. Thus, this is also a necessary condition for a system to be stable.

Note: A system that meets this condition is not always stable. For example, the system with $A(s) = 3s^3 + s^2 + 2s + 8$ is just an unstable system.

E. J. Routh and A. Hurwitz put forward their own additional conditions in 1877 and 1895, respectively, and their achievements have since been collectively referred to as the Routh–Hurwitz criterion by later generations, that is:

The necessary and sufficient condition that $A(s)$ is a Hurwitz polynomial is that signs of elements in the first column in the Routh array are the same, otherwise, the change times of the signs are the number of positive real roots of $A(s) = 0$.

Thus, the necessary and sufficient condition to determine the system stability includes:

(1) There is no missing term in $A(s)$, or the coefficients of all terms are not zero;
(2) the plus or minus signs of all coefficients $a_n, a_{n-1}, \ldots, a_1, a_0$ in the $A(s)$ polynomial are the same;
(3) the plus or minus signs of the elements in the first column in the Routh–Hurwitz array are the same.

Note that if $A(s)$ is a second-order or first-order polynomial, the necessary and sufficient condition of system stability can be simplified as that all the coefficients a_2, a_1, a_0 (or a_1, a_0) exist and have the same plus or minus signs.

The arrangement of the Routh array is introduced in the following. For an nth power equation with real coefficients,

$$A(s) = a_n s^n + a_{n-1} s^{n-1} + \cdots + a_1 s + a_0 = 0 \, ,$$

the Routh array is of the form

$$
\begin{array}{llll}
\text{first row:} & a_n & a_{n-2} & a_{n-4} & \cdots \\
\text{second row:} & a_{n-1} & a_{n-3} & a_{n-5} & \cdots \\
\text{third row:} & b_{n-1} & b_{n-3} & b_{n-5} & \cdots \\
\text{fourth row:} & c_{n-1} & c_{n-3} & c_{n-5} & \cdots
\end{array}
$$

where

$$b_{n-1} = -\frac{1}{a_{n-1}} \begin{vmatrix} a_n & a_{n-2} \\ a_{n-1} & a_{n-3} \end{vmatrix}$$

$$b_{n-3} = -\frac{1}{a_{n-1}} \begin{vmatrix} a_n & a_{n-4} \\ a_{n-1} & a_{n-5} \end{vmatrix}$$

$$b_{n-5} = -\frac{1}{a_{n-1}} \begin{vmatrix} a_n & a_{n-6} \\ a_{n-1} & a_{n-7} \end{vmatrix}$$

$$\vdots$$

$$c_{n-1} = -\frac{1}{b_{n-1}} \begin{vmatrix} a_{n-1} & a_{n-3} \\ b_{n-1} & b_{n-3} \end{vmatrix}$$

$$c_{n-3} = -\frac{1}{b_{n-1}} \begin{vmatrix} a_{n-1} & a_{n-5} \\ b_{n-1} & b_{n-5} \end{vmatrix}$$

$$c_{n-5} = -\frac{1}{b_{n-1}} \begin{vmatrix} a_{n-1} & a_{n-7} \\ b_{n-1} & b_{n-7} \end{vmatrix}$$

$$\vdots$$

The array is arranged sequentially in this method until all zero elements appear in one row, and it can be generally arranged to row $n + 1$. Usually, any new element in the array can be obtained by two elements in the first column of two rows above the unknown element, and two elements in the column of the top right of the unknown element can form a determinant. Then this determinant is divided by the first element in the row above close to the unknown element, and finally, a negative sign is added to this determinant. Letting $a_{j,k}$ denote the element in row j and column k, we have

$$a_{j,k} = -\frac{1}{a_{j-1,1}} \begin{vmatrix} a_{j-2,1} & a_{j-2,k+1} \\ a_{j-1,1} & a_{j-1,k+1} \end{vmatrix}$$

Example 7.2-3. Determine the system stability of $A(s) = s^4 + 7s^3 + 17s^2 + 17s + 6$.

Solution. The Routh array is arranged as follows:

$$
\begin{array}{lccc}
s^4 & 1 & 17 & 6 \\
s^3 & 7 & 17 & \\
s^2 & 14.58 & 6 & \\
s^1 & 14.12 & & \\
s^0 & 6 & &
\end{array}
$$

The first column element signs are all positive, and the coefficients of $A(s)$ are not missing, so the system is stable.

Example 7.2-4. With $A(s) = s^5 - 2s^4 + 2s^3 + 4s^2 - 11s - 10$ known, find the number of roots of the characteristic equation which have positive real part.

Solution. Obviously, such a system with $A(s)$ is unstable because there are coefficients with different signs in $A(s)$. Next, the number of positive real part roots in it can be found by the Routh–Hurwitz criterion. The Routh array is arranged as follows:

$$
\begin{array}{llll}
s^5 & 1 & 2 & -11 \\
s^4 & -2 & 4 & -10 \\
s^3 & 4 & -16 & \\
s^2 & -4 & -10 & \\
s^1 & -26 & & \\
s^0 & -10 & &
\end{array}
$$

It can be seen that element signs in the first column change three times between positive and negative, and therefore, $A(s)$ has three positive real port roots. From the solution for $A(s) = 0$, we have $A(s) = (s+1)^2(s-2)(s-1+j2)(s-1-j2) = 0$. We can see it has three roots with positive real parts, such as 2, $1 + j2$ and $1 - j2$, so the Routh–Hurwitz criterion is verified.

Some special problems should appear while a Routh array is arranged, and the corresponding solutions are concluded in the following.

First case: The first element of a row is zero, but the remaining elements of the row are not all zero. Now, the element in the first column of the next row will be uncertain. Considering $A(s) = s^5 + 2s^4 + 2s^3 + 4s^2 + s + 1$, its Routh array is of the form

$$
\begin{array}{llll}
s^5 & 1 & 2 & 1 \\
s^4 & 2 & 4 & 1 \\
s^3 & 0 & \frac{1}{2} & 0 \\
s^2 & -\frac{1}{0} & &
\end{array}
$$

Because the 0 appears in the denominator of the element in row 4, the array cannot be continued. There are ways to deal with this problem; one of them is to use an arbitrarily small number δ to replace the 0 in the first column in row 3, and then continue to arrange the array by normal steps. The term containing δ^2 can be ignored unless the corresponding coefficient is an uncertainty value. In fact, the total changing times of signs in first column elements are independent of the sign of δ.

Continuing to arrange the array, we have

$$
\begin{array}{llll}
s^5 & 1 & 2 & 1 \\
s^4 & 2 & 4 & 1 \\
s^3 & \delta & \frac{1}{2} & 0 \\
s^2 & 4 - \frac{1}{\delta} & 1 & 0 \\
s^1 & \frac{1}{2} - \frac{\delta^2}{4\delta - 1} & 0 & 0
\end{array}
$$

Neglecting the δ^2 term, we have $\frac{\delta^2}{4\delta-1} = 0$. Because δ is far less than 1, $4 - \frac{1}{\delta} \approx -\frac{1}{\delta}$.
Then the array becomes

$$
\begin{array}{cccc}
s^5 & 1 & 2 & 1 \\
s^4 & 2 & 4 & 1 \\
s^3 & \delta & \frac{1}{2} & 0 \\
s^2 & -\frac{1}{\delta} & 1 & 0 \\
s^1 & \frac{1}{2} & 0 & 0 \\
s^0 & 1 & & \\
\end{array}
$$

It can be seen that regardless of whether the sign of δ is positive or negative, the element signs in the first column are changed twice, so $A(s)$ has two zeros in the right half-plane.

Another way to arrange is to reverse the order of the coefficients, namely from a_0,

$$
\begin{array}{ccc}
a_0 & a_2 & a_4 & \cdots \\
a_1 & a_3 & a_5 & \cdots \\
\end{array}
$$

For this example, we have

$$
\begin{array}{cccc}
s^5 & 1 & 4 & 2 \\
s^4 & 1 & 2 & 1 \\
s^3 & 2 & 1 & 0 \\
s^2 & \frac{3}{2} & 1 & 0 \\
s^1 & -\frac{1}{3} & 0 & 0 \\
s^0 & 1 & & \\
\end{array}
$$

The changing times of the signs in the first column are the same as previously.

Second case: The elements in a row are all zero. This occurs when the element numbers of two adjacent rows are the same, and the corresponding elements are proportional. Now, the Routh array need not be arranged sequentially, we can assert that the equation $A(s) = 0$ has some roots on the imaginary axis or in the right half-plane, the system is either boundary stable nor unstable.

Example 7.2-5. Find the range of the gain K to make the system shown in ▶ Figure 7.26 stable.

Solution. The transfer function of the system is $H(s) = \frac{K}{s^3+6s^2+8s+K}$, and the Routh array is

$$
\begin{array}{cc}
1 & 8 \\
6 & K \\
\frac{48-K}{6} & 0 \\
K & 0 \\
\end{array}
$$

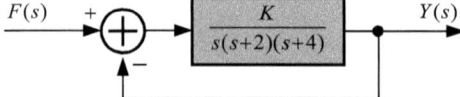

Fig. 7.26: E7.2-5.

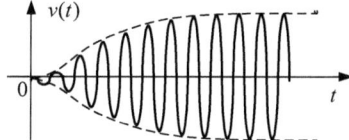

Fig. 7.27: Unstable system output waveform.

It can be seen that only if $0 < K < 48$, the plus or minus signs of the first column elements are all the same, and the system is stable.

An unstable electric system subjected to a disturbance, no matter how small the disturbance, will produce a natural response that increases with time, and the response will become infinite in theory. If the response becomes very large, the system will be ruined. However, the amplitude growth of the response can usually cause changes of other parameters in the system, so that it will limit the continuous growth of the response and impair the instability of the system. Specifically, changes in the parameters cause make the poles in the right half-plane to move left onto the $j\omega$ axis, and the natural response grows to a certain value and is maintained, which will result in a persistent oscillation waveform, as shown in ▶ Figure 7.27.

Obviously, an unstable system cannot be used to amplify or process signals, because the response of the system to a signal will be submerged by gradually increasing natural response components. For example, there is a sound amplifying system shown in ▶ Figure 7.28; when we sing a song into a microphone, if the acoustic waves from the loudspeaker have also entered the microphone and are loud enough (and have satisfied the requirements for phases), the loudspeaker will produce howling. This phenomenon is called self-excitation, and the sound system becomes unstable. However, an unstable system is also useful. For example, it can be used as a sinusoidal signal generator.

An unstable system must have a feedback branch in structure, which can lead the output signal into the input port of the system. According to its effect, the feedback can be classified as positive and negative feedback. The system whose feedback signal offsets the original input signal is considered negative feedback, otherwise, it is positive feedback. Of course, only positive feedback can cause the system to be unstable. In the above sound amplifying system, the acoustic waves from the loudspeaker are the

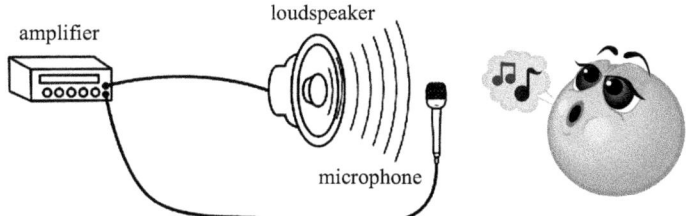

Fig. 7.28: The sound amplifying system schematic diagram.

output signal, but the microphone is the input port for the human voice signal. Once the acoustic waves from the loudspeaker are fed back to the microphone and result in the positive feedback effect, self-excitation is more likely to be produced.

7.3 Controllability and observability of a system

In the control technology field, in addition to being stable, a system should also be controllable. In other words, it is necessary to analyze the controllability of the system.

Controllability refers to whether a system can be effectively controlled to reach the desired state by means of the input signal. Observability is another issue that should be studied; that is, whether the relevant information about the state of the system can be obtained in time through observation and measurement of the system's output signal, so that the system can be controlled by adjusting the input signal. The objective of the analysis of the two problems, which are the two basic concepts in modern control theory, is to achieve optimal control.

The usual requirement of a control system from the human is that the output and the input of the system are the same as much as possible.

As a result, people would like to adopt different input signals to test and analyze a system to achieve the control indexes enacted by them. The unit step signal is chosen as the most representative testing signal, because not only many properties of system can be reflected by using it, but also the responses produced by other testing signals can be deduced by its differential or integral. The following example will illustrate this point.

Example 7.3-1. A feedback system is given in ▶ Figure 7.29. The gain of the amplifier is k, $H_O(s) = \frac{1}{s(s+4)}$ and $H_B(s) = 1$.
(1) Judge the stability of the system.
(2) If the input is a unit step signal $\varepsilon(t)$, what is the appropriate value of k to make the output a constant value within a short adjusting time and with good stability?

Solution. (1) The system function is

$$H(s) = \frac{kH_O(s)}{1 + kH_O(s)H_B(s)} = \frac{kH_O(s)}{1 + kH_O(s)} = \frac{k}{s^2 + 4s + k}.$$

The poles are $\lambda_{1,2} = \frac{-4 \pm \sqrt{16-4k}}{2}$. No matter what k is a real number greater than zero, these two poles will locate on the left half-plane, and the system is stable.

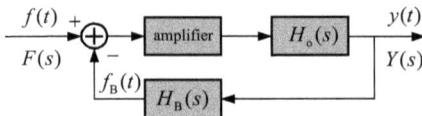

Fig. 7.29: E7.3-1 (1).

(2) The Laplace transform of the step response is

$$G(s) = F(s)H(s) = \frac{1}{s} \cdot \frac{k}{s^2 + 4s + k} = \frac{k}{s(s^2 + 4s + k)} .$$

It can be seen that $G(s)$ has a pole located at the original point, and that the other two poles are located in the left half-plane, so increasing values of the step response over time necessarily reach a stable value. Three cases based on different values of k will be discussed.

(a) $k > 4$, where $\lambda_{1,2} = -2 \pm j\sqrt{k-4}$ are a pair of the conjugate complex roots. If $k = 40$, $\lambda_{1,2} = -2 \pm j6$, and then

$$G(s) = \frac{40}{s(s-\lambda_1)(s-\lambda_2)} = \frac{1}{s} - \frac{\sqrt{5}}{3\sqrt{2}}e^{-j18.4°}\frac{1}{s-\lambda_1} - \frac{\sqrt{5}}{3\sqrt{2}}e^{j18.4°}\frac{1}{s-\lambda_2} .$$

Finding the inverse Laplace transform of $G(s)$, we have

$$g(t) = \varepsilon(t) - \frac{\sqrt{5}}{3\sqrt{2}}e^{-2t}[e^{j6t}e^{-j18.4°} + e^{-j6t}e^{-j18.4°}]\varepsilon(t)$$

$$= \varepsilon(t) - \frac{\sqrt{10}}{3}e^{-2t}\cos(6t + 18.4°)\varepsilon(t) .$$

(7.3-1)

(b) $k = 4$, where $\lambda_{1,2} = -2$ are the real repeated roots. Then

$$G(s) = \frac{4}{s(s-\lambda_1)(s-\lambda_2)} = \frac{4}{s(s+2)^2} = \frac{1}{s} + \frac{-1}{s+2} + \frac{-2}{(s+2)^2} .$$

Finding the inverse Laplace transform of $G(s)$, we have

$$g(t) = \varepsilon(t) - (1+2t)e^{-2t}\varepsilon(t) .$$

(7.3-2)

(c) $k < 4$. Suppose $k = 2$, so $\lambda_{1,2} = -2 \pm \sqrt{2}$ are two real roots, then

$$G(s) = \frac{2}{s(s-\lambda_1)(s-\lambda_2)} = \frac{1}{s} + \frac{-1.2}{s+2-\sqrt{2}} + \frac{0.2}{s+2+\sqrt{2}} .$$

Finding the inverse Laplace transform of $G(s)$, we have

$$g(t) = \varepsilon(t) - (1.2e^{-0.586t} - 0.2e^{-3.414t})\varepsilon(t) .$$

(7.3-3)

The $g(t)$ in equations (7.3-1)–(7.3-3), are plotted in ▶ Figure 7.30.
From the observation of the three waveforms, we can conclude that:

(1) The step response includes an oscillation stage when $k > 4$, but it can reach a stable value after attenuation in an interim time t_s. This circumstance is called the underdamping state. The changing time of the response from the initial zero value to the maximum value is called the peak time t_p. The overshoot of the step response is defined by

$$\sigma = \frac{g(t_p) - g(\infty)}{g(\infty)}100\% .$$

(7.3-4)

Usually, we require the interim and peak times to be short, and the overshoot smaller or equal to zero, so that the difference between the response and the excitation is minimized.

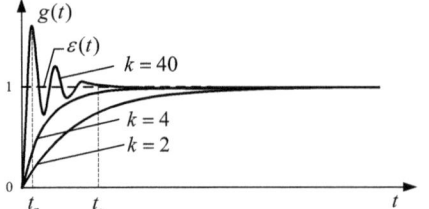

Fig. 7.30: E7.3-1 (2).

(2) There is no oscillation stage in the step response when $k = 4$, so the step response at that time can be called the critical damping state.

(3) There is no oscillation stage in the step response when $k < 4$, but compared with $k = 4$, the transient time t_s becomes longer, and the circumstance of the step response can be called the overdamping state.

(4) Obviously, at the critical damping state, the track feature of the system or the consistency between the input and the output is better. In practical applications, to reduce the transient time t_s, the system should often be allowed to be in the underdamping state with a bit overshoot.

From the above discussions, transient time t_s, peak time t_p and overshoot σ can be considered as three important indicators to describe the control properties of a system. We can regulate k to satisfy different requirements and achieve the optimal control result in practice.

When we analyze a control system using an external analysis method, generally, we do not need to discuss controllable and observable problems; as long as the zeros and poles of the system function do not counteract with each other, the system is controllable and observable, otherwise, it is not controllable and observable.

The usage of knowledge from the signals and systems topic in Automation Control Systems is briefly introduced here, with the aim of helping readers better understand the automation control principle and laying the foundation for further research or further study.

7.4 Solved questions

Question 7-1. A causal feedback LTI system is shown in ▶ Figure Q7-1 (1). The system 1 is represented by a differential equation $y'(t) - y(t) = \varepsilon(t)$, the system function of the feedback path is $F(s) = K/(s + 2)$, and K can be adjusted for any real number.

(1) Find the system function $H_1(s)$ of system 1, and work out the zeros, poles and ROC of it, and judge the stability of system 1.

(2) Find an adjustable range of K in $F(s)$ to make the feedback system stable.

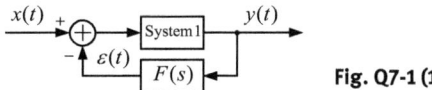

Fig. Q7-1 (1)

Solution. (1) The system function and the ROC of system 1 are, respectively, $H_1(s) = \frac{1}{s-1}$ and ROC : $\sigma > 1$.

A first-order pole $p = 1$, and the ROC does not include imaginary axis; they are plotted in ▶ Figure Q7-1 (2). According to the concept of stability, the system is not stable.

(2) The system function of the feedback system in ▶ Figure Q7-1 (1) is

$$H(s) = \frac{H_1(s)}{1 + H_1(s)F(s)} = \frac{\frac{1}{s-1}}{1 + \frac{K}{(s-1)(s+2)}} = \frac{s+2}{s^2 + s + (K-2)} \ .$$

Obviously, $A(s)$ is the second-order polynomial, if $k - 2 > 0$, the system is stable, so, $K > 2$.

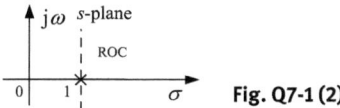

Fig. Q7-1 (2)

Question 7-2. For the system shown as ▶ Figure Q7-2 $H(s) = \frac{U_2(s)}{U_1(s)} = 4$ is known.
(1) Find the subsystem function $H_2(s)$.
(2) Make the subsystem $H_2(s)$ stable and find the range of k.

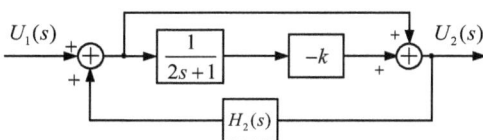

Fig. Q7-2

Solution. (1) According to ▶ Figure Q7-2, we have

$$U_1(s) + U_2(s)H_2(s)\left(1 - \frac{k}{2s+1}\right) = U_2(s) \ .$$

Because $H(s) = \frac{U_2(s)}{U_1(s)} = 4$, we can conclude that $H_2(s) = \frac{6s+3+k}{4(2s+1-k)}$.

(2) To make $H_2(s)$ stable, we obtain $k < 1$.

Question 7-3. The system equation $y''(t) + 3y'(t) + 2y(t) = 2f'(t) + f(t)$. Work out the flow graph of the system in direct form.

Solution. Based on the zero-state conditions, with the Laplace transform for this differential equation, we obtain

$$s^2 Y_f(s) + 3s Y_f(s) = 2sF(s) + F(s) .$$

So, the system function is

$$H(s) = \frac{2s + 1}{s^2 + 3s + 1} .$$

The flow graph of the system in direct form is depicted in ▸ Figure Q7-3.

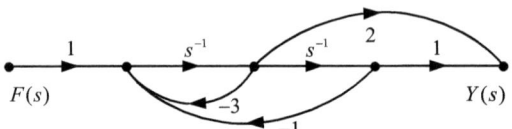

Fig. Q7-3

Question 7-4. The differential equation of an LTI causal continuous system

$$y''(t) + 4y'(t) + 3y(t) = 4f'(t) + 2f(t) .$$

(1) Find the impulse response $h(t)$ of the system.
(2) Judge whether the system is stable.

Solution. (1) With the Laplace transform on both sides of the equation, we have

$$s^2 Y_f(s) + 4s Y_f(s) + 3 Y_f(s) = 4sF(s) + 2F(s) .$$

Changing the form of the above equation, yields

$$H(s) = \frac{Y_f(s)}{F(s)} = \frac{4s + 2}{s^2 + 4s + 3} = \frac{4s + 2}{(s + 1)(s + 3)} = \frac{5}{s + 3} - \frac{1}{s + 1} .$$

With the inverse Laplace transform, the impulse response is

$$h(t) = (5e^{-3t} - e^{-t})\varepsilon(t) .$$

(2) From the denominator polynomial of $H(s)$, the system function has two poles, such as $\lambda_1 = -1, \lambda_2 = -3$, which are located on the left half-plane, so the system is stable.

Question 7-5. The differential equation of a LTI causal continuous system $y''(t) + 5y'(t) + 6y(t) = 2f'(t) + f(t)$. known $f(t) = e^{-t}\varepsilon(t)$, $y(0_-) = 1$, $y'(0_-) = 1$. Try to find in s domain:

(1) zero-input and zero-state responses $y_x(t)$ and $y_f(t)$, the total response $y(t)$.
(2) the system function $H(s)$ and impulse response $h(t)$, and judge system stability.

Solution. (1) With the Laplace transform on both sides of the differential equation,

$$s^2 Y(s) - sy(0_-) - y'(0_-) + 5sY(s) - 5y(0_-) + 6Y(s) = (2s+1)F(s) .$$

So,

$$Y(s) = \frac{sy(0_-) + y'(0_-) + 5y(0_-)}{s^2 + 5s + 6} + \frac{2s+1}{s^2 + 5s + 6}F(s) .$$

Two parts on the right side of the equation are representations of zero-input and zero-state responses in the s domain; the total response can be obtained from their inverse Laplace transform.

Since $y(0_-) = 1$, $y'(0_-) = 1$, and so the representation of the zero-input response in the s domain is

$$Y_x(s) = \frac{sy(0_-) + y'(0_-) + 5y(0_-)}{s^2 + 5s + 6} = \frac{s+6}{s^2 + 5s + 6} = \frac{4}{s+2} + \frac{-3}{s+3} ,$$

and so, the zero-input response is

$$y_x(t) = 4e^{-2t} - 3e^{-3t}, \quad t \geq 0 .$$

The representation of zero-state response in the s domain is

$$Y_f(s) = \frac{2s+1}{s^2 + 5s + 6}F(s) = \frac{2s+1}{(s^2 + 5s + 6)(s+1)} = \frac{-1/2}{s+1} + \frac{3}{s+2} + \frac{-5/2}{s+3} ,$$

so the zero-state response

$$y_f(t) = \left(-\frac{1}{2}e^{-t} + 3e^{-2t} - \frac{5}{2}e^{-3t}\right)\varepsilon(t) .$$

The total response

$$y(t) = y_x(t) + y_f(t) = -\frac{1}{2}e^{-t} + 7e^{-2t} - \frac{11}{2}e^{-3t}, \quad t \geq 0 .$$

(2) The system function

$$H(s) = \frac{Y_f(s)}{F(s)} = \frac{2s+1}{s^2 + 5s + 6} = \frac{-3}{s+2} + \frac{5}{s+3} ,$$

and so, the impulse response

$$h(t) = \left(-3e^{-2t} + 5e^{-3t}\right)\varepsilon(t) .$$

From $H(s)$, both of the two poles, -2 and -3, are located on the left half-plane, so the system is stable.

7.5 Learning tips

The graphical system model has incomparable advantages over the analytic formula. Furthermore, besides the relationship between the response and the excitation, stability is an important consideration in system analysis. Please pay attention to the following points.

(1) Besides the mathematical model, a system can also be simulated more intuitively and concisely by the block diagram and the flow graph.

(2) The stability of a system can be seen from the distribution of the pole-zero of the system function which is also a type of the mathematical model.

(3) The Rose–Hurwitz criterion.

7.6 Problems

Problem 7-1. Two systems are shown in ▶ Figure P7-1, find the system functions and prove that the two block diagrams correspond to the same system.

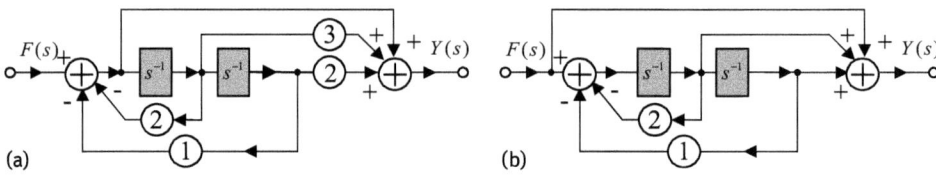

(a) (b)

Fig. P7-1

Problem 7-2. Make simulation diagrams in direct form for the following systems:

(1) $H(p) = \dfrac{1}{p^3 + 3p + 2}$

(2) $H(p) = \dfrac{p^2 + 2p}{p^3 + 3p^2 + 3p + 2}$

(3) $H(s) = \dfrac{2s + 3}{(s + 2)^2(s + 3)}$

(4) $H(s) = \dfrac{s^2 + 4s + 5}{(s + 1)(s + 2)(s + 3)}$

Problem 7-3. Find the system function $H(s) = \dfrac{Y(s)}{F(s)}$ for each system shown in ▶ Figure P7-3.

Problem 7-4. For the following system functions, try to draw their simulation block diagrams in direct, cascade and parallel forms.

(1) $H(s) = \dfrac{5s + 5}{s(s + 2)(s + 5)}$, (2) $H(s) = \dfrac{5s^2 + s + 1}{s^3 + s^2 + s}$, (3) $H(s) = \dfrac{3s}{s^3 + 4s^2 + 6s + 4}$.

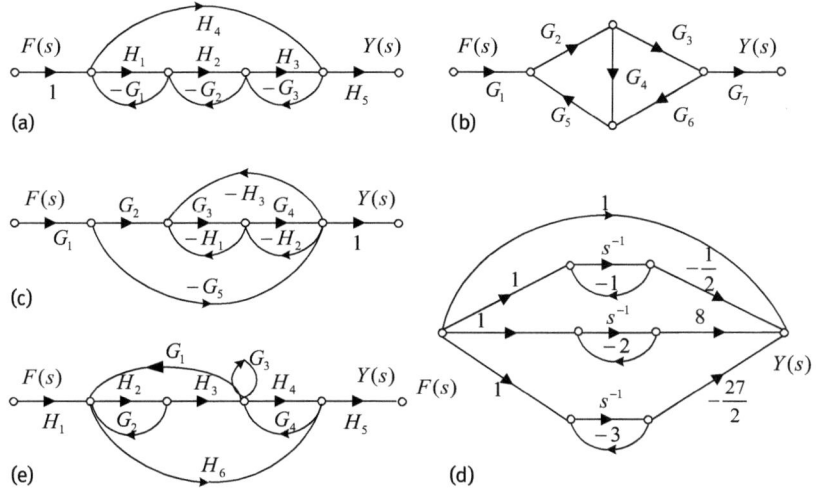

Fig. P7-3

Problem 7-5. Find $H(s)$ of the system shown in ▶ Figure P7-5 and its flow graphs in direct form.

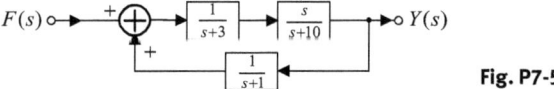

Fig. P7-5

Problem 7-6. A flow graph in parallel form is shown in ▶ Figure P7-6. Find $H(s) = \frac{Y(s)}{F(s)}$ and draw its flow graphs in cascade form and in direct form.

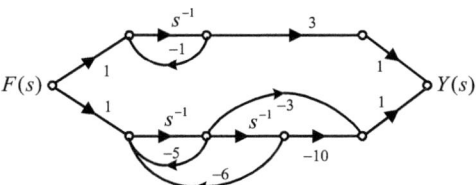

Fig. P7-6

Problem 7-7. A system flow graph is shown in ▶ Figure P7-7, and $f(t) = e^{-2t}\varepsilon(t)$ is known. Find the zero-state response $y_f(t)$ of the system.

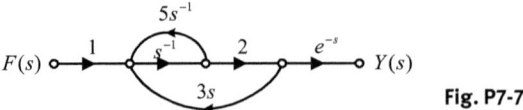

Fig. P7-7

Problem 7-8. Find system functions of circuits in ▶ Figure P7-8 and their pole-zero plots.

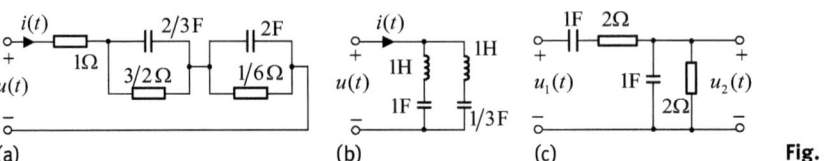

(a)   (b)   (c)   **Fig. P7-8**

Problem 7-9. The poles of a system function are $\lambda_1 = 0$, $\lambda_2 = -1$, the zero is $z_1 = 1$, and the terminal value of the impulse response is $h(\infty) = -10$. Find
(1) the system function $H(s)$;
(2) the stable response $y_s(t)$ for the excitation $f(t) = 3\sin(3t)\varepsilon(t)$.

Problem 7-10. A circuit is shown in ▶ Figure P7-10a, the pole-zero diagram of the transfer function $H(s) = \frac{U_2(s)}{U_1(s)}$ is shown in ▶ Figure P7-10b, and $H(0) = 1$. Find L, R and C.

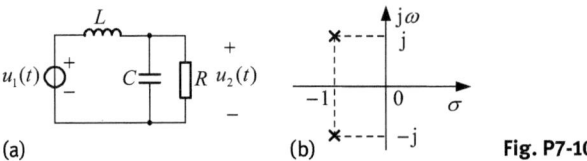

(a)   (b)   **Fig. P7-10**

Problem 7-11. The pole-zero diagram is shown in ▶ Figure P7-11, and $h(0_+) = 1$. If the excitation $f(t) = \cos t\varepsilon(t)$, find the zero-state response, the free response and the forced response.

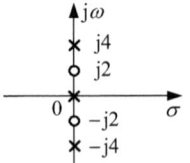

Fig. P7-11

Problem 7-12. Judge the stability of the following systems and point out the number of eigenvalues with a positive real part.
(1) The characteristic equation is $s^4 + 7s^3 + 17s^2 + 6 = 0$.
(2) The characteristic equation is $s^6 + 7s^5 + 16s^4 + 14s^3 + 25s^2 + 7s + 12 = 0$.
(3) The system function is $H(s) = \frac{4s^3+2s^2+3s+1}{s^5+2s^4+2s^3+4s^2+11s+4}$.
(4) The system function is $H(s) = \frac{s+1}{s^6+5s^5+11s^4+25s^3+36s^2+30s+36}$.
(5) The characteristic equation is $s^4 + 2s^3 + 7s^2 + 10s + 10 = 0$.

Problem 7-13. For the two feedback systems shown in ▶ Figure P7-13, find the value of k that can make the systems stable.

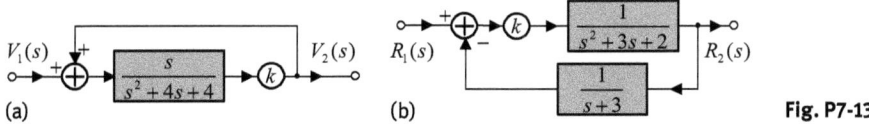

(a)   (b)   **Fig. P7-13**

Problem 7-14. A feedback system is shown in ▶ Figure P7-14; the impulse response of the subsystem $h_1(t) = (2e^{-2t} - e^{-t})\varepsilon(t)$.
(1) If the system is stable, what is the value of the real coefficient k?
(2) Under the condition of boundary stability, find the impulse response $h(t)$ of the whole system.

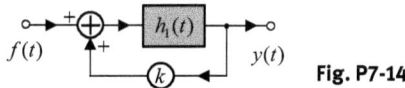

Fig. P7-14

Problem 7-15. A system is shown in ▶ Figure P7-15.
(1) What k can make the system is stable?
(2) Under the condition of boundary stability, find the impulse response $h(t)$.

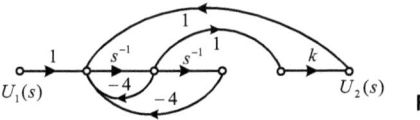

Fig. P7-15

A Reference answers

Chapter 1

Problem 1-1:

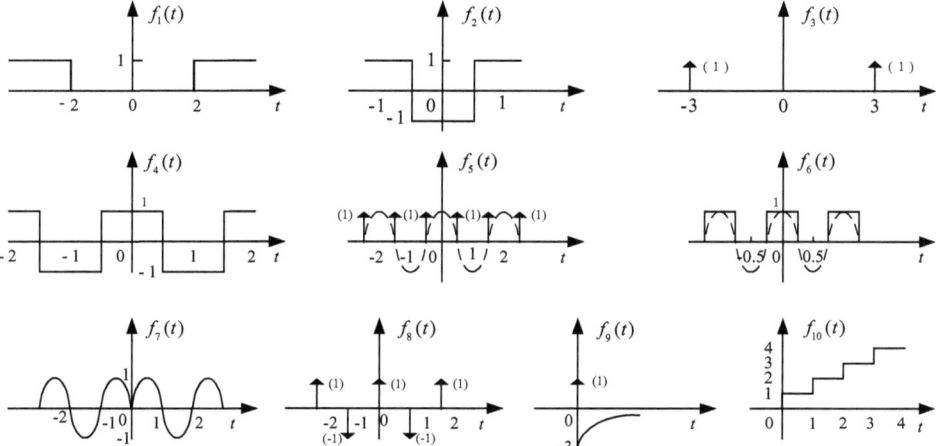

Fig. A1-1

Problem 1-2:

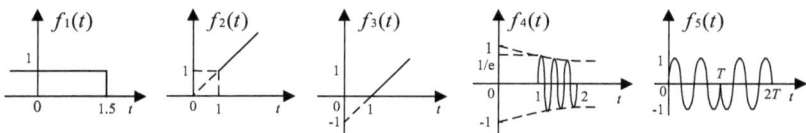

Fig. A1-2

Problem 1-3:

The answer is omitted.

https://doi.org/10.1515/9783110419535-app-008

Problem 1-4:

(1) $f(-t_0)$,

(2) $f(t_0)$,

(3) 1,

(4) $e^{-4} + 2e^{-2}$

(5) $-\frac{\sqrt{3}}{2}e^{-1}$,

(6) 2,

(7) 0,

(8) 0,

(9) $\frac{1}{2}$,

(10) $\frac{1}{4}[\delta(t) + \varepsilon(t)]$,

(11) 1,

(12) 0

Problem 1-5:

(1) $\delta'(t) - \delta(t)$,

(2) $\delta(t) - 2e^{-2t}\varepsilon(t)$,

(3) $\delta(t) - 2\varepsilon(t-1) - 2\delta(t-1)$

Problem 1-6:

(1) $f_1(t) = (1 + \cos \pi t)[\varepsilon(t) - \varepsilon(t - 2)]$,

(2) $f_2(t) = 2t[2\varepsilon(t) - \varepsilon(t + 1) - \varepsilon(t - 1)]$

(3) $f_3(t) = \sin(t)[\varepsilon(t) - \varepsilon(t - \pi)]$,

(4) $f_4(t) = \sin(\pi t) \cdot \text{sgn}(t)$

Problem 1-7:

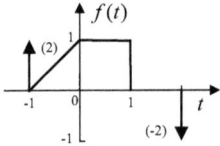

Fig. A1-7

Problem 1-8:

(1) energy

(2) power

(3) power

(4) energy

(5) energy

(6) no energy, no power

Problem 1-9:

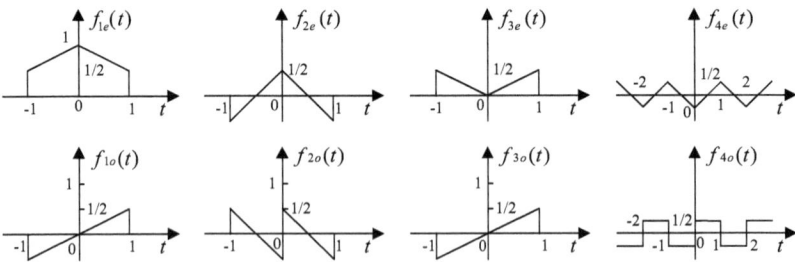

Fig. A1-9

Problem 1-10:

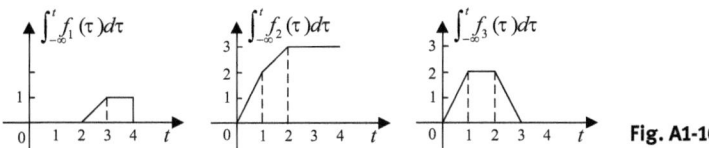

Fig. A1-10

Problem 1-11:

$$f_1(t)*f_2(t) = \int\limits_{-\infty}^{+\infty} f_1(\tau)f_2(t-\tau)d\tau = \int\limits_{-\infty}^{+\infty} e^{2\tau}e^{-(t-\tau)}\varepsilon(t-\tau)d\tau$$

$$= e^{-t}\int\limits_{-\infty}^{t} e^{3\tau}d\tau = \frac{1}{3}e^{-t}e^{3\tau}|_{-\infty}^{t} = \frac{1}{3}e^{-t}e^{3t} = \frac{1}{3}e^{2t} \quad -\infty < t < +\infty$$

Problem 1-12:

$$f_1(t) = e^{-t}\varepsilon(t), \qquad f_2(t) = \varepsilon(t) - \varepsilon(t-1).$$

Problem 1-13:

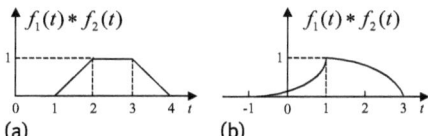

(a)   (b)   Fig. A1-13

(a) $f_1(t)*f_2(t) = \begin{cases} 0 & \text{else} \\ t-1 & 1 \le t < 2 \\ 1 & 2 \le t < 3 \\ 4-t & 3 \le t < 4 \end{cases}$

(b) $f_1(t)*f_2(t) = \begin{cases} 0 & \text{else} \\ \frac{1}{4}(t+1)^2 & -1 \le t < 1 \\ -\frac{1}{4}t^2 + \frac{1}{2}t + \frac{3}{4} & 1 \le t \le 3 \end{cases}$

Problem 1-14:

(1) $(1 - e^{-t})\varepsilon(t)$;

(2) $f_2(t) = \frac{1}{2\pi}[1 - \cos 2\pi t][\varepsilon(t) - \varepsilon(t-1)]$

Chapter 2

Problem 2-1:

(1) System does not meet zero input linear and zero state linear, is a nonlinear system.
(2) System does not meet decomposability, is a nonlinear system.
(3) System is a linear system.
(4) System does not meet zero input linear, is a nonlinear system.
(5) System is a nonlinear system.

Problem 2-2:

(1) System is a time invariant system.
(2) System is a time-variant system.
(3) System is a time variant system.
(4) System is a time variant system.

Problem 2-3:

(1) causal
(2) noncausal
(3) noncausal
(4) $b < 0$ noncausal; $b \geq 0$ causal
(5) causal

Problem 2-4:

(1) linear, time variant, causal
(2) nonlinear, time invariant, noncausal
(3) linear, time variant, causal
(4) linear, time invariant, causal
(5) linear, time-invariant, noncausal
(6) nonlinear, time variant, causal

Problem 2-5:

When $\tau \geq 0$, time variant, causal;   when $\tau < 0$, time variant, noncausal.

Problem 2-6:

$$\because f_2(t) = \varepsilon(t) - 2\varepsilon(t-1) + \varepsilon(t-2), \qquad \therefore y_2(t) = y_1(t) - 2y_1(t-1) + y_1(t-2)$$

$$\because f_3(t) = \int_{-\infty}^{t} f_2(\tau)d\tau, \qquad \therefore y_3(t) = \int_{-\infty}^{t} y_2(\tau)d\tau$$

$$\because f_4(t) = f_2'(t), \qquad \therefore y_4(t) = y_2'(t)$$

Problem 2-7:

$y_1(t) = y_x(t) + y_{f1}(t) = 3e^{-2t} + \sin 4t,$ $y_2(t) = y_x(t) + 2y_{f1}(t) = 4e^{-2t} + 2\sin 4t,$
$y_{f1}(t) = e^{-2t} + \sin 4t,$ $y_x(t) = 2e^{-2t},$
$y_3(t) = y_x(t) + 3y_{f1}(t) = 5e^{-2t} + 3\sin 4t$ $t > 0$

Problem 2-8:

If $x_1(0_-) = 1$, zero-input response is $y_{x1}(t)$; $x_2(0_-) = 1$, zero-input response is $y_{x2}(t)$;
$f(t) = \begin{cases} 1 & t > 0 \\ 0 & t < 0 \end{cases}$, zero-state response is $y_{f1}(t)$.

$\therefore y_{x1}(t) = te^{-t} + e^{-t},$ $y_{x2}(t) = te^{-t},$ $y_{f1}(t) = -te^{-t},$
$\therefore y_f(t) = 3y_{f1}(t) = -3te^{-t}$ $t > 0$

Problem 2-9:

$2u_C'(t) + 2u_C(t) = u_s(t)$

Problem 2-10:

$u_C''(t) + 7u_C'(t) + 6u_C(t) = 6\sin 2t\varepsilon(t)$

Problem 2-11:

$(p^2 + 3p + 3)u_1 = (2p + 2)i(t),$ $(p^2 + 3p + 3)u_2 = 2pi(t)$

Problem 2-12:

(1) $y''(t) + 7y'(t) + 12y(t) = f(t),$
(2) $y'''(t) + 4y''(t) + 10y'(t) + 3y(t) = f''(t) + 10f(t)$

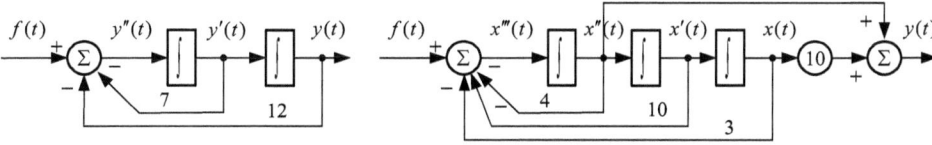

Fig. A2-12

Chapter 3

Problem 3-1:

$y(t) = \frac{1}{2}e^{-3t} + 1$ $t > 0$,   Natural response $\frac{1}{2}e^{-3t}\varepsilon(t)$,   forced response $\varepsilon(t)$.

Problem 3-2:

$y(t) = 4te^{-2t} - 3e^{-2t} + 6e^{-t}$ $t \geq 0$;   natural $4te^{-2t} - 3e^{-2t}$,   forced $6e^{-t}$ $t \geq 0$

Problem 3-3:

(1) $y'(0_-) = y'(0_+) = 2$, $y(0_-) = y(0_+) = 1$, $y(t) = 3e^{-t} - \frac{5}{2}e^{-2t} + \frac{1}{2}$ $t > 0$

(2) $y'(0_+) = y'(0_-) = 2$, $y(0_-) = y(0_+) = 1$, $y(t) = 4e^{-t} - 3e^{-2t} - te^{-2t}$ $t > 0$

Problem 3-4:

$\because y''(t)$ includes $\delta(t)$, $y'(t)$ includes $\varepsilon(t)$, $y(t)$ includes $t\varepsilon(t)$,

$\therefore \Delta y'(0) = 1$ $\Delta y(0) = 0$, $y'(0_+) = y'(0_-) + \Delta y'(0) = 2 + 1 = 3$, $y(0_+) = y(0_-) + \Delta y(0) = 1 + 0 = 1$,

$y(t) = 2e^{-t} - \frac{5}{2}e^{-2t} + \frac{3}{2}$ $t > 0$, $y_c(t) = 2e^t - \frac{5}{2}e^{-2t}$, $y_p(t) = \frac{3}{2}$.

$\because y_x(t) = c_3 e^{-t} + c_4 e^{-2t}$, $y_x(0_+) = y_x(0_-) = y(0_-) = 1$, $y_x'(0_+) = y'(0_-) = 2$,

$\therefore y_x(t) = 4e^{-t} - 3e^{-2t}$,

$\therefore y_f(t) = y(t) - y_x(t) = 2e^{-t} - \frac{5}{2}e^{-2t} + \frac{3}{2} - 4e^{-t} + 3e^{-2t} = -2e^{-t} + \frac{1}{2}e^{-2t} + \frac{3}{2}$

Problem 3-5:

$y_x(t) = c_1 e^{(-1-2j)t} + c_2 e^{(-1+2j)t}$ $\begin{cases} y_x(0_-) = c_1 + c_2 = 2 \\ y_x'(0_-) = (-1 - 2j)c_1 + (-1 + 2j)c_2 = -2 \end{cases}$,

$y_x(t) = 2e^{-t}\cos 2t$ $t \geq 0$

Problem 3-6:

$y_x(t) = c_1 e^{-3t} + c_2 e^{-t}$ $\begin{cases} c_1 + c_2 = 1 \\ -3c_1 - c_2 = 2 \end{cases}$,

$y_x(t) = -\frac{3}{2}e^{-3t} + \frac{5}{2}e^{-t}$

Problem 3-7:

$y_f(t) = 5e^{-t} - 5e^{-2t} + e^{-3t}$ $t \geq 0$

Problem 3-8:

(1) $u_C(0_+) = 10\,\text{V}$, $\qquad i(0_+) = \frac{20-u_C(0_+)}{R} = 5\,\text{A}$;

(2) $u_C(t) = 20 - 10e^{-2t}, t \geq 0_+$

Problem 3-9:

$i_{Lx}(t) = \frac{1}{2}e^{-t} + \frac{1}{2}e^{-2t} \quad t \geq 0, \qquad i_{Lf}(t) = -\frac{3}{2}e^{-t} + \frac{1}{2}e^{-2t} + 1 \quad t \geq 0$

Problem 3-10:

(a) $H(p) = \dfrac{2(p+10)}{(p+5)(p+6)}$; $\qquad\qquad$ (b) $H(p) = \dfrac{10(p+1)}{(p+5)(p+6)}$

Problem 3-11:

(1) $y_x(t) = 5e^{-t} - 3e^{-2t} \quad t \geq 0$

(2) $y_x(t) = e^{-t}[A_1 \cos t + A_2 \sin t] = e^{-t}[\cos t + 3 \sin t] \quad t \geq 0$

(3) $y_x(t) = 1 - e^{-t} - te^{-t} \quad t \geq 0$

Problem 3-12:

$h(t) = 0.5e^{-2t}\varepsilon(t)$

Problem 3-13:

$h(t) = e^{t-1}\varepsilon(3-t)$

Problem 3-14:

$y_{f1}(t) = \frac{1}{\pi}(1 - \cos \pi t)[\varepsilon(t) - \varepsilon(t-2)]$

Problem 3-15:

$y_f(t) = t[\varepsilon(t) - \varepsilon(t-T)] - (t-2T)[\varepsilon(t-T) - \varepsilon(t-2T)]$

Problem 3-16:

$y_{f_1}(t) = \dfrac{df_1(t)}{dt} * g(t) = g(t) - 2g(t-2) + g(t-3)$

$y_{f_2}(t) = \{[\varepsilon(t) - \varepsilon(t-1)] - \delta(t-1)\} * (2e^{-2t} - 1)\varepsilon(t)$

$\qquad = (1 - t - e^{-2t})\varepsilon(t) - \left[1 - t + e^{-2(t-1)}\right]\varepsilon(t-1)$

Problem 3-17:

$h(t) = [h_1(t) + h_2(t) * h_1(t) * h_3(t)] * h_4(t) = 3 [\varepsilon(t) - \varepsilon(t-1)]$

$g(t) = 3\varepsilon(t) * [\varepsilon(t) - \varepsilon(t-1)] = 3t [\varepsilon(t) - \varepsilon(t-1)] + 3\varepsilon(t-1)$

Problem 3-18:

$h(t) = \varepsilon(t) + \varepsilon(t-1) + \varepsilon(t-2) - \varepsilon(t-3) - \varepsilon(t-4) - \varepsilon(t-5)$

Problem 3-19:

$g(t) = 4e^{-2t}\varepsilon(t), \qquad h(t) = 4\delta(t) - 8e^{-2t}\varepsilon(t)$

Problem 3-20:

$h(t) = e^{-t} \cos t \varepsilon(t)$

Chapter 4

Problem 4-1:

(a) only include $a_n \cos n\omega_0 t$.
(b) only include odd harmonic components.

Problem 4-2:

(a) $F_n = \dfrac{1}{T}, \quad f_1(t) = \dfrac{1}{T} \displaystyle\sum_{n=-\infty}^{+\infty} e^{jn\omega_0 t},$

(b) $a_0 = \dfrac{E}{2}, \quad a_n = \begin{cases} 0 & (n = 2, 4, \ldots) \\ -\dfrac{4E}{(n\pi)^2} & (n = 1, 3, \ldots) \end{cases}, \quad b_n = 0$

$f_2(t) = -\dfrac{4E}{\pi^2} \left(\cos \omega_0 t + \dfrac{1}{3^2} \cos 3\omega_0 t + \dfrac{1}{5^2} \cos 5\omega_0 t + \ldots + \dfrac{1}{n^2} \cos n\omega_0 t + \ldots \right)$
$(n = 1, 3, 5, \ldots)$

(c) $F_0 = \dfrac{1}{\pi}, \quad F_1 = \dfrac{1}{4}e^{-j\frac{\pi}{2}}, \quad F_{-1} = \dfrac{1}{4}e^{j\frac{\pi}{2}}, \quad F_n = \dfrac{-\cos^2 \frac{n\pi}{2}}{\pi(n^2 - 1)} \ (|n| > 1)$

Problem 4-3:

(1) $a_0 = \dfrac{1}{4}, \quad a_n = \dfrac{1}{(n\pi)^2}(\cos n\pi - 1), \quad b_n = -\dfrac{1}{n\pi} \cos n\pi$

$$f_1(t) = \frac{1}{4} + \frac{1}{\pi^2} \sum_{n=1}^{\infty} \frac{\cos n\pi - 1}{n^2} \cos n\omega_0 t - \frac{1}{\pi} \sum_{n=1}^{\infty} \frac{\cos n\pi}{n} \sin n\omega_0 t$$

(2) $$f_2(t) = \frac{1}{4} + \frac{1}{\pi^2} \sum_{n=1}^{\infty} \frac{1 - \cos n\pi}{n^2} \cos n\omega_0 t - \frac{1}{\pi} \sum_{n=1}^{\infty} \frac{1}{n} \sin n\omega_0 t$$

$$f_3(t) = \frac{1}{4} + \frac{1}{\pi^2} \sum_{n=1}^{\infty} \frac{1 - \cos n\pi}{n^2} \cos n\omega_0 t + \frac{1}{\pi} \sum_{n=1}^{\infty} \frac{1}{n} \sin n\omega_0 t$$

$$f_4(t) = \frac{1}{2} + \frac{2}{\pi^2} \sum_{n=1}^{\infty} \frac{1 - \cos n\pi}{n^2} \cos n\omega_0 t$$

Problem 4-4:

(a) $$f_1(t) = \frac{1}{2} - \frac{1}{\pi} \sum_{n=1}^{\infty} \frac{1}{n} \sin n\omega_0 t$$

(b) $$f_2(t) = \frac{1}{2} + \frac{2}{\pi} \sum_{n=1}^{\infty} \frac{1}{n} \sin n\pi t, \quad n = 1, 3, 5 \dots$$

Problem 4-5:

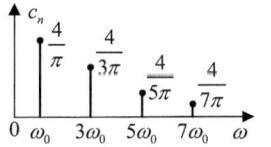 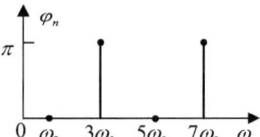

Fig. A4-5 (1)

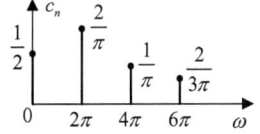 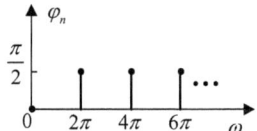

Fig. A4-5 (2)

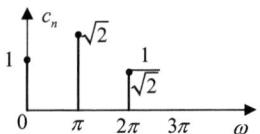 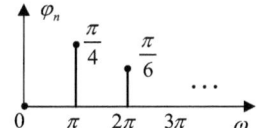

Fig. A4-5 (3)

Problem 4-8:

$$a_0 = \frac{E}{\pi}, \quad b_n = 0, \quad a_n = \begin{cases} \frac{E}{2} & (n = 1) \\ 0 & (n = 3, \dots) \\ \frac{2E}{(1-n^2)\pi} \cos \frac{n\pi}{2} & (n = 2, 4, \dots) \end{cases}$$

$$f(t) = \frac{E}{\pi} + \frac{E}{2}(\cos \omega_0 t + \frac{4}{3\pi} \cos 2\omega_0 t - \frac{4}{15\pi} \cos 4\omega_0 t + \dots)$$

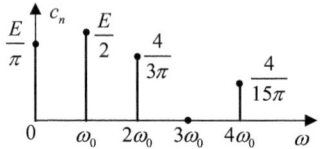 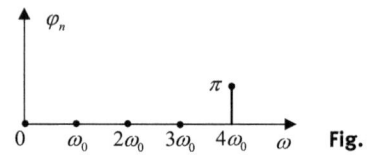

Fig. A4-8

Problem 4-9:

$$f(t) = \frac{4}{\pi}\left(\frac{1}{2} + \frac{1}{3}\cos 2\omega_0 t - \frac{1}{15}\cos 4\omega_0 t + \dots - \frac{\cos \frac{n\pi}{2}}{n^2 - 1}\cos n\omega_0 t + \dots\right)$$

$$(n = 2, 4, 6, \dots)$$

Problem 4-10:

$$u_c(t) = 6 + 8\cos\left(10^3 t - 36.9°\right) + 3.33\cos\left(2 \times 10^3 t - 56.3°\right) V$$

Problem 4-11:

$$u_2(t) = \frac{2}{\pi}[0.104 \sin(\pi t + 84°) + 0.063 \sin(3\pi t - 79°)] V$$

Problem 4-12:

$$i(t) = 0.5 + 0.450\cos(\omega_0 t - 45°) + 0.067\cos(3\omega_0 t + 108.4°) A$$

Chapter 5

Problem 5-1:

$$F_1(j\omega) = \frac{\frac{A}{\tau} - \frac{A}{\tau}e^{-j\omega\tau} - jA\omega e^{-j\omega\tau}}{(j\omega)^2}; \quad F_2(j\omega) = \tau Sa^2\left(\frac{\omega\tau}{2}\right);$$

$$F_3(j\omega) = 4\tau Sa(\omega\tau) + 2\tau Sa\left(\frac{1}{2}\omega\tau\right)\cos(\omega\tau);$$

$$F_4(j\omega) = e^{-j(2\omega-\frac{\pi}{2})}[Sa(\omega + \pi) - Sa(\omega - \pi)];$$

$$F_5(j\omega) = \frac{1}{2}\left[Sa^2\left(\frac{\omega + \omega_0}{2}\right)e^{-j(\omega+\omega_0)} + Sa^2\left(\frac{\omega - \omega_0}{2}\right)e^{-j2(\omega-\omega_0)}\right]$$

Problem 5-2:

(1) $F_1(j\omega) = \dfrac{1}{2}\left[1 - \dfrac{|\omega|}{4\pi}\right][\varepsilon(\omega + 4\pi) - \varepsilon(\omega - 4\pi)]$;

(2) $F_2(j\omega) = \dfrac{\pi}{100}e^{-j3\omega}g_{100}(\omega)$;

(3) $F_3(j\omega) = \pi e^{-2|\omega|}$

Problem 5-3:

(1) $F_1(j\omega) = \dfrac{Sa\left(\frac{\omega}{2}\right)e^{-j2.5\omega} - e^{-j4\omega}}{j\omega}$;

(2) $F_2(j\omega) = \dfrac{2}{j\omega}\left[Sa\left(\dfrac{\omega T}{2}\right) - \cos\left(\dfrac{\omega T}{2}\right)\right]$;

(3) $F_3(j\omega) = \dfrac{2E}{\omega}Sa\left[\dfrac{\omega(\tau_2 - \tau_1)}{4}\right]\cdot \sin\left[\dfrac{\omega(\tau_2 + \tau_1)}{4}\right]$

Problem 5-4:

(1) $\frac{1}{2}F\left(j\frac{\omega}{2}\right)e^{-j2.5\omega}$

(2) $\frac{1}{5}F\left(j\frac{-\omega}{5}\right)e^{-j\frac{3}{5}\omega}$

(3) $\frac{j}{2}F'\left(j\frac{\omega}{2}\right)$

(4) $\frac{j}{2}F'\left(-j\frac{\omega}{2}\right) - 2F\left(j\frac{-\omega}{2}\right)$

(5) $-F(j\omega) - \omega F'(j\omega)$;

(6) $F[j2(\omega - 4)]e^{j6(\omega-4)} + F[j2(\omega + 4)]e^{j6(\omega+4)}$

(7) $F(j\omega)e^{-j3\omega} - \frac{1}{2}\left\{F\left[j(\omega - 4)e^{-j3(\omega-4)}\right] + F\left[j(\omega + 4)e^{-j3(\omega+4)}\right]\right\}$

Problem 5-5:

(1) $\frac{\pi}{\alpha}$ 　　　　　　(2) $\frac{2}{3}\pi a^3$ 　　　　　　(3) $\frac{\pi}{\omega_0}$

Problem 5-6:

(1) $-\dfrac{1}{2}|t|$; 　　　　　　(3) $\dfrac{1}{2\pi(\alpha + jt)}$;

(2) $\dfrac{1}{\pi j}\sin(100t)$; 　　　　(4) $e^{2(t-1)}\varepsilon(t - 1) - e^{-3(t-1)}\varepsilon(t - 1)$

Problem 5-7:

(a) $f(t) = \dfrac{A\omega_0}{\pi} Sa(\omega_0 t - 1);$

(c) $f_3(t) = \dfrac{1}{j\pi t^2}(\sin t - \sin 2t + t\cos 3t)$

(b) $f(t) = \dfrac{-2A}{\pi t}\sin \tfrac{1}{2}\omega_0 t \cdot \cos \tfrac{1}{2}\omega_0 t;$

Problem 5-8:

$y_f(t) = \left(e^{-3t} + e^{-t} - e^{-2t}\right)\varepsilon(t)$

Problem 5-9:

$y_x(t) = 7e^{-2t} - 5e^{-3t}, \qquad y_f(t) = \left(-\tfrac{1}{2}e^{-t} + 2e^{-2t} - \tfrac{3}{2}e^{-3t}\right)\varepsilon(t)$

$y(t) = \left(-\tfrac{1}{2}e^{-t} + 9e^{-2t} - \tfrac{13}{2}e^{-3t}\right)\varepsilon(t)$

Problem 5-10:

$y_f(t) = \left[1 - e^{-2(t-1)}\right]\varepsilon(t-1) - \left[1 - e^{-2(t-2)}\right]\varepsilon(t-2)$

Problem 5-11:

$u_C(t) = (3 - 3e^{-t} - t)\varepsilon(t) + (t - 3)\varepsilon(t - 3)$

Problem 5-12:

(1) $i_0(t) = \tfrac{1}{8}e^{-\frac{5}{8}t}\varepsilon(t)$ A;

(3) $i_0(t) = \tfrac{1}{3}\left(e^{-\frac{5}{8}t} - e^{-t}\right)\varepsilon(t)$ A

(2) $i_0(t) = \tfrac{1}{5}\left(1 - e^{-\frac{5}{8}t}\right)\varepsilon(t)$ A;

Problem 5-13:

$h(t) = \tfrac{1}{2}e^{-2t}\varepsilon(t) \quad i_f(t) = \left(5e^{-t} - 5.5e^{-2t} + \tfrac{1}{2}\right)\varepsilon(t)$A

Problem 5-14:

$H(j\omega) = \dfrac{1}{(j\omega)^2 + 6j\omega + 8}$

$h(t) = \tfrac{1}{2}\left[e^{-2t} - e^{-4t}\right]\varepsilon(t)$

$y_f(t) = \left(-\dfrac{1}{8}e^{-4t} + \dfrac{1}{8}e^{-2t} - \dfrac{1}{4}te^{-2t} + \dfrac{1}{4}t^2 e^{-2t}\right)\varepsilon(t)$

Chapter 6

Problem 6-1:

(1) $\dfrac{6}{s^2+4}+\dfrac{2s}{s^2+9};$

(5) $\dfrac{e^{-s}}{s+1};$

(9) $\dfrac{\pi}{s^2+\pi^2}+\dfrac{\pi e^{-2s}}{s^2+\pi^2};$

(2) $2-\dfrac{1}{s+1};$

(6) $\dfrac{e}{s+1};$

(10) $\dfrac{e^{-2s-1}(2s+3)}{(s+1)^2};$

(3) $\dfrac{1}{2}\left(\dfrac{1}{s}+\dfrac{s}{s^2+16}\right);$

(7) $\dfrac{1}{s^2}-\dfrac{e^{-s}}{s^2}-\dfrac{e^{-s}}{s};$

(11) $\dfrac{e^{-s}}{s^3}(2+2s+s^2);$

(4) $e^{-\alpha}\dfrac{s+1}{(s+1)^2+\omega^2};$

(8) $\dfrac{e^s}{s^2};$

(12) $\dfrac{2s^3-6s}{(s^2+1)^3}$

Problem 6-2:

(1) $f(4t)\leftrightarrow\dfrac{1}{4}F\left(\dfrac{s}{4}\right)=\dfrac{1}{4}\dfrac{1}{\left(\frac{s}{4}\right)^2+3\frac{s}{4}-5}=\dfrac{4}{s^2+12s-80}$

$e^{-t}f(4t)\leftrightarrow\dfrac{4}{(s+1)^2+12(s+1)-80}=\dfrac{4}{s^2+14s-67}$

(2) $f(2t)\leftrightarrow\dfrac{1}{2}F\left(\dfrac{s}{2}\right)$

$f(2t-4)\leftrightarrow\dfrac{1}{2}F\left(\dfrac{s}{2}\right)e^{-2s}=\dfrac{e^{-2s}}{2}\dfrac{1}{\left(\frac{s}{2}\right)^2+3\left(\frac{s}{2}\right)-5}=\dfrac{2e^{-2s}}{s^2+6s-20}$

(3) $f''(t)\leftrightarrow s^2F(s),\qquad tf''(t)\leftrightarrow-\left[s^2F(s)\right]'=-\left(\dfrac{s^2}{s^2+3s-5}\right)'=\dfrac{10s-3s^2}{(s^2+3s-5)^2}$

(4) $f(t)\sin 2t=f(t)\dfrac{e^{j2t}-e^{-j2t}}{2j}=\dfrac{1}{2j}\left[f(t)e^{j2t}-f(t)e^{-j2t}\right]\leftrightarrow\dfrac{1}{2j}\left[F(s-2j)-F(s+2j)\right]=$

$\dfrac{1}{2j}\left[\dfrac{1}{(s-2j)^2+3(s-2j)-5}-\dfrac{1}{(s+2j)^2+3(s+2j)-5}\right]$

(5) $f(\tau)e^\tau\leftrightarrow\dfrac{1}{(s-1)^2+3(s-1)-5}=\dfrac{1}{s^2+s-7}\qquad\left[\displaystyle\int_0^t f(\tau)e^\tau d\tau\right]=\left[\displaystyle\int_{-\infty}^t f(\tau)e^\tau d\tau\right]=$

$\dfrac{1}{s}\cdot\dfrac{1}{s^2+s-7},\qquad\left[\displaystyle\int_0^{t-2}f(\tau)e^\tau d\tau\right]\leftrightarrow\dfrac{e^{-2s}}{s\left(s^2+s-7\right)}$

(6) $\left[\dfrac{1}{t}f(t)\right]\leftrightarrow\displaystyle\int_s^\infty\dfrac{1}{s^2+3s-5}ds$

Problem 6-3:

(a) $f_1(t)=2\varepsilon(t)-\varepsilon(t-1)-\varepsilon(t-2)\leftrightarrow\dfrac{2}{s}-\dfrac{e^{-s}}{s}-\dfrac{e^{-2s}}{s}=\dfrac{2-e^{-s}-e^{-2s}}{s}$

(b) $f_2(t) = \frac{1}{T}t\,[\varepsilon(t) - \varepsilon(t - T)] = \frac{t}{T}\varepsilon(t) - \frac{t-T}{T}\varepsilon(t-T) - \varepsilon(t-T) \leftrightarrow \frac{1}{T}\frac{1}{s^2} - \frac{1}{T}\frac{1}{s^2}\frac{e^{-Ts}}{s^2} - \frac{e^{-Ts}}{s} =$

$\dfrac{1 - e^{-Ts} - Tse^{-Ts}}{Ts^2}$

(c) $f_3(t) = \sin(\omega t)\left[\varepsilon(t) - \varepsilon\left(t - \dfrac{T}{2}\right)\right] = \sin(\omega t)\cdot\varepsilon(t) - \sin(\omega t)\varepsilon\left(t - \dfrac{T}{2}\right) = \sin(\omega t)\varepsilon(t) +$

$\sin\left[\omega\left(t - \dfrac{T}{2}\right)\right]\varepsilon\left(t - \dfrac{T}{2}\right) \leftrightarrow \dfrac{\omega}{s^2 + \omega^2} + \dfrac{\omega e^{-\frac{T}{2}s}}{s^2 + \omega^2}$

Problem 6-4:

$f(t) = f(t)\varepsilon(t) = f_1(t) + f_1(t - T) + \cdots = \displaystyle\sum_{T=0}^{\infty} f_1(t - nT)$

$F(s) = [f(t)] = F_1(s) + e^{-Ts}F_1(s) + e^{-2Ts}F_1(s) + \cdots + e^{-nTs}F_1(s) = \dfrac{F_1(s)}{1 - e^{-Ts}}$

Problem 6-5:

(a) $f_1(t) = E\left[\varepsilon(t) - \varepsilon\left(t - \dfrac{T}{2}\right)\right] \leftrightarrow E\left[\dfrac{1}{s} - \dfrac{e^{-\frac{T}{2}s}}{s}\right]$

$F_1(s) = \dfrac{1}{1 - e^{-sT}}E\left(\dfrac{1 - e^{-\frac{T}{2}s}}{s}\right) = \dfrac{E}{s}\dfrac{1}{1 + e^{-\frac{T}{2}s}}$

(b) $f_2(t) = E\sin(\omega t)\,[\varepsilon(t) - \varepsilon(t - T)] \leftrightarrow \dfrac{E\omega}{s^2 + \omega^2}\left(1 - e^{-Ts}\right)$

$F_2(s) = \dfrac{1}{1 - e^{-Ts}}\dfrac{E\omega}{s^2 + \omega^2}\left(1 - e^{-Ts}\right) = \dfrac{E\omega}{s^2 + \omega^2}$

(c) $f_3(t) = \delta(t) + \delta(t - 2) + \delta(t - 4) + \cdots + \delta(t - 2n) + \cdots \leftrightarrow 1 + e^{-2s} + e^{-4s} + \cdots + e^{-2ns} + \cdots =$

$\dfrac{1}{1 - e^{-2s}}$

(d) $f'_{4T}(t) = \varepsilon(t) - 2\varepsilon(t - 1) + \varepsilon(t - 2) \leftrightarrow \dfrac{1}{s} - \dfrac{2}{s}e^{-s} + \dfrac{1}{s}e^{-2s}$

$f_{4T}(t) \leftrightarrow \dfrac{1}{s^2}\left(1 - 2e^{-s} + e^{-2s}\right)$

$f_4(t) \leftrightarrow \dfrac{1}{s^2\left(1 - e^{-s^2}\right)}\left(1 - 2e^{-s} + e^{-2s}\right) = \dfrac{1}{s^2\left(1 - e^{-2s}\right)}\left(1 - e^{-s}\right)^2 = \dfrac{1 - e^{-s}}{s^2\left(1 + e^{-s}\right)}$

Problem 6-6:

(1) $f(0_+) = \displaystyle\lim_{s\to\infty} s\dfrac{s + 3}{(s + 1)(s + 2)^2} = 0,\qquad f(\infty) = \displaystyle\lim_{s\to 0} s\dfrac{s + 3}{(s + 1)(s + 2)^2} = 0$

(2) $F(s) = s + \dfrac{-2s}{s^2 + 6s + 8} = s + F_1(s)$

$f(0_+) = -\displaystyle\lim_{s\to\infty}\dfrac{2s^2}{s^2 + 6s + 8} = -2,\qquad f(\infty) = \displaystyle\lim_{s\to 0}\dfrac{s^4 + 6s^3 + 6s^2}{s^2 + 6s + 8} = 0$

(3) $f(0_+) = \lim\limits_{s \to \infty} s \cdot \dfrac{s^2 + 2s + 3}{(s+1)(s^2+2)} = 1,$ $f(\infty)$ no exist, $p = \pm 2j$

(4) $f(0_+) = 0,$ $f(\infty) = \lim\limits_{s \to 0} s \cdot \dfrac{2s+1}{s^3 + 3s^2 + 2s} = \dfrac{1}{2}$

(5) $f(0_+) = \lim\limits_{s \to \infty} s \cdot \dfrac{1 - e^{-2s}}{s(s^2+4)} = 0,$ $f(\infty)$ no exist.

(6) $f(0_+) = \lim\limits_{s \to \infty} \dfrac{1}{1 + e^{-s}} = 1,$ $f(\infty) = \lim\limits_{s \to 0} \dfrac{1}{1 + e^{-s}} = \dfrac{1}{2}$

Problem 6-7:

(1) $f(t) = \left[-5e^{-3t} + 9e^{-5t}\right] \varepsilon(t),$

(2) $f(t) = \delta'(t) - 3\delta(t) - 5e^{-t}\varepsilon(t) + 13e^{-2t}\varepsilon(t),$

(3) $f(t) = e^{-t}\varepsilon(t) + 2e^{-2t}t\varepsilon(t) - e^{-2t}\varepsilon(t),$

(4) $f(t) = \varepsilon(t) - e^{-t}\cos 2t\varepsilon(t),$

(5) $f(t) = 2e^{t}\varepsilon(t) + 2e^{-t}(\cos t - 2\sin t)\varepsilon(t),$

(6) $f(t) = e^{-t}t^2\varepsilon(t) - e^{-t}t\varepsilon(t) + 2e^{-t}\varepsilon(t) - e^{-2t}\varepsilon(t)$

Problem 6-8:

(1) $2e^{-2t}\varepsilon(t) - e^{-2(t-3)}\varepsilon(t-3),$

(2) $\cos \pi t \cdot \varepsilon(t) + \cos \pi(t-T) \cdot \varepsilon(t-T),$

(3) $t\varepsilon(t) + 2(t-2)\varepsilon(t-2) + (t-4)\varepsilon(t-4),$

(4) $\frac{1}{4}\left[\varepsilon(t) - \varepsilon(t-1)\right] - \frac{1}{4}\cos t \cdot \varepsilon(t) + \frac{1}{4}\cos(t-1)\varepsilon(t-1),$

(5) $\cos 3t \cdot \varepsilon(t) + 2\sin 3(t-1) \cdot \varepsilon(t-1),$

(6) $\sum_{n=0}^{\infty} (-1)^n \delta(t-n),$

(7) $\sum_{n=0}^{+\infty} (-1)^n \varepsilon(t-n),$

(8) $\sum_{n=0}^{\infty} \delta(t-n),$

(9) $\sum_{n=0}^{\infty} \varepsilon(t-n),$

(10) $f(t) = -\frac{1}{t}\varepsilon(t) + \frac{1}{t}e^{-9t}\varepsilon(t)$

Problem 6-9:

(1) $h(t) = \left(e^{-2t} - e^{-3t}\right)\varepsilon(t).$

Problem 6-10:

$$H(s) = 1 - \frac{11}{s+10}, \qquad Y_f(s) = \frac{1}{s+10} - \frac{11}{(s+10)^2}, \qquad F(s) = \frac{1}{s+10},$$

$$f(t) = e^{-10t}\varepsilon(t)$$

Problem 6-11:

(1) $Y_f(s) = \dfrac{s}{s^2+3s+2}\dfrac{10}{s} = \dfrac{10}{s+1} - \dfrac{10}{s+2}, \qquad y_f(t) = \left(10e^{-t} - 10e^{-2t}\right)\varepsilon(t)$

(2) $Y_f(s) = \dfrac{s}{s^2+3s+2}\cdot\dfrac{10}{s^2+1} = \dfrac{-5}{s+1} + \dfrac{4}{s+2} + \dfrac{1}{2}(1+3j)\dfrac{1}{s+j} + \dfrac{1}{2}(1-3j)\dfrac{1}{s-j}$

$y_f(t) = \left(4e^{-2t} - 5e^{-t} + \cos t + 3\sin t\right)\varepsilon(t)$

Problem 6-12:

$$I_s(s) = \frac{U_L(s)}{3} + \frac{U_L(s) + 1}{\frac{1}{2}s}, \qquad U_L(s) = \frac{3sI_s(s)}{s+6} - \frac{6}{s+6} = \frac{3s}{s+6}\cdot\frac{5}{s^2} - \frac{6}{s+6} = \frac{-8.5}{s+6} + \frac{2.5}{s}$$

$$u_L(t) = 2.5\varepsilon(t) - 8.5e^{-6t}\varepsilon(t)$$

Problem 6-13:

$$u_C(t) = \tfrac{1}{2}\left[te^{-t} - e^{-t} + 1\right]\varepsilon(t)$$

Problem 6-14:

$$\left(\frac{1}{5} + \frac{1}{s}\right)I_1(s) - \frac{1}{5}I_2(s) = \frac{15}{s}, \qquad -\frac{1}{5}I_1(s) + \left(\frac{1}{2}s + \frac{6}{5}\right)I_2(s) = 2$$

$$I_1(s) = \frac{-57}{s+3} + \frac{136}{s+4}, \qquad i_1(t) = \left(-57e^{-3t} + 136e^{-4t}\right)\varepsilon(t)$$

Problem 6-15:

$$Y_1(s) = \frac{-3}{s+1} = Y_x(s) + Y_{f_1}(s) = Y_x(s) + H(s),$$

$$Y_2(s) = \frac{1}{s} - \frac{5}{s+1} = Y_x(s) + Y_{f_2}(s) = Y_x(s) + \frac{1}{s}H(s),$$

$$H(s) = \frac{1}{s+1}, \qquad Y_x(s) = \frac{-4}{s+1}, \qquad y_x(t) = -4e^{-t}\varepsilon(t),$$

$$Y_{f_3}(s) = H(s)\cdot\frac{1}{s^2} = \frac{1}{s+1} + \frac{1}{s^2} - \frac{1}{s}, \qquad y_{f_3}(t) = \left(e^{-t} + t - 1\right)\varepsilon(t),$$

$$y_3(t) = y_x(t) + y_{f_3}(t) = t - 1 - 3e^{-t} \qquad t \geq 0$$

Problem 6-16:

$$Y_1(s) = Y_x(s) + Y_{f_1}(s) = 1 + \frac{1}{1+s}, \qquad Y_2(s) = Y_x(s) + Y_{f_2}(s) = \frac{3}{1+s}, \qquad H(s) = \frac{s}{s+1},$$

$$Y_x(s) = \frac{2}{s+1}, \qquad y_x(t) = 2e^{-t}, \qquad Y_{f_3}(s) = \frac{1-e^{-s}}{s^2} \cdot \frac{s}{s+1} = \frac{1}{s(s+1)} - \frac{e^{-s}}{s(s+1)},$$

$$f_3(t) = t\left[\varepsilon(t) - \varepsilon(t-1)\right] + \varepsilon(t-1) = t\varepsilon(t) - (t-1)\varepsilon(t-1), \qquad F_3(s) = \frac{1}{s^2} - \frac{e^{-s}}{s^2},$$

$$y_{f_3}(t) = \varepsilon(t) - \varepsilon(t)e^{-t} - \varepsilon(t-1)e^{-(t-1)},$$

$$y_3(t) = y_x(t) + y_{f_3}(t) = \left(1 + e^{-t}\right)\varepsilon(t) - \left[1 - e^{-(t-1)}\right]\varepsilon(t-1)$$

Problem 6-17:

$$Y(s) = \frac{s+3}{s^2+3s+2}F(s) + \frac{s+5}{s^2+3s+2}; \qquad y_x(t) = \left(4e^{-t} - 3e^{-2t}\right)\varepsilon(t)$$

$$y_f(t) = \left(e^{-t} - e^{-2t}\right)\varepsilon(t); \qquad y(t) = \left(5e^{-t} - 4e^{-2t}\right)\varepsilon(t), \text{ no forced response.}$$

Problem 6-18:

$$sY(s) - y(0_-) + 2Y(s) = F(s), \qquad Y(s) = \frac{F(s)}{s+2} + \frac{y(0_-)}{s+2}, \qquad F(s) = \frac{2}{s^2+4}$$

$$Y(s) = \frac{2}{(s+2)(s^2+4)} + \frac{1}{s+2} = \frac{5}{4}\frac{1}{s+2} + \left(-\frac{1}{8} - \frac{1}{8}j\right)\frac{1}{s-2j} + \left(-\frac{1}{8} + \frac{1}{8}j\right)\frac{1}{s+2j} \leftrightarrow$$

$$\frac{5}{4}e^{-2t} + \frac{1}{4}(\sin 2t - \cos 2t) \qquad (t \geq 0)$$

Problem 6-19:

$$s^2 Y(s) - sy(0_-) - y'(0_-) + 4sY(s) - 4y(0_-) + 3y(s) = 3F(s)$$

$$Y(s) = \frac{3}{s^2+4s+3}F(s) + \frac{sy(0_-) + y'(0_-) + 4y(0_-)}{s^2+4s+3} = \frac{1}{s} + \frac{1}{s+1}, \, y(t) = (1 + e^{-t})\varepsilon(t)$$

Problem 6-20:

$$s^2 Y(s) - sy(0_-) - y'(0_-) + 5sY(s) - 5y(0_-) + 6Y(s) = 6F(s)$$

$$Y(s) = \frac{6}{s^2+5s+6}F(s) + \frac{sy(0_-) + y'(0_-) + 5y(0_-)}{s^2+5s+6}$$

$$y_f(t) = 6\left(e^{-t} + 2e^{-3t} - 2e^{-2t}\right)\varepsilon(t), \qquad y_x(t) = \left(e^{-2t} - e^{-3t}\right)\varepsilon(t)$$

Problem 6-21:

$$s^2 Y(s) - sy(0_-) - y'(0_-) + 2sY(s) - 2y(0_-) + Y(s) = sF(s)$$

$$Y(s) = \frac{s}{s^2+2s+1}F(s) + \frac{s+4}{s^2+2s+1}; \qquad y_x(t) = (1 + 3t)e^{-t}\varepsilon(t); \qquad y_f(t) = \left(t - \frac{t^2}{2}\right)e^{-t}\varepsilon(t)$$

Chapter 7

Problem 7-1:

(a) and (b) $H(s) = \dfrac{s^2 + 3s + 2}{s^2 + 2s + 1}$

Problem 7-2:

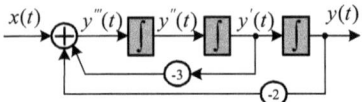

Fig. A7-2 (1)

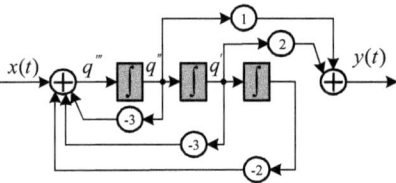

Fig. A7-2 (2)

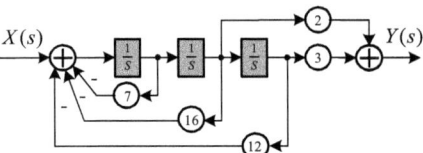

Fig. A7-2 (3)

Problem 7-3:

(a) $H(s) = \dfrac{H_1 H_2 H_3 H_5 + H_4 H_5(1 + G_2 H_2)}{1 + G_1 H_1 + G_2 H_2 + G_3 H_3 + G_1 G_2 G_3 H_4 + G_1 G_3 H_1 H_3}$,

(b) $H(s) = \dfrac{G_1 G_2 G_3 G_7}{1 - G_2 G_3 G_5 G_6 - G_2 G_4 G_5}$,

(c) $H(s) = \dfrac{G_1 G_2 G_3 G_4 - G_1 G_5 (1 + G_3 H_1)}{1 + (G_3 H_1 + G_4 H_2 + G_2 G_4 H_3)}$,

(d) $H(s) = \dfrac{s^3}{s^3 + 6s^2 + 11s + 6}$,

(e) $H(s) = \dfrac{H_1 H_2 H_3 H_4 H_5 + H_1 H_5 H_6 (1 - G_3)}{1 - (H_2 G_2 + H_4 G_4 + H_2 H_3 G_1 + G_3 + H_6 G_1 G_4) + H_2 G_2 (G_3 + H_4 G_4)}$

Problem 7-4:

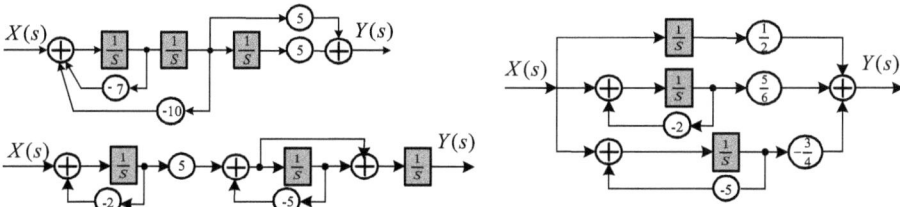

Fig. A7-4 (1)

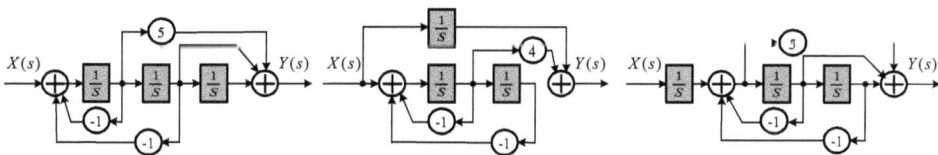

Fig. A7-4 (2)

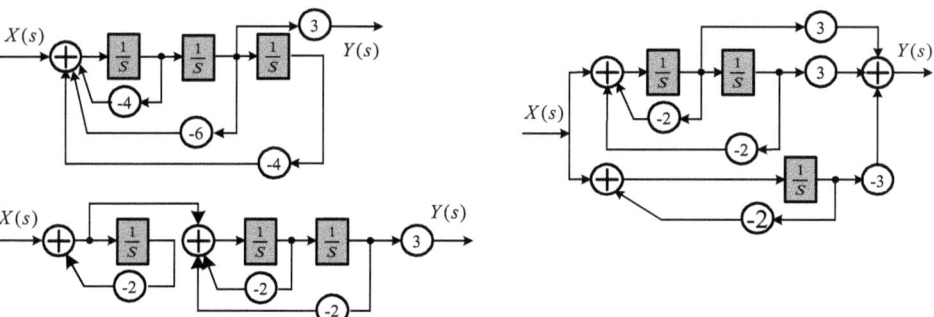

Fig. A7-4 (3)

Problem 7-5:

$$H(s) = \frac{s^2 + s}{s^3 + 14s^2 + 42s + 30}$$

Problem 7-6:

$$H(s) = \frac{2s + 8}{s^3 + 6s^2 + 11s + 6}$$

Problem 7-7:

$$y_f(t) = -\frac{2}{5}\left[\frac{2}{5}\cos(t-1) + \frac{1}{5}\sin(t-1) - \frac{2}{5}e^{-2(t-1)}\right]\varepsilon(t-1)$$

Problem 7-8:

(a) $H(s) = \dfrac{(s+2)(s+4)}{(s+1)(s+3)}$; (b) $H(s) = \dfrac{(s^2+1)(s^2+3)}{2s(s^2+2)}$; (c) $H(s) = \dfrac{2s}{4s^2 + 6s + 1}$

Problem 7-9:

(1) $H(s) = 10\dfrac{s-1}{s(s+1)}$; (2) $y_s(t) = 6\sin 3t - 8\cos 3t$

Problem 7-10:

$R = 2\,\Omega$, $L = 2\,\text{H}$, $C = 0.25\,\text{F}$

Problem 7-11:

$y_f(t) = \left(\dfrac{1}{5}\sin t + \dfrac{4}{5}\sin 4t\right)\varepsilon(t)$, no free response.

Problem 7-12:

(1) unstable, negative real roots 4, characteristic equation misses term s;
(2) unstable, positive real roots 2;
(3) critical, negative real roots 2;
(4) critical, negative real roots 2;
(5) unstable, positive real roots 2, negative real roots 3.

Problem 7-13:

(1) $k < 4$;
(2) $0 < k < 60$

Problem 7-14:

(1) $k < 3$;
(2) when $k = 3$, critically stable, $h(t) = \cos \sqrt{2}t\varepsilon(t)$

Problem 7-15:

(1) $k < 4$;
(2) when $k = 4$, critically stable, $h(t) = 4\cos 2t\varepsilon(t)$

Bibliography

Weigang Zhang, Wei-Feng Zhang. Signals and Systems (Chinese edition). Beijing, Tsinghua university press, 2012.

Weigang Zhang. Circuits Analysis (Chinese edition). Tsinghua university press, January 2015.

Alin V. Oppenheim, Alin S. Willsky, S. Hamid Nawab. Signals and Systems (Second Edition). Beijing, Electronic industry press, February 2004.

Bernd Girod, Rudolf Rabenstein, Alexander Strenger. Signals and Systems. Beijing, Tsinghua university press, March 2003.

Charles L. Phillips, John M. Parr, Eve A. Riskin. Signals, Systems and Transforms (Third Edition). Beijing, Mechanical industry press, January 2004.

M. J. Roberts. Signals and Systems – Analysis Using Transform Methods and MATLAB. McGrawHill Companies, 2004.

Weigang Zhang, Lina Cao. Communication Principles (Chinese Edition). Beijing, Tsinghua university press, 2016.

https://doi.org/10.1515/9783110419535-009

Index

https://doi.org/10.1515/9783110419535-010